Aircraft Design Projects

Dedications

To Jessica, Maria, Edward, Robert and Jonothan – in their hands rests the future.

To my father, J. F. Marchman, Jr, for passing on to me his love of airplanes and to my teacher, Dr Jim Williams, whose example inspired me to pursue a career in education.

Aircraft Design Projects

for engineering students

Lloyd R. Jenkinson
James F. Marchman III

BUTTERWORTH
HEINEMANN

OXFORD   AMSTERDAM   BOSTON   LONDON   NEW YORK   PARIS
SAN DIEGO   SAN FRANCISCO   SINGAPORE   SYDNEY   TOKYO

Butterworth-Heinemann
An imprint of Elsevier Science
Linacre House, Jordan Hill, Oxford OX2 8DP
200 Wheeler Road, Burlington MA 01803

First published 2003

British Library Cataloguing in Publication Data
A catalogue record for this book is available from the British Library

Library of Congress Cataloguing in Publication Data
A catalogue record for this book is available from the Library of Congress

ISBN 0 7506 5772 3

Typeset by Newgen Imaging Systems (P) Ltd., India
Printed and bound in Great Britain by Biddles Ltd. *www.biddles.co.uk*

For information on all Butterworth-Heinemann publications
visit our website at www.bh.com

Contents

Preface

There are many excellent texts covering aircraft design from a variety of perspectives.[1] Some of these are aimed at specific audiences ranging from practising aerospace engineers, to engineering students, to amateur airplane builders. Others cover specialized aspects of the subject such as undercarriage or propulsion system design. Some of these are quite detailed in their presentation of the design process while others are very general in scope. Some are overviews of all the basic aeronautical engineering subjects that come together in the creation of a design.

University faculty that teach aircraft design courses often face difficult choices when evaluating texts or references for their students' use. Many texts that are suitable for use in a design class are biased toward particular classes of aircraft such as military aircraft, general aviation, or airliners. A text that gives excellent coverage of design basics may prove slanted toward a class of aircraft different from that year's project. Alternatively, those that emphasize the correct type of vehicle may treat design fundamentals in an unfamiliar manner. The situation may be further complicated in classes that have several teams of students working on different types of designs, some of which 'fit' the chosen text while others do not.

Most teachers would prefer a text that emphasizes the basic thought processes of preliminary design. Such a text should encourage students to seek an understanding of the approaches and constraints appropriate to their design assignment before they venture too far into the analytical processes. On the other hand, students would like a text which simply tells them where to input their design objectives into a 'black-box' computer code or generalized spreadsheet, and preferably, where to catch the final design drawings and specifications as they are printed out. Faculty would like their students to begin the design process with a thorough review of their previous courses in aircraft performance, aerodynamics, structures, flight dynamics, propulsion, etc. Students prefer to start with an Internet search, hoping to find a solution to their problem that requires only minimal 'tweaking'.

The aim of this book is to present a two pronged approach to the design process. It is expected to appeal to both faculty and students. It sets out the basics of the design thought process and the pathway one must travel in order to reach an aircraft design goal for any category of aircraft. Then it presents a variety of design case studies. These are intended to offer examples of the way the design process may be applied to conceptual design problems typical of those actually used at the advanced level in academic and other training curricula. It does not offer a step-by-step 'how to' design guide, but shows how the basic aircraft preliminary design process can be successfully applied to a wide range of unique aircraft. In so doing, it shows that each set of design objectives presents its own peculiar collection of challenges and constraints. It also shows how the classical design process can be applied to any problem.

Case studies provide both student and instructor with a valuable teaching/learning tool, allowing them to examine the way others have approached particular design challenges. In the 1970s, the American Institute of Aeronautics and Astronautics (AIAA) published an excellent series of design case studies[2] taken from real aircraft project developments. These provided valuable insights into the development of several, then current, aircraft. Some other texts have employed case studies taken from industrial practice. Unfortunately, these tend to include aspects of design that are beyond the conceptual phase, and which are not covered in academic design courses. While these are useful in teaching design, they can be confusing to the student who may have difficulty discerning where the conceptual aspects of the design process ends and detailed design ensues. The case studies offered in this text are set in the preliminary design phase. They emphasize the thought processes and analyses appropriate at this stage of vehicle development.

Many of the case studies presented in this text were drawn from student projects. Hence, they offer an insight into the conceptual design process from a student perspective. The case studies include design projects that won top awards in national and international design competitions. These were sponsored by the National Aeronautics and Space Administration (NASA), the US Federal Aviation Administration (FAA), and the American Institute of Aeronautics and Astronautics (AIAA).

The authors bring a unique combination of perspectives and experience to this text. It reflects both British and American academic practices in teaching aircraft design to undergraduate students in aeronautical and aerospace engineering. Lloyd Jenkinson has taught aircraft design at both Loughborough University and Southampton University in the UK and Jim Marchman has taught both aircraft and spacecraft design at Virginia Tech in the US. They have worked together since 1997 in an experiment that combines students from Loughborough University and Virginia Tech in international aircraft design teams.[3] In this venture, teams of students from both universities have worked jointly on a variety of aircraft design projects. They have used exchange visits, the Internet and teleconference communications to work together progressively, throughout the academic year, on the conceptual design of a novel aircraft.

In this book, the authors have attempted to build on their experience in international student teaming. They present processes and techniques that reflect the best in British and American design education and which have been proven to work well in both academic systems. Dr Jenkinson also brings to this text his prior experience in the aerospace industry of the UK, having worked on the design of several successful British aircraft. Professor Marchman's contribution to the text also reflects his experiences in working with students and faculty in Thailand and France in other international design team collaborative projects.

The authors envision this text as supplementing the popular aircraft design textbooks, currently in use at universities around the world. Books such as those reviewed by Mason[1] could be employed to present the detailed aspects of the preliminary design process. Working within established conceptual design methodology, this book will provide a clearer picture of the way those detailed analyses may be adapted to a wide range of aircraft types.

It would have been impossible to write this book without the hard work and enthusiasm shown by many of our students over more years than we care to remember. Their continued interest in aircraft design project work and the smoothing of the difficulties they sometimes experienced in progressing through the work was our inspiration. We have also benefited from the many colleagues and friends who have been generous in sharing their encouragement and knowledge with us. Aircraft design educators seem to be a special breed of engineers who selflessly give their effort and time to inspire anyone who wants to participate in their common interest. We are fortunate to count them as our friends.

References

1 Bill Mason's web page: www.aoe.vt.edu/Mason/ACinfoTOC.html.
2 AIAA web page: www.aiaa.org/publications/index.
3 Jenkinson, L. R., Page, G. J., Marchman, J. F., 'A model for international teaming in aircraft design education', *Journal of Aircraft Design*, Vol. 3, No. 4, pp. 239–247, Elsevier, December 2000.

Acknowledgements

To all the students and staff at Loughborough and Southampton Universities who have, over many years, contributed directly and indirectly to my understanding of the design of aircraft, I would like to express my thanks and appreciation. For their help with proof reading and technical advice, I thank my friends Paul Eustace and Keith Payne. Our gratitude to all those people in industry who have provided assistance with the projects. Finally, to my wife and family for their support and understanding over the time when my attention was distracted by the writing of the book.

Lloyd Jenkinson

I would like to acknowledge the work done by the teams of Virginia Tech and Loughborough University aircraft design students in creating the designs which I attempted to describe in Chapters 7 and 10 and the contributions of colleagues such as Bill Mason, Nathan Kirschbaum, and Gary Page in helping guide those students in the design process. Without these people these chapters could not have been written.

Jim Marchman

Introduction

It is tempting to title this book 'Flights of Fancy' as this captures the excitement and expectations at the start of a new design project. The main objective of this book is to try to convey this feeling to those who are starting to undertake aircraft conceptual design work for the first time. This often takes place in an educational or industrial training establishment. Too often, in academic studies, the curiosity and fascination of project work is lost under a morass of mathematics, computer programming, analytical methods, project management, time schedules and deadlines. This is a shame as there are very few occasions in your professional life that you will have the chance to let your imagination and creativity flow as freely as in these exercises. As students or young engineers, it is advisable to make the most of such opportunities.

When university faculty or counsellors interview prospective students and ask why they want to enter the aeronautics profession, the majority will mention that they want to design aircraft or spacecraft. They often tell of having drawn pictures of aeroplanes since early childhood and they imagine themselves, immediately after graduation, producing drawings for the next generation of aircraft. During their first years in the university, these young men and women are often less than satisfied with their basic courses in science, mathematics, and engineering as they long to 'design' something. When they finally reach the all-important aircraft design course, for which they have yearned for so long, they are often surprised. They find that the process of design requires far more than sketching a pretty picture of their dream aircraft and entering the performance specifications into some all-purpose computer program which will print out a final design report.

Design is a systematic process. It not only draws upon all of the student's previous engineering instruction in structures, aerodynamics, propulsion, control and other subjects, but also, often for the first time, requires that these individual academic subjects be applied to a problem concurrently. Students find that the best aerodynamic solution is not equated to the best structural solution to a problem. Compromises must be made. They must deal with conflicting constraints imposed on their design by control requirements and propulsion needs. They may also have to deal with real world political, environmental, ethical, and human factors. In the end, they find they must also do practical things like making sure that their ideal wing will pass through the hangar door!

An overview of the book

This book seeks to guide the student through the preliminary stages of the aircraft design process. This is done by both explaining the process itself (Chapters 1 and 2) and by providing a variety of examples of actual student design projects (Chapters 3 to 10). The projects have been used as coursework at universities in the UK and the US. It should be noted that the project studies presented are not meant to provide a 'fill in the blank' template to be used by future students working on similar design problems but to provide insight into the process itself. Each design problem, regardless of how similar it may appear to an earlier aircraft design, is unique and requires a thorough and systematic investigation. The project studies presented in this book merely serve as examples of how the design process has been followed in the past by other teams faced with the task of solving a unique problem in aircraft design.

It is impossible to design aircraft without some knowledge of the fundamental theories that influence and control aircraft operations. It is not possible to include such information in this text but there are many excellent books available which are written to explain and present these theories. A bibliography containing some of these books and other sources of information has been added to the end of the book. To understand the detailed calculations that are described in the examples it will be necessary to use the data and theories in such books. Some design textbooks do contain brief examples on how the analytical methods are applied to specific aircraft. But such studies are mainly used to support and illustrate the theories and do not take an overall view of the preliminary design process.

The initial part of the book explains the preliminary design process. Chapter 1 briefly describes the overall process by which an aircraft is designed. It sets the preliminary design stages into the context of the total transformation from the initial request for proposal to the aircraft first flight and beyond. Although this book only deals with the early stages of the design process, it is necessary for students to understand the subsequent stages so that decisions are taken wisely. For example, aircraft design is by its nature an iterative process. This means that estimates and assumptions have sometimes to be made with inadequate data. Such 'guesstimates' must be checked when more accurate data on the aircraft is available. Without this improvement to the fidelity of the analytical methods, subsequent design stages may be seriously jeopardized.

Chapter 2 describes, in detail, the work done in the early (conceptual) design process. It provides a 'route map' to guide a student from the initial project brief to the validated 'baseline' aircraft layout. The early part of the chapter includes sections that deal with 'defining and understanding the problem', 'collecting useful information' and 'setting the aircraft requirements'. This is followed by sections that show how the initial aircraft configuration is produced. Finally, there are sections illustrating how the initial aircraft layout can be refined using constraint analysis and trade-off studies. The chapter ends with a description of the 'aircraft type specification'. This report is commonly used to collate all the available data about the aircraft. This is important as the full geometrical description and data will be needed in the detailed design process that follows.

Chapter 3 introduces the seven project studies that follow (Chapters 4 to 10). It describes each of the studies and provides a format for the sequence of work to be followed in some of the studies. The design studies are not sequential, although the earlier ones are shown in slightly more detail. It is possible to read any of the studies separately, so a short description of each is presented.

Chapters 4 to 10 inclusive contain each of the project studies. The projects are selected from different aeronautical applications (general aviation, civil transports, military aircraft) and range from small to heavy aircraft. For conciseness of presentation the detailed calculations done to support the final designs have not been included in these chapters but the essential input values are given so that students can perform their own analysis. The projects are mainly based on work done by students on aeronautical engineering degree courses. One of the studies is from industrial work and some have been undertaken for entry to design competitions. Each study has been selected to illustrate a different aspect of preliminary design and to illustrate the varied nature of aircraft conceptual design.

The final chapter (11) offers guidance on student design work. It presents a set of questions to guide students in successfully completing an aircraft design project. It includes some observations about working in groups. Help is also given on the writing of technical reports and making technical presentations.

Engineering units of measurement

Experience in running design projects has shown that students become confused by the units used to define parameters in aeronautics. Some detailed definitions and conversions are contained in Appendix A at the end of the book and a quick résumé is given here.

Many different systems of measurement are used throughout the world but two have become most common in aeronautical engineering. In the US the now inappropriately named 'British' system (foot, pound and second) is widely used. In the UK and over most of Europe, System International (SI) (metres, newton and second) units are standard. It is advised that students only work in one system. Confusion (and disaster) can occur if they are mixed. The results of the design analysis can be quoted in both types of unit by applying standard conversions. The conversions below are typical:

1 inch = 25.4 mm	1 foot = 0.305 metres
1 sq. ft = 0.0929 sq. m	1 cu. ft = 28.32 litres
1 US gal = 3.785 litres	1 Imp. gal = 4.546 litres
1 US gal = 0.833 Imp. gal	1 litre = 0.001 cubic metres
1 statute mile = 1.609 km	1 nautical mile = 1.852 km
1 ft/s = 0.305 m/s	1 knot = 0.516 m/s
1 knot = 1.69 ft/s	1 knot = 1.151 mph
1 pound force = 4.448 newtons	1 pound mass = 0.454 kilogram
1 horsepower = 745.7 watts	1 horsepower = 550 ft lb/s

To avoid confusing pilots and air traffic control, some international standardization of units has had to be accepted. These include:

Aircraft altitude – feet	Aircraft forward speed – knots*
Aircraft range – nautical miles	Climb rate – feet per minute

(* Be extra careful with the definition of units used for aircraft speed as pilots like to use airspeed in IAS (indicated airspeed as shown on their flight instruments) and engineers like TAS (true airspeed, the speed relative to the ambient air)).

Fortunately throughout the world, the International Standard Atmosphere (ISA) has been adopted as the definition of atmospheric conditions. ISA charts and data can be found in most design textbooks. In this book, which is aimed at a worldwide readership, where possible both SI and 'British' units have been quoted. Our apologies if this confuses the text in places.

English – our uncommon tongue

Part of this book grew out of the authors' collaboration in a program of international student design projects over several years. As we have reported our experiences from that program, observers have often noted that one thing that makes our international collaboration easier than some others is the common language. On the other hand, one thing we and our students have learned from this experience is that many of the aspects of our supposedly common tongue really do not have much in common.

Pairing an Englishman and an American to create a textbook aimed at both the US, British and other markets is an interesting exercise in spelling and language skills. While (or is it whilst?) the primary language spoken in the United Kingdom and the United States grows from the same roots, it has very obviously evolved somewhat differently. An easy but interesting way to observe some of these differences is to take a page of text from a British book and run it through an American spelling check program. Checking an American text with an 'English' spell checker will produce similar surprises. We spell many words differently, usually in small ways. Is it 'color' or 'colour'; do we 'organize' our work or 'organise' it? In addition, do we use double (") or single (') strokes to indicate a quote or give emphasis to a word or phrase? Will we hold our next meeting at 9:00 am or at 9.00 am? (we won't even mention the 24 hour clock!).

There are also some obvious differences between terminology employed in the US and UK. Does our automobile have a 'bonnet' and a 'boot' or a 'hood' and a 'trunk' and does its engine run on 'gasoline' or 'petrol'? American 'airplanes' have 'landing gear' while British 'aeroplanes/airplanes or aircraft' have 'undercarriages', does it have 'reheat' or an 'afterburner'. Fortunately, most of us have watched enough television shows and movies from both countries to be comfortable with these differences.

As we have pieced together this work we have often found ourselves (and our computer spell checkers) editing each other's work to make it conform to the conventions in spelling, punctuation, and phraseology, assumed to be common to each of our versions of this *common* language. The reader may find this evident as he or she goes from one section of the text to another and detects changes in wording and terminology which reflect the differing conventions in language use in the US and UK. It is hoped that these variations, sometimes subtle and sometimes obvious, will not prove an obstacle to the reader's understanding of our work but will instead make it more interesting.

Finally

All aircraft projects are unique, therefore, it is impossible to provide a 'template' for the work involved in the preliminary design process. However, with knowledge of the detail steps in the preliminary design process and with examples of similar project work, it is hoped that students will feel freer to concentrate on the innovative and analytical aspects of the project. In this way they will develop their technical and communication abilities in the absorbing context of preliminary aircraft design.

1

Design methodology

The start of the design process requires the recognition of a 'need'. This normally comes from a 'project brief' or a 'request for proposals (RFP)'. Such documents may come from various sources:

- Established or potential customers.
- Government defence agencies.
- Analysis of the market and the corresponding trends from aircraft demand.
- Development of an existing product (e.g. aircraft stretch or engine change).
- Exploitation of new technologies and other innovations from research and development.

It is essential to understand at the start of the study where the project originated and to recognise what external factors are influential to the design before the design process is started.

At the end of the design process, the design team will have fully specified their design configuration and released all the drawings to the manufacturers. In reality, the design process never ends as the designers have responsibility for the aircraft throughout its operational life. This entails the issue of modifications that are found essential during service and any repairs and maintenance instructions that are necessary to keep the aircraft in an airworthy condition.

The design method to be followed from the start of the project to the nominal end can be considered to fall into three main phases. These phases are illustrated in Figure 1.1.

The preliminary phase (sometimes called the conceptual design stage) starts with the project brief and ends when the designers have found and refined a feasible baseline design layout. In some industrial organisations, this phase is referred to as the 'feasibility study'. At the end of the preliminary design phase, a document is produced which contains a summary of the technical and geometric details known about the baseline design. This forms the initial draft of a document that will be subsequently revised to contain a thorough description of the aircraft. This is known as the aircraft 'Type Specification'.

The next phase (project design) takes the aircraft configuration defined towards the end of the preliminary design phase and involves conducting detailed analysis to improve the technical confidence in the design. Wind tunnel tests and computational fluid dynamic analysis are used to refine the aerodynamic shape of the aircraft. Finite element analysis is used to understand the structural integrity. Stability and control analysis and simulations will be used to appreciate the flying characteristics. Mass and balance estimations will be performed in increasingly fine detail. Operational factors (cost, maintenance and marketing) and manufacturing processes will be investigated

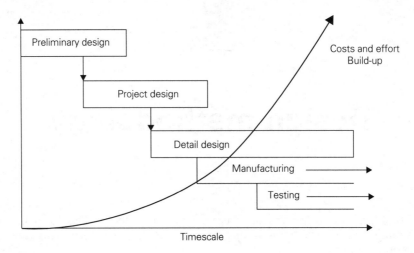

Fig. 1.1 The design process

to determine what effects these may have on the final design layout. All these invest-igations will be done so that the company will be able to take a decision to 'proceed to manufacture'. To do this requires knowledge that the aircraft and its novel features will perform as expected and will be capable of being manufactured in the timescales envisaged. The project design phase ends when either this decision has been taken or when the project is cancelled.

The third phase of the design process (detail design) starts when a decision to build the aircraft has been taken. In this phase, all the details of the aircraft are translated into drawings, manufacturing instructions and supply requests (subcontractor agree-ments and purchase orders). Progressively, throughout this phase, these instructions are released to the manufacturers.

Clearly, as the design progresses from the early stages of preliminary design to the detail and manufacturing phases the number of people working on the project increases rapidly. In a large company only a handful of people (perhaps as few as 20) will be involved at the start of the project but towards the end of the manufacturing phase several thousand people may be employed. With this build-up of effort, the expenditure on the project also escalates as indicated by the curved arrow on Figure 1.1.

Some researchers[1] have demonstrated graphically the interaction between the cost expended on the project, the knowledge acquired about the design and the resulting reduction in the design freedom as the project matures. Figure 1.2 shows a typical distribution. These researchers have argued for a more analytical understanding of the requirement definition phase. They argue that this results in an increased understand-ing of the effects of design requirements on the overall design process. This is shown on Figure 1.2 as process II, compared to the conventional methods, process I. Under-standing these issues will increase design flexibility, albeit at a slight increase in initial expenditure. Such analytical processes are particularly significant in military, multi-role, and international projects. In such case, fixing requirements too firmly and too early, when little is known about the effects of such constraints, may have considerable cost implications.

Much of the early work on the project is involved with the guarantee of technical competence and efficiency of the design. This ensures that late changes to the design

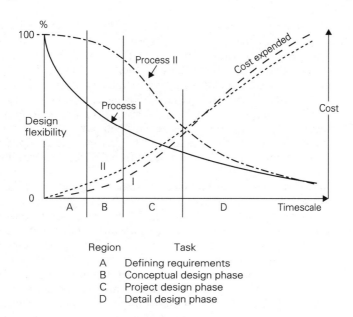

Region	Task
A	Defining requirements
B	Conceptual design phase
C	Project design phase
D	Detail design phase

Fig. 1.2 Design flexibility

layout are avoided or, at best, reduced. Such changes are expensive and may delay the completion of the project. Managers are eager to validate the design to a high degree of confidence during the preliminary and project phases. A natural consequence of this policy is the progressive 'freezing' of the design configuration as the project matures. In the early preliminary design stages any changes can (and are encouraged to) be considered, yet towards the end of the project design phase only minor geometrical and system modifications will be allowed. If the aircraft is not 'good' (well engineered) by this stage then the project and possibly the whole company will be in difficulty. Within the context described above, the preliminary design phase presents a significant undertaking in the success of the project and ultimately of the company.

Design project work, as taught at most universities, concentrates on the preliminary phase of the design process. The project brief, or request for proposal, is often used to define the design problem. Alternatively, the problem may originate as a design topic in a student competition sponsored by industry, a government agency, or a technical society. Or the design project may be proposed locally by a professor or a team of students. Such design project assignments range from highly detailed lists of design objectives and performance requirements to rather vague calls for a 'new and better' replacement for existing aircraft. In some cases student teams may even be asked to develop their own design objectives under the guidance of their design professor.

To better reflect the design atmosphere in an industry environment, design classes at most universities involve teams of students rather than individuals. The use of multi-disciplinary design teams employing students from different engineering disciplines is being encouraged by industry and accreditation agencies.

The preliminary design process presented in this text is appropriate to both the individual and the team design approach although most of the cases presented in later chapters involved teams of design students. While, at first thought, it may appear that the team approach to design will reduce the individual workload, this may not be so.

The interpersonal dynamics of working in a team requires extra effort. However, this greatly enhances the design experience and adds team communications, management and interpersonnel interaction to the technical knowledge gained from the project work.

It is normal in team design projects to have all students conduct individual initial assessments of the design requirements, study comparable aircraft, make initial estimates for the size of their aircraft and produce an initial concept sketch. The full team will then begin its task by examining these individual concepts and assessing their merits as part of their team concept selection process. This will parallel the development of a team management plan and project timeline. At this time, the group will allocate various portions of the conceptual design process to individuals or small groups on the team.

At this point in this chapter, a word needs to be said about the role of the computer in the design process. It is natural that students, whose everyday lives are filled with computer usage for everything from interpersonal communication to the solution of complex engineering problems, should believe that the aircraft design process is one in which they need only to enter the operational requirements into some supercomputer and wait for the final design report to come out of the printer (Figure 1.3).

Indeed, there are many computer software packages available that claim to be 'aircraft design programs' of one sort or another. It is not surprising that students, who have read about new aircraft being 'designed entirely on the computer' in industry, believe that they will be doing the same. They object to wasting time conducting all of the basic analyses and studies recommended in this text, and feel that their time would be much better spent searching for a student version of an all-encompassing aircraft design code. They believe that this must be available from Airbus or Boeing if only they can find the right person or web address.

While both simple aircraft 'design' codes and massive aerospace industry CAD programs do exist and do play important roles, they have not yet replaced the basic processes outlined in this text. Simple software packages which are often available freely at various locations on the Internet, or with many modern aeronautical engineering texts, can be useful in the specialist design tasks if one understands the assumptions and limitations implicit in their analysis. Many of these are simple computer codes based on

Fig. 1.3 Student view of design

Fig. 1.4 The 'real' design process

the elementary relationships used for aircraft performance, aerodynamics, and stability and control calculations. These have often been coupled to many simplifying assumptions for certain categories of aircraft (often home-built general aviation vehicles). The solutions which can be obtained from many such codes can be obtained more quickly, and certainly with a much better understanding of the underlying assumptions, by using directly the well-known relationships on which they are based. In our experience, if students spent half the time they waste searching for a design code (which they expect will provide an instant answer) on thinking and working through the fundamental relationships with which they are already supposedly familiar, they would find themselves much further along in the design process.

The vast and complex design computer programs used in the aerospace industry have not been created to do preliminary work. They are used to streamline the detail design part of the process. Such programs are not designed to take the initial project requirements and produce a final design. They are used to take the preliminary design, which has followed the step-by-step processes outlined in this text, and turn it into the thousands of detailed CAD drawings needed to develop and manufacture the finished vehicle.

It is the task of the aircraft design students to learn the processes which will take them from first principles and concepts, through the conceptual and preliminary design stages, to the point where they can begin to apply detailed design codes (Figure 1.4).

At this point in time, it is impossible to envisage how the early part of the design process will ever be replaced by off-the-shelf computer software that will automatically design novel aircraft concepts. Even if this program were available, it is probably not a substitute for working steadily through the design process to gain a fundamental understanding of the intricacies involved in real aircraft design.

Reference

1 Mavris, D. *et al.*, 'Methodology for examining the simultaneous impact of requirements, vehicle characteristics and technologies on military aircraft design', ICAS 2000, Harrogate UK, August 2000.

2

Preliminary design

Conceptual design is the **organised** application of **innovation** to a **real problem** to produce a **viable** product for the **customer**.

(Anon.)

As previously described, the preliminary design phase starts with the recognition of need. It continues until a satisfactory starting point for the conceptual design phase has been identified. The aircraft layout at the end of the phase is referred to as the 'baseline' configuration. Between these two milestones there are a number of distinctive, and partially sequential, stages to be investigated. These stages are shown in Figure 2.1 and described below:

2.1 Problem definition

For novice aircraft designers the natural tendency when starting a project is to want to design aircraft. This must be resisted because when most problems are originally presented they do not include all the significant aspects surrounding the problem. As a lot of time and effort will be spent on the design of the aircraft, it is important that all the criteria, constraints and other factors are recognised before starting, otherwise a lot of work and effort may be wasted. For this reason, the first part of the conceptual design phase is devoted to a thorough understanding of the problem.

The definition of conceptual design quoted above raises a number of questions that are useful in analysing the problem.

For example (in reverse order to the above definition):

1. Who are the **customers**?
2. How should we assess if the product is **viable**?
3. Can we completely **define the problem** in terms that will be useful to the technical design process?
4. What are the **new/novel features** that we hope to exploit to make our design better than the existing competition and to build in flexibility to cater for future developments?
5. What is the best way to tackle the problem and how will this be **managed**?

These questions are used to gain more insight into the definition of the problem as explained below.

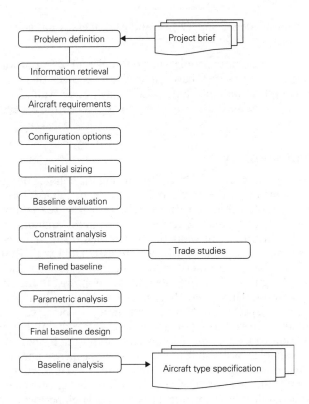

Fig. 2.1 The preliminary design flowchart

2.1.1 The customers

Who are your 'customers'? They are not only the purchasers of the aircraft; many groups of people and organisations will have an interest in the design and their expectations and opinions should be determined. For example, it would be technically straightforward to design a new supersonic airliner to replace Concorde. The operating and technical issues are now well understood. However, the environmental lobby (who want to protect the upper atmosphere from further contamination) and the airport noise abatement groups have such political influence as to render the project unfeasible at this time. For all new designs it is necessary to identify all the influential people and find out their views before starting the project.

Who are the influential people?

- Obviously at the top of the list are the clients (the eventual purchasers of the aircraft).
- Their customers (people who fly and use the aircraft, people who operate and maintain it, etc.).
- Your technical director, departmental head and line supervisor (these have a responsibility for the company and its shareholders to make a reasonable return on investments).
- Your sales team (they know the market and understand customers and they will eventually have to market the aircraft).

- As a student, your academic supervisors and examiners (what is it that they expect to see from the project work).

It is useful to make a list of those people who you think will be important to the project and then find out what views they have. In academic courses the available timescale and facility to accomplish this consultation fully may not be available. In this case, set up your own focus groups and role-play to try to appreciate the expected opinions of various groups.

2.1.2 Aircraft viability

It will be impossible to make rational decisions during the detailed design stages unless you can clearly establish how the product/aircraft is to be judged. Often this is easier said than done, as people will have various views on what are the important criteria (i.e. what you should use to make judgements). The aircraft manufacturing company and particularly its directors will want the best return on their investments (ROI). Unfortunately, so many non-technical issues are associated with ROI that it is too complicated to be used as a design criterion in the initial stages of the project. In the early days aircraft designers solved this dilemma by adopting aircraft mass (weight) as their minimising criteria. They knew that aircraft mass directly affected most of the performance and cost aspects and it had the advantage of being easy to estimate and control. Without any other information about design criteria, minimum mass is still a valid overall criterion to use. As more knowledge about the design and its operating regime becomes available it is possible to use a more appropriate parameter. For example, minimum direct operating cost (DOC) is frequently used for civil transport aircraft. For military aircraft, total life cycle cost (LCC), operational effectiveness (e.g. lethality, survivability, dependability, etc.) are more appropriate. High performance aircraft may be assessed by their operating parameters (e.g. maximum speed, turn rate, sink rate).
  Some time ago A. W. Bishop of British Aerospace observed:

> The message is clear – if everyone can agree beforehand on how to measure the effectiveness of the design, then the designer has a much simpler task. But even if everyone does not agree, the designer should still quantify his own ideas to give himself a sensible guide.

The procedure is therefore relatively simple – ask all those groups and individuals, who you feel are important to the project, how they would assess project effectiveness. Add any weightings you feel are appropriate to these opinions and decide for yourself what criteria should be adopted (or get the project group to decide if you are not working alone). Remember that the criteria must be capable of being quantified and related to the design parameters. Criteria such as 'quality', 'goodness' and 'general effectiveness' are of no use unless such a description can be translated into meaningful design parameters. For example, the effectiveness of a fighter aircraft may be judged by its ability to manoeuvre and launch missiles quicker than an opponent.

2.1.3 Understanding the problem

It is unusual if the full extent of the problem is included in the initial project brief. Often the subtlety of the problem is not made clear because the people who draft the problem are too familiar with the situation and incorrectly assume that the design team will be equally knowledgeable. It is also found that the best solution to a problem is always found by considering the circumstances surrounding the problem in as broad a manner as possible. This procedure has been called 'system engineering'. In this approach, the aircraft is considered only as one component in the total operating environment. The design of the aircraft is affected by the design of all the components in the whole system. For example, a military training aircraft is only one element in the airforce flight/pilot training process. There are many other parts to such a system including other aircraft, flight simulators and ground schools. The training aircraft is also part of the full operational activity of the airforce and cannot be divorced from other aircraft in the service, the maintenance/service sector, the flight operations and other airport management activities. On the other hand, the training aircraft itself can be considered as a total system including airframe, flight control, engine management, weapon on sensor systems, etc. All of these systems will interact to influence the total design of the aircraft.

Such considerations may lead to conflicts in the realisation of the project. For example, although the airforce may have a particular view of the aircraft, the manufacturers may have a different perspective. The airforce will only be focused on their aircraft but the manufacturers will want the aircraft to form part of a family of aircraft, which will have commercial opportunities beyond the supply to the national airforce. Within this context the aircraft may not be directly optimised for a particular role. The best overall configuration for the aircraft will be a compromise between, sometimes competing, requirements. It is the designer's responsibility to consider the layout from all the different viewpoints and to make a choice on the preferred design. He therefore needs to understand all aspects of the overall system in which the aircraft will operate. Some of the most notable past failures in aircraft projects have arisen due to designs initially being specified too narrowly. Conversely, successful designs have been shown to have considerable flexibility in their design philosophy.

Part of the problem definition task is to identify the various constraints to which the aircraft must conform. Such constraints will arise from performance and operational requirements, airworthiness requirements, manufacturing considerations, and limitation on resources. There will also be several non-technical constraints that must be recognised. These may be related to political, social, legal, economic, and commercial issues. However, it is important that the problem is not overconstrained as this may lead to no feasible solution existing. To guard against this it is necessary to be forceful in only accepting constraints that have been fully justified and their consequences understood. For technical constraints (e.g. field performance, climb rate, turn performance, etc.) there will be an opportunity to assess their influences on the design in the later stages (a process referred to as constraint analysis). Non-technical restrictions are more difficult to quantify and therefore must be examined carefully.

In general, the problem definition task can be related to the following questions:

- Has the problem been considered as broadly as possible? (i.e. have you taken a systems approach?)
- Have you identified all the 'real' constraints to the solution of the problem?
- Are all the constraints reasonable?
- Have you thoroughly examined all the non-technical constraints to determine their suitability? (Remember that such constraints will remain unchallenged after this time.)

2.1.4 Innovation

The design and development of a new aircraft is an expensive business. The people who invest in such an enterprise need to be confident that they will get a safe and profitable return on their outlay. The basis for confidence in such projects lies in the introduction and exploitation of new technologies and other innovations. Such developments should give an operational and commercial advantage to the new design to make it competitive against existing and older products. Innovation is therefore an essential element in new aircraft design. The downside of introducing new technology is the increase in commercial risk. The balancing of risk against technical advantage is a fundamental challenge that must be accepted by the designers. Reduction of technological risk will be a high priority within the total design process. Empirical tests and analytical verification of the effects of innovative features are the designer's insurance policy.

Innovation does not just apply to the introduction of new technology. Novel business and commercial arrangements and new operational practices may be used to provide a commercial edge to the new design. Whatever is planned, the designer must be able to identify it early so that he can adjust the baseline design accordingly.

The designers should be able to answer the following questions:

- What are the new technologies and other innovations that will be incorporated into the design?
- How will such features provide an advantage over existing/competing aircraft?
- If the success of the innovation is uncertain, how can the risk to the project be mitigated?

2.1.5   Organising the design process

Gone are the days, if they ever existed, of a project being undertaken by an individual working alone in a back room. Modern design practice is the synthesis of many different skills and expertise. Such combination of talent, as in an orchestra, requires organisation and management to ensure that all players are using the same source of information. The establishment of modern computer assisted design (CAD) software and other information technology (IT) developments allows disparate groups of specialists and managers to be working on the same design data (referred to in industry as 'concurrent engineering').

The organisation of such systems demands careful planning and management. Design-build teams are sometimes created to take control of specific aircraft types within a multi-product company. The design engineer is central to such activity and therefore a key team player. It is essential for him to know the nature of the team structure, the design methods to be adopted, the standards to be used, the facilities to be required, and not least, the work schedules and deadlines to be met. Such considerations are particularly significant in student project work, as there are many other demands on team members. All students will have to personally time-manage all their commitments.

Whether the team is selected by an advising faculty member or is self-selected, students will face numerous challenges during the course of a design project. In most student design projects the organisation of the work is managed by the 'design team'. Good team organisation and an agreed management structure are both essential to success. These issues are discussed in detail in Chapter 11, with particular emphasis to teaming issues in sections 11.2 and 11.3 respectively. When working in a team environment, students are advised to consult these sections before attempting to proceed with the preliminary design.

2.1.6 Summary

The descriptions above indicate that there is a lot of work and effort to be exerted before it is possible to begin the laying-out of the aircraft shape. Each project is different so it is impossible to produce a template to use for the design process. The only common factor is that if you start the design without a full knowledge of the problem then you will, at best, be wasting your time but possibly also making a fool of yourself. Use the comments and questions above to gain a complete understanding of the problem. Write out a full description of the problem in a report to guide you in your subsequent work.

An excellent way for design teams to begin this process of understanding the design problem is the use of the process known as 'brainstorming'. This is discussed in more detail in section 11.2.5. Brainstorming is essentially a process in which all members of a team are able to bring all their ideas about the project to the table with the assurance that their ideas, no matter how far-fetched they may at first appear, are considered by the team. Without such an open mind, a team rarely is able to gain a complete understanding of the problem.

2.2 Information retrieval

Later stages of the design process will benefit from knowledge of existing work published in the area of the project. Searching for such information will involve three tasks:

1. Finding data on existing and competitive aircraft.
2. Finding technical reports and articles relating to the project area and any advanced technologies to be incorporated.
3. Gathering operational experience.

2.2.1 Existing and competitive aircraft

The first of these searches is relatively straightforward to accomplish. There are several books and published surveys of aircraft that can be easily referenced. The first task is to compile a list of all the aircraft that are associated with the operational area. For example, if we are asked to design a new military trainer we would find out what training aircraft are used by the major air forces in the world. This is published in the reviews of military aircraft, in magazines like *Flight International* and *Aviation Week*.

Systematically go through this list, progressively gathering information and data on each aircraft. A spreadsheet is the best way of recording numerical values for common parameters (e.g. wing area, installed thrust, aircraft weights (or masses), etc.). A database is a good way to record other textural data on the aircraft (e.g. when first designed and flown, how many sold and to whom, etc.). The geometrical and technical data can be used to obtain derived parameters (e.g. wing loading, thrust to weight ratio, empty weight fraction, etc.). Such data will be used to assist subsequent technical design work. It is possible, using the graph plotting facilities of modern spreadsheet programs, to plot such parameters for use in the initial sizing of the aircraft. For instance, a graph showing wing loading against thrust loading for all your aircraft will be useful in selecting specimen aircraft to be used in comparison with your design. Such a plot also allows operational differences between different aircraft types to be identified. Categories of various aircraft types can be identified.

2.2.2 Technical reports

As there are so many technical publications available, finding associated technical reports and articles can be time consuming. A good search engine on a computer-based information retrieval system is invaluable in this respect. Unfortunately, such help is not always available but even when it is, the database may not contain recent articles. Older, but still quite relevant, technical articles might also be easily missed when a search relies on computer search and retrieval systems. All computer search systems are very dependent on the user's ability to choose key words which will match those used by whoever catalogued the material in the search system database. Success with such systems is often both difficult and incomplete as the user and the computer try to match an often quite different set of key words to describe a common subject. It becomes somewhat of a game, in which two people with different backgrounds try to describe the same physical object based on their own experiences. Often, a manual search of shelves in a library will product far better results in less time. Manual search is more laborious but such effort is greatly rewarded when appropriate material is found. This makes subsequent design work easier and it provides extra confidence to the final design proposal.

An excellent place to start a technical search is with the reference section at the end of each chapter in your preferred textbooks. Start with a text with which you are already familiar and track down relevant references. Do this either by using computer methods, or in a manual search of the library shelves. This can rapidly lead to an expanding array of background material as subsequent reference lists, in the newly found reports (etc.), are also interrogated.

2.2.3 Operational experience

One of the best sources of information, data and advice comes from the existing area of operation appropriate to your project. People and organisations that are currently involved with your study area are often very willing to share their experiences. However, treat such opinions with due caution as individual responses are sometimes not representative of the overall situation.

The best advice on information retrieved is to collect as much as you can in the time available and to keep your lines of enquiry open so that new information can be considered as it becomes available throughout the design process.

2.3 Aircraft requirements

From the project brief and the first two stages of the design process it is now possible to compile a statement regarding the requirements that the aircraft should meet. Such requirements can be considered under five headings:

1. Market/Mission
2. Airworthiness/other standards
3. Environment/Social

4. Commercial/Manufacturing
5. Systems and equipment

The detail to be considered under each of these headings will naturally vary depending on the type of aircraft. Some general advice for each section is offered below but it will also be necessary to consider specific issues relating to your design.

2.3.1 Market and mission issues

The requirements associated with the mission will generally be included in the original project brief. Such requirements may be in the form of point performance values (e.g. field length, turn rates, etc.), as a description of the mission profile(s), or as operational issues (e.g. payload, equipment to be carried, offensive threats, etc.). The market analysis that was undertaken in the problem definition phase might have produced requirements that are associated with commonality of equipment or engines, aircraft stretch capability, multi-tasking, costs and timescales.

2.3.2 Airworthiness and other standards

For all aircraft designs, it is essential to know the airworthiness regulations that are appropriate. Each country applies its own regulations for the control of the design, manufacture, maintenance and operation of aircraft. This is done to safeguard its population from aircraft accidents. Many of these national regulations are similar to the European Joint Airworthiness Authority (JAA) and US-Federal Aviation Administration (FAA) rules.[1,2] Each of these regulations contains specific operational requirements that must be adhered to if the aircraft is to be accepted by the technical authority (ultimately the national government from which the aircraft will operate). Airworthiness regulations always contain conditions that affect the design of the aircraft (e.g. for civil aircraft the minimum second segment climb gradient at take-off with one engine failed). Although airworthiness documents are not easy to read because they are legalistic in form, it is important that the design team understands all the implications relating to their design. Separate regulations apply to military and civil aircraft types and to different classes of aircraft (e.g. very light aircraft, gliders, heavy aircraft, etc.). It is also important to know what operational requirements apply to the aircraft (e.g. minimum number of flight crew, maintenance, servicing, reliability, etc.). The purchasers of the aircraft may also insist that particular performance guarantees are included in the sales contract (e.g. availability, timescale, fuel use, etc.). Obviously all the legal requirements are mandatory and must be met by the aircraft design. The design team must therefore be fully conversant with all such conditions.

2.3.3 Environmental and social issues

Social implications on the design and operation of the aircraft arise mainly from the control of noise and emissions. For civil aircraft such regulations are vested in separate operational regulations.[3] For light aircraft, some airfields have locally applied operation restrictions to avoid noise complaints from adjacent communities. Such issues are becoming increasingly significant to aircraft design.

2.3.4 Commercial and manufacturing considerations

Political issues may affect the way in which the aircraft is to be manufactured. Large aircraft projects will involve a consortium of companies and governments (e.g. Airbus). This will directly dictate the location of design and manufacturing activity. Such influence may also extend to the supply of specific systems, engines and components to be used on the aircraft. If such restrictions are to be applied, the design team should be aware of them as early as possible in the design process.

2.3.5 Systems and equipment requirements

Aircraft manufacture is no longer just concerned with the supply of a suitable airframe. All aircraft/engine and other operational systems have a significant influence in the overall design philosophy. Today many aircraft are not technically viable without their associated flying and control systems. Where such integration is to be adopted the design team must include this in the aircraft requirements. This aspect is particularly significant for the design of military aircraft that rely on weapon and other sensor systems to function effectively (e.g. stealth). Regulations for military aircraft usually fully describe the systems that the airframe must support.

2.4 Configuration options

With a fully described set of regulations, knowledge of existing aircraft data and a complete understanding of the problem, it is now possible to start the technical design tasks. Many project designers regard this stage as the best part of all the design processes. The question to be answered is simply this: Starting with a completely clear mind, what configurational options can you suggest that may solve the problem? For example, a two-seat light touring aircraft could be: side-by-side or tandem seating, high or low wing, tractor or pusher engine, canard or tail stabilised, nose or tail wheeled, conventional or novel planform (e.g. box wing, joined wing, delta, tandem), etc.

The following stage of the design process will sort through the 'weird and wonderful' configurations to eliminate the unfeasible and uncompetitive layouts. At this point in the layout process a quantity of ideas is needed and a judgement on their suitability will be left until later. With this in mind it is unnecessary to elaborate on an option past the point at which its characteristics can be appreciated. A good starting point for this work is to list the configurations that past and existing aircraft of this type have adopted. A brief synopsis of the strength and weaknesses of each option may be written so that improvements to the designs can be identified. Such analysis will also help in the concept-filtering phase that will follow.

In the conceptual design stage, designers have two options available for their choice of engines. Namely a 'fixed' (i.e. a specified/existing or manufacturers' projected engine), or an 'open' design (in which the engine parameters are not known). In most cases, and definitely at later stages in the design process, the size and type of engine will have been determined. The aircraft manufacturer will prefer that more than one engine supplier is available for his project. In this way he can be more competitive on price and supply deadlines. For design studies in which the engine choice is open, it is possible to adopt what is known as a 'rubber' engine. Obviously, such engines do not exist in practice. The type and initial size of the rubber engine can be based on existing or

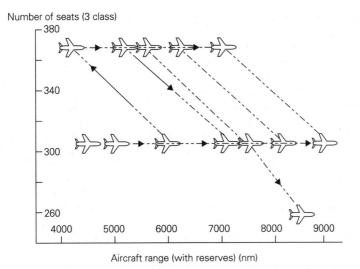

Number of seats (3 class)

Aircraft range (with reserves) (nm)

Fig. 2.2 Aircraft development programme (Boeing 777)

engine manufacturers' projected engine designs. Using a rubber engine, the aircraft designer has the opportunity to scale the engine to exactly match the optimum size for his airframe. Such optimisations enable the designer to identify the best combination of airframe and engine parameters. If an engine of the preferred size is not available, in the timescale of the project, the designer will need to reconfigure the airframe to match an available engine. Rubber engine studies show the best combination of airframe and engine parameters for a design specification and can be used to assess the penalties of selecting an available engine.

Aircraft and engine configuration and size is often compromised at the initial design stage to allow for aircraft growth (either by accidental weight growth or by intent (aircraft stretch)). Such issues must be kept in mind when considering the various options. Most aircraft projects start with a single operational purpose but over a period of time develop into a family of aircraft. Figure 2.2 shows the development originally envisaged by Boeing for their B777 airliner family. For military aircraft such developments are referred to as multi-role (e.g. trainer, ground support, etc.). It is important that designers appreciate future developments at an early design stage and allow for such flexibility, if desired.

2.5 Initial baseline sizing

At the start of this stage you will have a lot of design options available together with a full and detailed knowledge of the problem. It would be impossible and wasteful to start designing all of these options so the first task is to systematically reduce the number. First, all the obviously unfeasible and crazy ideas should be discarded but be careful that potentially good ideas are not thrown out with the rubbish. Statements and comments in the aircraft regulations and the problem definition reports will help to filter out uneconomic, weak and ineffective options. The object should be to reduce the list to a single preferred option but sometimes this is not possible and you may need to take another one or two into the next design stage. Obviously, the workload will be increased in the next stages if more options are continued. Eventually it will be necessary to choose a single aircraft configuration. This will mean that all the work on the rejected options may be wasted.

This can be a very difficult part of the design process for a student design team. At this point, it is common for each member of the team to have invested a lot of time and energy into his or her own proposed design concept. It is often difficult to get team members to release their emotional ties to their own proposals and begin to embrace those of others or even to find a viable compromise. To get through this stage of the process both good team management and an effective means of comparing and evaluating all proposed concepts are required. Some of these difficulties are discussed in Chapter 11 (section 11.2). All proposed solutions to the design objective need to be given a fair and impartial assessment during the selection of the final concept. Obviously, a compromise solution which draws upon key elements of every team member's contributions will result in a happier set of team players. On the other hand, it is important that the selected concept embodies the best design elements that the team has developed. These must be chosen for the benefit of the overall design and not just to keep each member of the team happy.

Once decisions have been made on the configuration(s) to be further considered it is necessary to size the aircraft. A three-view general arrangement scale drawing for each aircraft configuration will be required. Little detail will be known at this stage about the aircraft parameters (wing size, engine thrust, and aircraft weight) so some crude estimates have to be made. This is where data from previous/existing aircraft designs will be useful. Although the new design will be different from previous aircraft, such inconsistencies can be ignored at this stage. Use representative values from one or a small group of the specimen aircraft for wing loading, thrust loading and aircraft take-off weight. It is also possible to use a representative wing shape and associated tail sizes.

The design method that follows is an iterative process that usually converges on a feasible configuration quickly. The initial general arrangement drawing, produced to match existing aircraft parameters, provides the starting point for this process. Even though your design is relatively crude at this stage it is important to draw it to scale making approximations for the relative longitudinal position of the wing and fuselage and the location of tail surfaces and landing gear.

Most aircraft layouts start with the drawing of the fuselage. For many designs the geometry of the fuselage can be easily proportioned as it houses the payload and cockpit/flight deck. These parameters are normally specified in the project brief. They can be sized using design data from other aircraft. The non-fuselage components (e.g. wing, tail, engines and landing gear) are added as appropriate. With a reasonable first guess at the aircraft configuration, the aircraft can be sized by making an initial estimate of the aircraft mass. Once this is completed it is possible to more accurately define the aircraft shape by using the predicted mass to fix the wing area and engine size.

2.5.1   Initial mass (weight) estimation

The first step is to make a more accurate prediction of the aircraft maximum (take-off) mass/weight. (Note: if SI units are used for all calculations it is appropriate to consider aircraft mass (kilograms) in place of aircraft weight (Newtons).)

Aircraft design textbooks[4,5,6] show that the aircraft take-off mass can be found from:

$$M_{\text{TO}} = \frac{M_{\text{UL}}}{1 - (M_{\text{E}}/M_{\text{TO}}) - (M_{\text{F}}/M_{\text{TO}})}$$

where M_{TO} = maximum take-off mass
M_{UL}^* = mass of useful load (i.e. payload, crew and operational items)
M_{E}^* = empty mass
M_{F} = fuel mass

(*When using the above equation it is important not to double account for mass components. If aircraft operational mass is used for M_{E}, the crew and operational items in M_{UL} would not be included. One of the main difficulties in the analysis at this stage is the variability of definitions used for mass components in published data on existing aircraft. Some manufacturers will regard the crew as part of the useful load but others will include none or just the minimum flight crew in their definition of empty/operational mass. Such difficulties will be only transitional in the development of your design, as the next stage requires a more detailed breakdown of the mass items.)

The three unknowns on the right-hand side of the equation can be considered separately:

(a) *Useful load*
The mass components that contribute to M_{UL} are usually specified in the project brief and aircraft requirement reports/statements.
(b) *Empty mass ratio*
The aircraft empty mass ratio ($M_{\text{E}}/M_{\text{TO}}$) will vary for different types of aircraft and for different operational profiles. All that can be done to predict this value at the initial sizing stage is to assume a value that is typical of the aircraft and type of operation under consideration. The data from existing/competitor aircraft collected earlier is a good source for making this prediction. Figure 2.3 shows how the data might be viewed. Alternatively, aircraft design textbooks often quote representative values for the ratio for various aircraft types.

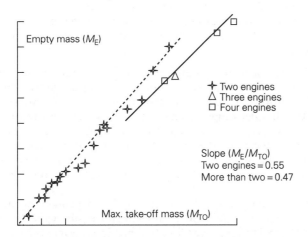

Fig. 2.3 Analysis of existing aircraft data (example)

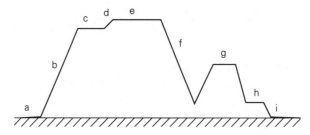

a – take-off, b – climb, c – cruise, d – step climb, e – continued cruise,
f – descent, g – diversion, h – hold, i – landing at alternate airstrip.

Fig. 2.4 Mission profile (civil aircraft example)

(c) *Fuel fraction*
For most aircraft the fuel fraction (M_F/M_{TO}) can be crudely estimated from the modified Brequet range equation:

$$\frac{M_F}{M_{TO}} = (\text{SFC}) \cdot \frac{1}{(L/D)} \cdot (\text{time})$$

where   (SFC) = engine specific fuel consumption (kg/N/hr)
   (L/D) = aircraft lift to drag ratio
   (time) = hours at the above conditions

The mission profile will have been specified in the project brief. Figure 2.4 illustrates a hypothetical profile for a civil aircraft.

This shows how the mission profile consists of several different segments (climb, cruise, etc.). The fuel fraction for each segment must be determined and then summed. Reserve fuel is added to account for parts of the mission not calculated. For example:

(a) for the fuel used in the warm-up and taxi manoeuvres,
(b) for the effects on fuel use of non-standard atmospheric conditions (e.g. winds),
(c) for the possibility of having to divert and hold at alternative airfield when landing.

The last item above is particularly significant for civil operations. In such applications designers sometimes convert the actual range flown to an equivalent still air range (ESAR) using a multiplying factor that accounts for all of the extra (to cruise) fuel.

When using the Brequet range equation it must be remembered that both engine (SFC) and aircraft (L/D) will be different for different flight conditions. These variations arise because the aircraft speed, altitude, weight and engine setting will be different for each flight segment. Typical values for (SFC) can be found in engine data books[7] or from aircraft and engine textbooks[4,8] for the type of engine to be used.

The aircraft lift to drag ratio (L/D) will vary and be dependent on aircraft geometry (particularly wing angle of attack). Such values are not easily available for the aircraft in the initial design stage. However, we know that previous designers have tried to achieve a high value in the principal flight phase (e.g. cruise). We can use the fact that in cruise 'lift equals weight' and 'drag equals thrust'. We can therefore transpose (L/D) into (W/T). Both aircraft weight and engine thrust (at cruise) could be estimated from our specimen aircraft data. This value will be close to the maximum (L/D) and relate only to the cruise condition. At flight conditions away from this point the value of (L/D) will reduce. It must be stressed that the engine thrust level in cruise will be substantially less than the take-off condition due to reduced engine thrust setting and the effect of altitude and speed. This reduction in thrust is referred to as 'lapse rate'. Engine specific fuel consumption will also change with height and speed. Values for (L/D) vary over a wide range depending on the aircraft type and configuration. Typical values range from 30 to 50 for gliders, 15 to 20 for transport/civil aircraft, 12 to 15 for smaller aircraft with reasonable aspect ratio and less than 10 for military aircraft with short span delta wing planforms. Aircraft design textbooks are a source of information on aircraft (L/D) if the values cannot be estimated from the engine cruise conditions and aircraft weight.

(Time) is usually easy to specify as each of the mission segments is set out in the project brief (mission profiles). Alternatively, it can be found by dividing the distance flown in a segment by the average speed in that segment.

2.5.2 Initial layout drawing

Obviously, all the above calculations require a lot of 'guesstimation' but at least at the end we will have a better estimate of the aircraft maximum take-off mass than previously. This value can then be used in conjunction with the previously assumed values for wing and thrust loading to refine the size of the wing and engine(s). The original concept drawing can be modified to match these changes. This drawing becomes the initial 'baseline' aircraft configuration.

2.6 Baseline evaluation

The methods used up to this point to produce the baseline aircraft configuration have been based mainly on data from existing aircraft and engines. In the next stage of the design process it is necessary to conduct a more in-depth and aircraft focused analysis. This will start with a detailed estimation of aircraft mass. This is followed by detailed aerodynamic and propulsion estimates. With aircraft mass, aerodynamic and engine parameters better defined it is then possible to conduct more accurate performance estimations. The baseline evaluation stage ends with a report that defines a modified baseline layout to match the new data. A brief description of each analysis conducted in this evaluation stage is given below.

2.6.1 Mass statement

Since the geometrical shape of each part of the aircraft is now specified, it is possible to make initial estimates for the mass of each component. This may be done by using empirical equations, as quoted in many design textbooks, or simply by assuming a value for the component as a proportion of the aircraft maximum or empty mass. Such ratios are also to be found in design textbooks or could match values for similar aircraft types, if known. The list below is typical of the detail that can be achieved.

Generating a mass statement like this one is the first task in the baseline evaluation phase.

Wing (M_W)
Tail (M_T)
Body (M_B)
Nacelle (M_N)
Landing gear (M_U)
Control surfaces (M_{CS})
$\sum$ **total aircraft structure** (M_{ST})

Engine basic (dry)
Engine systems
Induction (intakes)
Nozzle (exhaust)
Installation
$\sum$ **total propulsion system** (M_P)

Aircraft systems and equipment (M_{SE})
$\sum$ **aircraft empty mass** $= M_E = M_{ST} + M_P + M_{SE}$

Operational items (M_{OP})
$\sum$ **aircraft operational empty mass** $(M_{OE}) = M_E + M_{OP}$

Crew* (M_C)
Payload (M_{PL})
Fuel (M_F)
$\sum \sum$ **aircraft take-off mass** $(M_{TO}) = M_{OE} + M_C + M_{PL} + M_F$

(*For some military aircraft mass statements, the crew are considered to form part of the operational items and their mass is added to aircraft OEM.)

The main structural items in the list above (e.g. wing, fuselage, engine, etc.) can be estimated using statistically determined formulae which can be found in most aircraft design textbooks. (Note: if you are working in SI units be careful to convert mass values from historical reports, journals, and current US textbooks to kilograms (1 kg = 2.205 lb).) Many of these mass items are dependent on M_{TO}, therefore estimations involve an iterative process that starts with the assumed value of M_{TO}, as estimated in the initial sizing stage. Spreadsheet 'solver' methods will be useful when performing this analysis.

At the early design stages, the estimation of mass for some of the less significant (and smaller) components may be too time consuming to calculate in detail (e.g. tail, landing gear, flight controls, engine systems and components, etc.). To speed up the evaluation process, these can be estimated by assuming typical percentage values of M_{TO}, as mentioned above. Such values can be found from existing aircraft mass breakdowns, if available, or from aircraft design textbooks.

At the final stages of the conceptual phase an aircraft mass will be selected which is slightly higher than the estimated value of M_{TO}. This higher weight is known as the 'aircraft design mass'. All the structural and system components will be evaluated using the value for the aircraft design weight as this provides an insurance against weight growth in subsequent stages of the design process. For aircraft performance estimation, the mass to be used may be either the M_{TO} value shown above or something less (e.g. military aircraft manoeuvring calculations are frequently associated with the aircraft operational empty mass plus defensive weapons and half fuel load only).

2.6.2 Aircraft balance

With the mass of each component estimated and with a scale layout drawing of the aircraft it is possible, using educated guesses, to position the centre of mass for each component. This will allow the centre of gravity of the aircraft in various load conditions (i.e. different combinations of fuel or payload) to be determined. It is common practice to estimate the extreme positions (forward and aft) so that the trim loads on the control surfaces (tail/canard) and the reaction loads on the undercarriage wheels can be assessed.

Up to this point in the design process, the longitudinal position of the wing along the fuselage has been guessed. As part of the determination of the aircraft centre(s) of gravity, it is possible to check this position and, by iteration, to reposition it to suit the aircraft lift and inertia force (i.e. mass × acceleration) vectors. This process is referred to as 'aircraft balancing'. As moving the wing will affect the position of the aircraft centre of gravity and the wing lift aerodynamic centre from the datum, several iterations may be required. There are several methods that can be used to reduce the complications inherent in this iteration. The simplest method sets the position of the aircraft operational empty mass relative to a chosen point (per cent chord aft of the wing leading edge) on the wing mean aerodynamic chord line. To start the process the aircraft operational empty mass components are divided into two separate groups:

(a) Wing mass group (M_{WG}) (and associated components) – this will include the wing structure, fuel system (if the fuel is housed in the wing), main landing gear unit (even if it is structurally attached to the fuselage), wing mounted engines and all wing attached systems.
(b) Fuselage mass group (M_{FG}) (and associated components) – this group will include the fuselage structure, equipment, cockpit and cabin furnishings and systems, operational items, airframe services, crew, tail structure, nose landing gear and fuselage mounted engines and systems.

Note: if the position of wing mounted engines is linked to internal fuselage layout requirements (e.g. propeller plane be in line with non-passenger areas) then these masses should be transferred to the fuselage group.

Obviously all the aircraft components relating to the aircraft operational empty mass must be included in either of the above groups (i.e. $M_{OE} = M_{WG} + M_{FG}$). It is important to check that none of the component masses has been omitted before starting the balancing process.

It is possible to determine the centres of mass separately for each of the two mass groups above. The distance of the wing group centre of mass from the leading edge of the wing mean aerodynamic chord (MAC) is defined as X_{WG} (see Figure 2.5a).

The next stage is to select a suitable location for the centre of gravity of the aircraft operational empty weight, on the wing mean aerodynamic chord. If the centre of gravity is too far aft or forward then the balancing loads from the tail (or canard) will be high. This will result in a requirement for larger tail surfaces and thereby increased aircraft mass and trim drag. For most conventional aircraft configurations, a centre of gravity position coincident with the 25 per cent MAC position behind the wing leading edge is considered a good starting position. If it is known that loading the aircraft from the operational empty mass will progressively move the aircraft centre of gravity forward, then a 35 per cent MAC position would be a better starting point. Such cases arise on civil airliners with rear fuselage mounted engines. Conversely, a 20 per cent MAC would be chosen for designs with mainly aft centre of gravity movements. For aircraft flying at supersonic speed the centre of lift will be at about the 50 per cent MAC position. This must be carefully allowed for when selecting the operational mass position. The location of the chosen operational empty mass location with respect to the leading edge of the wing mean aerodynamic chord is defined as X_{OE}.

It is possible to take moments of the aircraft masses shown in Figure 2.5a. By rearranging the moment equation, the position of the fuselage group mass relative to line XX can be calculated. The resulting equation is shown below:

$$X_{FG} = X_{OE} + (X_{OE} - X_{WG})(M_{WG}/M_{FG})$$

Overlays of the separate wing and fuselage layouts provide the best method of fixing the wing relative to fuselage. On a plan view of the wing, determine the position of the wing MAC and its intersection with the wing leading edge (line XX). Also, on this drawing show the position of the wing group centre of mass, see Figure 2.5b.

Measure the distance X_{WG} from this drawing and use it in the formula above together with the selected value of X_{OE} and the calculated wing and fuselage group masses (M_{WG} and M_{FG}), to evaluate the distance X_{FG}. On a plan view of the fuselage, determine the position of the fuselage group mass centre (using any convenient datum plane) then draw a line XX at a distance X_{FG} forward of this position, as shown in Figure 2.5c.

Overlay the wing and fuselage diagram lines XX. This is the correct location of wing and fuselage to give the aircraft operational centre of gravity at the previously selected position on the wing MAC. It is not unusual to discover by this process that the originally assumed position of the wing relative to the fuselage, on the aircraft layout drawing, is incorrect and must be changed.

With the aircraft balanced, it is now possible to determine the range of aircraft centre of gravity movement about the operational empty position and to assess the effect of this on the tail sizing. Obviously, it is preferable to design for small movements of the aircraft centre of gravity to ensure the control forces are small. To do this, the disposable items of mass (fuel and payload) should be centred close to the aircraft operational empty centre of gravity position as practical.

At this stage in the development of the aircraft geometry it is possible to position the undercarriage units. The process involves geometric and load calculations associated with the aircraft mass and centre of gravity range. The main units must allow for adequate rotation of the aircraft on take-off and in the landing attitude. When the aircraft is in the maximum tail down attitude, the aircraft rearmost centre of gravity position must be forward of the wheel reactions. This will ensure the aircraft does not stay in this position. The loads on the main and nose units can be determined by simple mechanics. Make sure that the nose wheel load is not excessive as this will require a large tailplane force to lift the nose on take-off. On the other hand, if the load is too small on the nose wheel it will not generate an effective steering force. The forces determined for each unit will dictate the tyre size commensurate with the allowable tyre pressure and runway point-load capability. Several aircraft design textbooks include undercarriage layout guidelines.

2.6.3 Aerodynamic analysis

At the same time as the mass and balance estimation is made, or sequentially after if you are working alone, it is possible to make the initial estimations for the baseline aircraft aerodynamic characteristics (drag and lift). The aircraft drag estimation, like mass,

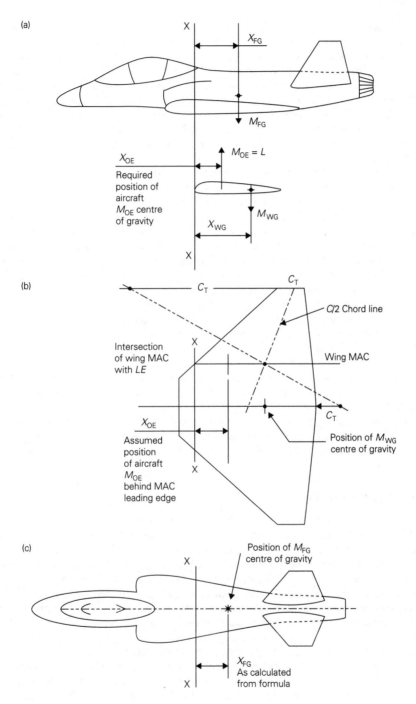

Fig. 2.5 Aircraft balance methodology (diagrams a, b and c)

can be broken down into individual components (e.g. wing, body, tail, etc.) and then summed. Allowance for interference effects between components must also be added to the value. Textbooks on aerodynamics and aircraft design provide several different methods for performing such calculations. The drag of the aircraft will eventually be used in the performance estimations, therefore it will be necessary to determine values at different flight conditions (e.g. take-off, climb, cruise, etc.). These calculations will involve the aircraft in different configurations with regard to the deployment of landing gear and flap extensions. The aircraft will also be at different speeds and altitudes for each condition. This affects the Reynolds number used in the drag calculations and other parameters.

You may find it useful to do the drag calculations during this stage in terms of 'drag area' rather than in the coefficient form. This effectively 'dimensionalises' the drag of each component, and ultimately the whole aircraft, in terms of the area of a flat plate that would have an equivalent drag to the component. As might be expected, this method is sometimes referred to as 'the equivalent flat plate area'. Drag area gives a better visualisation of the effectiveness (or otherwise) of the various components and their contributions to the total aircraft drag. It also provides an indication of the influence of the geometrical parameters of the component to its drag. In the early design stages the selection of aircraft gross wing area (i.e. reference area) is very tentative, as it has not been checked against the performance requirements. Using it as a reference area in drag coefficient form may be regarded as premature. On the other hand, in the determination of aircraft lift many of the established methods are based on the manipulation of lift coefficients. It is therefore impossible to avoid the potential wing area confusions for the estimation of lift.

As with drag estimation, it is necessary to determine lift coefficients at different operating conditions (i.e. various flap deflection angles – e.g. take-off and landing settings). Use design data from existing aircraft to initially set values for flap deflections and wing planform (flap span ratio) geometry. At later stages, when more detailed aerodynamic analysis of wing and other aircraft components has been completed, it will be necessary to select specific flap angles to suit your particular aircraft operational requirements.

2.6.4   Engine data

Before a detailed performance estimation can be made it is essential to have representative engine performance charts (or data) available. From the problem definition phase either the engine or the engine type may be known. The initial sizing work will have provided an estimate of the engine take-off thrust. To undertake aircraft performance calculations it is necessary to know what thrust (and SFC) the engine will give at thrust settings other than take-off (e.g. at continuous climb, cruise, etc.). It will also be necessary to know the effect of aircraft altitude and speed on the engine parameters. For some military aircraft it is also necessary to understand what effect the use of reheat (afterburning) will have on engine performance. For existing engines, data may be available from the engine manufacturers but sometimes it is difficult to obtain this data. Engine manufacturers are reticent to release technical detail for commercial reasons. It is also often impossible for them to provide the data in the form that students can use as their engine performance is held in extensive databases that require flight data as input. For many new aircraft projects a new engine is required, therefore manufacturers' data is not available. In these cases predictions based on similar engine types have to be made. Aircraft design and engine textbooks[4,8] often contain data on which to make such predictions.

2.6.5 Aircraft performance

With aircraft mass, drag, lift and engine characteristics known it is a relatively straight-forward process to make initial estimates of aircraft performance. This is done for each flight segment separately (climb, cruise, dash, loiter, descent, combat, etc.). The field performance (take-off and landing) is also required. Many textbooks are available on aircraft performance estimation.[4,5,6,9] These can be used, with appropriate simplifying assumptions, to estimate performance values.

The results from the performance estimates are compared to the aircraft require-ments. It is now that the original estimates for wing area and thrust are re-evaluated. Changes in these values are often necessary to obtain aircraft performance to meet the requirements. It is essential that new values for wing area and engine thrust are selected that allow such compliance but not too much in excess as this will make the design inefficient. As aircraft mass, drag, lift and engine characteristics are directly affected by changes in wing and engine size it will be necessary to repeat all the pre-vious initial estimates for the baseline aircraft. This is a laborious task but the use of modern spreadsheet methods does assist in such iterative processes.

2.6.6 Initial technical report

At the end of the baseline evaluation stage you should have a detailed knowledge of an aircraft configuration that will meet the original problem specifications. However, this configuration is unlikely to be 'optimum'.

It is now possible to produce a report which contains a scale drawing of the modified baseline configuration, a detailed mass breakdown, drag and lift assessments for each operational configuration, and engine and aircraft performance predictions for all flight segments. Some examples of these calculations are shown in the project studies that follow (Chapters 4 to 10). Subsequent stages in the conceptual design process are aimed at improving the aircraft configuration to make a more efficient design and to address non-technical factors.

2.7 Refining the initial layout

At this point in the design process we have an aircraft layout that has been based mainly on crude estimates taken from previous aircraft designs. We also have not assessed the overall problem definition with regard to any of the aircraft design parameters. It is now time to improve this situation and provide more confidence in the aircraft layout. From the previous stage, we have enough geometric and configurational details to make detailed estimates. These include the mass of aircraft components (giving a mass statement), aerodynamic coefficient assessments (lift and drag) and some knowledge of engine performance (thrust and fuel flow) at various operating conditions. With this data, it is possible to undertake a more detailed analysis of the aircraft design. The following studies will allow us to progressively adjust the aircraft geometrical and layout features to better match the problem constraints and to improve the aircraft effectiveness as judged by the overall assessment criteria. These studies will also allow us to test the sensitivity of the problem constraints against the aircraft configuration. Two design processes are used:

- Constraint analysis
- Aircraft trade studies

Although these methods are separately described in the following sections, they are often used in conjunction, and repeatedly, to refine the baseline layout and to assess the significance of the problem constraints.

2.7.1 Constraint analysis

From our earlier work on understanding the problem (section 2.1.3) we identified several constraints that must be satisfied by the aircraft design. Many of the specified constraints are related to the performance of the aircraft (for example, minimum speeds or climb gradients). The requirement will be written in the form:

> At a specified aircraft flight condition and configuration, the aircraft must demonstrate a performance no less than a specified value (e.g. climb gradient better than 0.024 with an engine failed on take-off with aircraft at max. weight).

The specified values are known as constraints. For most aircraft projects there are several constraints related to field performance (e.g. take-off, climb, balanced field length, stall speed, landing characteristics) and several linked to other mission segments (e.g. minimum cruise speed, climb rates/times, turn rates, specific excess power, loiter speed, dash speed, etc.). Each aircraft project will have a different set of such constraints. The design process known as constraint analysis is used to assess the relative significance of these constraints and their influence on the aircraft configuration.

It is common to display the constraint boundaries on a graph of aircraft thrust to weight ratio (T/W) against wing loading (W/S) as illustrated in Figure 2.6.

The area of the graph is called the design space. Constraint lines can be plotted to show the relationship of the two aircraft parameters at the constraint value. These lines represent boundaries between the unacceptable and feasible regions of the design space. It is usual to attach hatching to the line on the unacceptable side of the constraint. In this way, the design space is divided into the two regions by each constraint. A combination of constraints graphically identifies the feasible design space (shown shaded on Figure 2.6). Any combination of (T/W) and (W/S) values is possible within this region. The selection of a particular pair of values will dictate the size and nature of the aircraft layout. For example, given a reasonable estimate of the aircraft maximum weight (take-off mass), the engine thrust will be fixed by the T/W value and the gross wing area fixed by the W/S value.

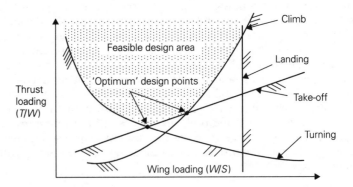

Fig. 2.6 Generalised constraint diagram

The original guesstimate of wing and thrust loading made in the initial sizing process can be checked to see if it lies in the feasible region. If not, the design point can be moved into the acceptable design space. The position of the design point in the feasible region, relative to the constraint boundaries, indicates the efficiency of the aircraft. It is desirable that the selected point lies close to the minimum value of T/W to reduce engine size and close to the maximum value of W/S to reduce wing size. Often such a point lies on the intersection of the constraint lines.

Although the constraint analysis will give guidance on the selection of the design point as described above, it is principally used to show the significance of the specified constraints to the aircraft configuration. For example, it may show that a constraint is too demanding (over-riding all other constraints) and significantly reduces the allowable design space. This will make the design excessively large and potentially not viable. In such cases, it is important to inform those who originally prescribed the constraint of the consequences. It is often possible to get a relaxation of the offending constraint to make the design more effective. The diagram will clearly show which of the constraints are inconsequential (i.e. falling well inside the unacceptable design space) and which are 'active' (i.e. forming the boundary of the feasible region). As described above, the 'optimum' design point will lie on the intersection of constraint lines, therefore the best set of constraints is that in which several intersect at roughly the same design point. This is sometimes referred to as a 'well-balanced' design.

As mentioned in the introduction to this section, the performance constraints are not the only ones to be imposed on the design. For example, for a naval aircraft operating from a carrier there will be geometrical limits that are set by the size and shape of the deck elevators. Such constraint may make it impossible to select the design point shown on the performance constraint diagram. However, the constraint analysis will provide a means of assessing the design penalty that has to be accepted for imposing the non-performance constraint.

The fundamental theory on which the constraint analysis is based involves the manipulation of the energy state (kinetic and potential) of the aircraft and its relationship to the available excess power, P_s:

$$P_s = [(T - D)/W]V \quad \text{(where weight} = W = Mg)$$

The excess power can be used to either climb (dh/dt) or accelerate ($[V/g] \cdot [dV/dt]$) the aircraft separately, or in combination within the limit of the excess power:

$$P_s = [(T/W) - (D/W)]V = dh/dt + ([V/g] \cdot [dV/dt])$$

This equation can be rearranged to provide a relationship between aircraft thrust and wing loadings (T/W, W/S). This is done by expanding the drag term ($D = qSC_D$) and the drag coefficient ($C_D = C_{DO} + k_1 C_L^2$) and writing C_L in terms of lift ($C_L = nW/qS$):

$$(T/W) = [qC_{DO}/(W/S) + k_1 \cdot n^2 \cdot (W/S)/q] + (1/V) \cdot dh/dt + ([1/g] \cdot [dV/dt])$$

where $q = 0.5\rho V^2$ and $n = L/W$

The thrust to weight ratio (T/W) can be normalised, for various different operating conditions, to the equivalent sea-level static thrust and maximum mass condition. This is done by the application of the following factors: thrust lapse rate $= \alpha = T/T_{SSL}$ and aircraft weight fraction $= \beta = W/W_{TO}$, where, T and W are the thrust and weight at the operating condition, T_{SSL} is the static sea-level total engine thrust and W_{TO} is the aircraft take-off weight. Rearranging the (T/W) equation above and introducing the parameters α and β we get:

$$(T_{SSL}/W_{TO}) = (\beta/\alpha)[\{(q/\beta)(C_{DO}/(W_{TO}/S)\} + \{[k_1 \cdot n^2 \cdot (W_{TO}/S)]/(q/\beta)\}]$$
$$+ (1/V) \cdot dh/dt + ([1/g] \cdot [dV/dt])$$

Several textbooks explain the constraint analysis process in detail.[6,8] Aircraft and engine characteristics in the form of drag coefficients (C_{DO} and k_1), weight (W), speed (V), altitude/density (ρ), normal acceleration (n) are set as constants for a particular flight/operation segment. Obviously, this may be a somewhat crude assumption as many of the constants are affected by the aircraft and engine size, and operating characteristics but it is possible to use iteration to reduce the errors once the critical constraints have been identified.

Each of the aircraft constraints can be analysed with the above equation (e.g. the constant speed climb rate of 2.4 per cent mentioned above would set the last term to zero and the penultimate one to 0.024). The resulting expression can be solved for (T/W) using increasing values of (W/S) as the variable input. The results can then be plotted to give a boundary to the feasible design region.

A graph of the form shown in Figure 2.7 can be drawn showing the extent of the specified constraints on a (W_{TO}/S) versus (T_{SSL}/W_{TO}) graph. The area above and between the constraint lines defines the feasible design space. Any combination of (W_{TO}/S) and (T_{SSL}/W_{TO}) in this area is allowable but the best design will lie towards the bottom and right-hand side of the diagram.

As mentioned above, constraint analysis is a crude analytical tool as it cannot easily take into account changes in the basic geometry of the aircraft (e.g. aspect ratio). It does, however, serve the principal purpose of identifying any constraint that is unduly

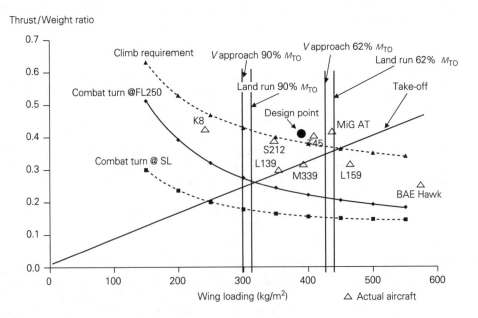

Fig. 2.7 Constraint diagram (example)

or harshly influencing the design configuration. For example, if the vertical line in Figure 2.7 (the approach speed constraint) had been positioned too far to the left of the diagram it would have eliminated much of the available design space. It would also have made the two lower constraints ineffective.

A secondary purpose for the constraint diagram is to use it in conjunction with the wing and thrust loading data from your specimen aircraft list as shown in Figure 2.7. Note: plotting existing data on the diagram drawn for your aircraft configuration is somewhat misleading as other designs are constrained differently but the diagram does hold some general interest. It illustrates how existing aircraft have been constrained. If existing aircraft lie outside your predicted design area this raises a further discussion with your clients about the values chosen for your constraints.

As stated above, many of the aircraft and engine characteristics are held as constant values in the constraint analysis. It is important to conduct trade (sensitivity) studies on any assumptions used to generate these values (e.g. drag or thrust). Those shown to be critical should be more accurately predicted. Be careful not to get too involved in external discussions before such validation work has been completed.

The outcome of the constraint analysis may be the selection of a different geometry for the aircraft than currently specified for the baseline configuration. When you are ready to proceed a little backtracking will be necessary to re-evaluate the new configuration using methods that have been developed in the previous design stages.

2.7.2   Trade-off studies

The outcome of the constraint analysis may be the selection of a different geometry for the aircraft than currently specified for the baseline configuration. Since it will be necessary to recalculate the baseline aircraft with these new values, it is worth considering other changes that might be beneficial to the design. For example, the wing aspect ratio may have been selected from arbitrary data from other aircraft. It would be appropriate to assess this decision as more detailed analysis of the design is now available. Such methods are referred to as trade-off or sensitivity studies. They generally investigate the variation of a single parameter while keeping all others constant. Multi-variable investigations are referred to as parametric studies, these are explained later in this chapter.

Trade studies can take many different forms. They can be used, as implied above, to assess the selection of aircraft geometrical features (aspect ratio, taper ratio, wing thickness/sweep combinations, etc.). They can be used to indicate the sensitivity of the design to assumptions made in the analysis (e.g. the affect of the assumed extent of laminar flow, the affect of high strength or composite materials, etc.). Alternatively, they may be used to show the effect of changing design requirements (e.g. field lengths, mass variations, engine characteristics, etc.). All of these have the objective of building confidence in the design predictions on which the aircraft configuration is based.

Some examples of trade studies based on the example aircraft (described in section 2.9) are shown in Figures 2.8 to 2.10. These figures are introduced here only as examples of the type of studies that are possible. The first example of a trade study (Figure 2.8) shows the sensitivity for the choice of wing taper ratio.

Various aircraft parameters are plotted which show variations with increasing wing taper ratio. A lower limit for taper ratio may be imposed to avoid wing tip stall resulting from low Reynolds number flow over the outer wing sections. (This provides a good example of considerations other than those of the trade-off study also being influential.) The second Figure (2.9) shows the effect of wing aspect ratio.

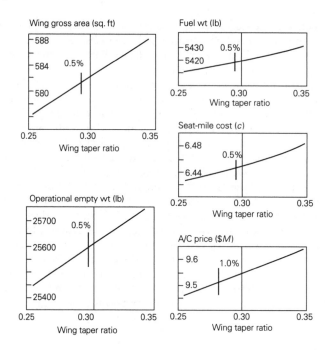

Fig. 2.8 Wing taper ratio sensitivity study (example)

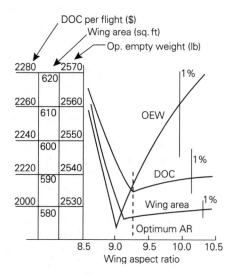

Fig. 2.9 Trade-off study of wing aspect ratio (example)

The 'kinks' in the plots show the effect of two separate constraint boundaries on the problem (i.e. the influence of span loading on climb performance and the activation of the WAT (weight, altitude and temperature) restrictions at low values of aspect ratio). For this civil airliner the optimum aspect ratio for minimum direct operating cost (DOC) is seen to be about 9.32.

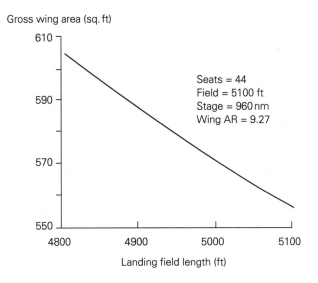

Gross wing area (sq. ft)

Seats = 44
Field = 5100 ft
Stage = 960 nm
Wing AR = 9.27

Landing field length (ft)

Fig. 2.10 Sensitivity graph – landing versus wing area (example)

Varying an aircraft geometrical parameter, like taper and aspect ratio, is relatively easy as these are inputs to the aerodynamic, mass and performance equations. Altering a main operational parameter (e.g. landing field length, LFL) involves much more iteration. It is possible to show the sensitivity of such parameters using trade study methods as illustrated in Figure 2.10 (LFL versus aircraft wing area).

2.8 Refined baseline design

Up to this point in the design sequence the aircraft has retained many of the features assumed from the initial configuration stage. For example, the wing planform shape will not have been carefully considered. It is at this stage in the development of the baseline layout that more rational decisions can be taken on the size and shape of all the aircraft components. One of the first decisions to be considered is the aircraft balance. As we now have a detailed mass statement for the aircraft that shows the mass of all the major components and a scale drawing of the aircraft layout, it is possible to locate the centres of each component mass. It is then possible to predict the positions of the aircraft centre of gravity more accurately than previously. (Note: 'positions' is plural as the aircraft can be loaded and flown at different payload or fuel masses.) The final objective of this analysis is to assess the longitudinal position of the wing relative to the fuselage to achieve an acceptable static margin. (Refer to textbooks on aircraft stability and control or aircraft design for the definition of static margin.)

As mentioned earlier (Section 2.6.2), it is acceptable to position the aircraft centre of gravity for subsonic aircraft just ahead of the wing mean aerodynamic quarter chord position. For supersonic flight the wing lift moves back to about half chord but remember that these aircraft also have to fly subsonically so a compromise or other technical measures (e.g. fuel management systems) may have to be introduced. The locations of the landing gear units is directly related to the aircraft centre of gravity, therefore repositioning of these units may also be considered at this time.

Detailed assessment of the 'packaging' of the aircraft is another task that can be considered at this point in the design process. An analysis of the space required and available for all components and systems must be made. This will include a check on the space available within the structure to hold all the fuel that is required to fly the mission. Will the landing gear have space for retraction? Will the pilot have adequate sight lines from the cockpit (these are mandatory)? Engine intake and exhaust flow requirements can be assessed to dictate the cross-sectional area distributions. Radar and other flight sensors must be suitably positioned. Will servicing maintenance and turn-round be easy to accomplish? In other words, it is possible at this stage to undertake a full review of all the installation and space aspects relating to the layout.

As mentioned earlier, the wing planform must be examined carefully. This will be reconsidered in work to be done later (e.g. parametric and trade-off studies) but for now some simple aerodynamic and mass analyses will provide evidence to make sensible judgements. For example, would an increase (or decrease) in wing aspect ratio be advisable. What about wing sweepback, thickness and taper? All these and other wing planform issues should be considered as thoroughly as technical methods, your ability and timescales allow.

Tail sizes can also be reassessed using typical values for tail volume coefficients, suitably adjusted to account for wing flaps, centre of gravity movement and aerodynamic efficiency. The planforms of the tail surfaces may also be altered to match the revised wing shape and rear fuselage geometry. It must be emphasised that using this method is relatively crude as each aircraft configuration produces different flow conditions over the tail surfaces. Also, some types may have uncharacteristic centre of gravity limits, offset thrust line, excessive aerodynamic moments and other features not representative of the general aircraft layouts. Such issues must be carefully taken into account when assessing a suitable value for the tail volume coefficients. As soon as enough detail is known it is important to conduct more precise control and stability calculations.

When all the geometric reviews have been completed it is necessary to recalculate the new aircraft configuration using the methods developed in the previous stage (baseline evaluation). However, it is now possible to incorporate more detail in the mass estimation methods used previously. Wherever possible the percentage M_{TO} assumptions for components should be replaced by detailed estimates using geometrical formulae or suppliers' data. Such information can be found in most aircraft design textbooks or in specialist technical papers on mass estimation (e.g. Society of Allied Weight Engineers, SAWE).

It is now possible to produce a report containing the revised baseline drawing and a summary of the mass, drag, lift and engine data, and aircraft performance predictions. This is sometimes regarded as the completion of the conceptual design phase. If deemed necessary it is possible to conduct more refined and extensive configurational studies using the methods described below.

2.9 Parametric and trade studies

The importance of clearly establishing the criteria upon which the effectiveness of the design will be judged was discussed earlier (section 2.1.2). A strong argument was made for this criteria to be used to direct the design process to arrive at the 'optimum' aircraft. This is done by the application of parametric and trade-off studies. For example, in many design projects aircraft costs figure highly in the assessment of aircraft effectiveness. The cost of buying the aircraft and operating it can be assessed against aircraft design variables (e.g. wing area, aspect ratio, engine bypass ratio, thrust, etc.) and values selected that give the best overall configuration. For some aircraft the criteria on which the aircraft is judged may be related to aircraft performance (e.g. for fighters/manoeuvrability, racing aircraft/circuit speed, aerobatic aircraft/handling and control characteristics). In such cases, as with costs, parametric and trade studies will show the most effective choice of the aircraft design variables.

It is unusual for an aircraft design to be centred on one precise specification. Even if the original problem definition seems tight, the manufacturing company will be hoping to extend the design into other operating environments (e.g. a military trainer may be developed into a close-air support aircraft and civil aircraft are often stretched (or shrunk) to suit different markets). For most aircraft types some form of multi-tasking or even simple re-engining will be envisaged. To account for such considerations the baseline aircraft configuration defined in the previous stage will be assessed for its suitability to fulfil other roles. This may force a change to the baseline configuration so that future developments are easier to achieve. For example, a new design may have a wing area that is too large for the initial specification in order to allow for future heavier weight derivatives without the need for a new wing to be designed. However, this strategy must not make the original design too ineffective in its primary role. The design team must draw a fine line between existing and future requirements for the aircraft. Parametric studies are used to explore how successful the design can be made to meet all the potential developments.

Parametric design studies are often performed by adopting the classical nine-point method (see Figure 2.11a). This shows the variation of the objective function (e.g. M_{TO} or direct operating cost) to changes in two of the aircraft parameters x and y (e.g. wing area and aspect ratio). The baseline aircraft, as developed in the previous stages of the design process, is often used as the centre point of the nine points. The parameters x and y are then taken as those for the baseline aircraft together with plus and minus steps each side of the central value (e.g. areas 54, 60, 66 and aspect ratios 7, 9, 11, where 60 and 9 are the values for the baseline design). The method can be extended to consider a third parameter (z) by producing a series of nine-point studies (see Figure 2.11b).

2.9.1 Example aircraft used to illustrate trade-off and parametric studies

To show how trade-off and parametric studies are employed in the design process, the following examples of typical studies are described. The design project used in these studies relates to a 40–50 seat regional jet operating over a 1000 nm single-stage length, flying out of a 5300 ft field on a hot day (ISA +25°). The full study resulted in the selection of an aircraft configuration that was regarded as offering the minimum direct operating cost for the design mission. This configuration (the baseline aircraft layout) is shown in Figure 2.12 and some technical details are listed below:

Overall dimensions (m)	*Wing geometry*	*Mass* (kg)	
Overall length 24.5	Gross area 55.0 sq. m	Maximum take-off	18 730
Overall width 22.6	Span 22.6 m	Maximum zero fuel	17 006
Overall height 7.9	Aspect ratio 9.32	Payload	4790
	Wing t/c 15% root, 11% tip		
	Sweepback 20° at quarter chord		

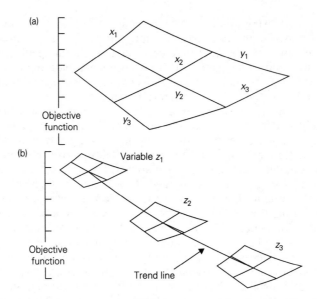

Fig. 2.11 Classical nine-point carpet plots (diagrams a and b)

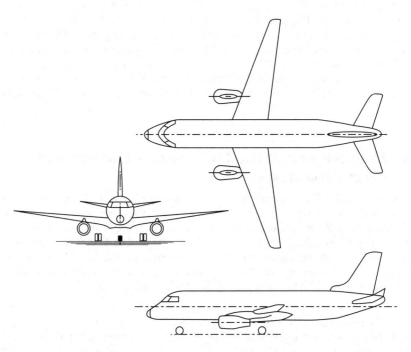

Fig. 2.12 Example aircraft layout

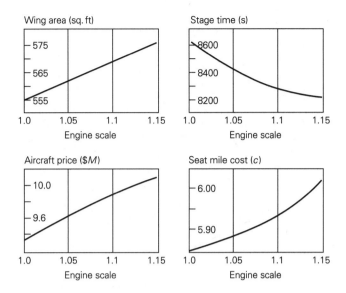

Fig. 2.13 Parameter study – engine size (example)

The results presented in the graphs (Figures 2.13 to 2.19) and described below are taken from a muli-variable optimisation (MVO) study in which all the aircraft design parameters were allowed to vary within predetermined limits. The aircraft parameter under investigation (e.g. wing taper ratio) was fixed at a selected value and the aircraft optimised for minimum aircraft direct operating cost. This process was repeated for other values of the parameter. The resulting optimised values of the design variables (e.g. wing area, aircraft empty mass, fuel mass, seat mile cost, etc.) were recorded for each value of the study parameter. The results when plotted show the sensitivity of the aircraft configuration to changes in the study parameter. As mentioned above, if a high sensitivity is indicated, more care must be taken in the selection of the parameter value for the final design configuration (and vice versa). The process described here is sometimes conducted without involving an MVO programme. This is less accurate as it requires some of the design variables to be held constant. It is easier and quicker to perform and, providing that the range of parameter variation is kept narrow, it is sufficiently accurate for the initial design stages. To avoid the complications associated with aircraft price and cost estimation it is possible to simplify the studies by adopting aircraft mass as the design objective function.

During the earlier description of the design process mention was made of the adoption of a 'rubber' engine in the aircraft to determine optimum engine size. Figure 2.13 shows the results of such a study. The penalties for including an oversize engine in the initial design for this project are quantified in this study. Of the parameters investigated, only stage-time benefits from the installation of the larger engine.

To study the stretch potential of the original baseline aircraft two parametric nine-point studies were completed. Figure 2.14 shows the effect on aircraft wing area and Figure 2.15 shows the effect on maximum take-off weight, for an aircraft with 56 seats (20 per cent increase) at various field length and stage distances.

With a newly designed 60-passenger baseline layout, further stretch potential was investigated by conducting a series of nine-point studies. Figure 2.16 shows the consequential effect on required engine size (still using the rubber engine).

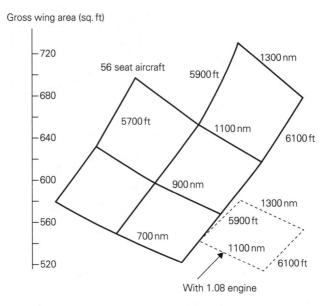

Fig. 2.14 Parameter study – wing area (example)

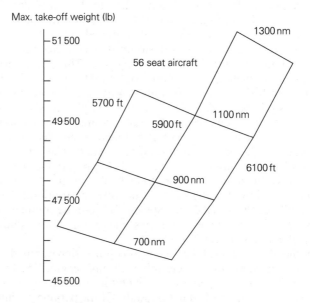

Fig. 2.15 Parameter study – take-off mass (example)

Figure 2.17 shows the required wing area (indicative of aircraft size). Figure 2.18 shows the effect of stretch on aircraft maximum take-off mass.

Finally, the effect of all these changes on the aircraft seat mile cost (SMC) is shown in Figure 2.19. The diminishing improvement of SMC with aircraft stretch is clearly shown. Only the advantage of the initial stretch to 70 passengers (PAX) looks attractive.

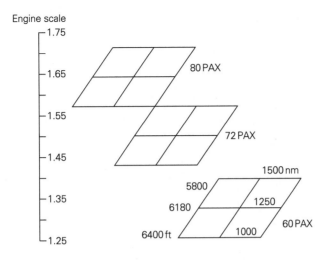

Fig. 2.16 Operational design study – engine size (example)

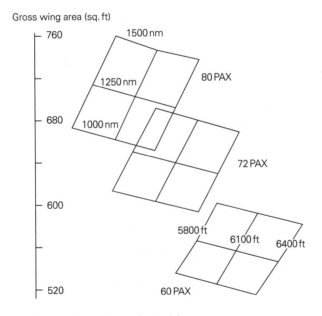

Fig. 2.17 Operational design study – wing area (example)

The examples described above are only a brief selection of the types of investigation that can be conducted using parametric methods. Each aircraft project will raise specific types of study that are significant. By the time the project is developed to this stage the designers will be aware of the nature of the parametric studies that are of interest to them.

Max. take-off weight (lb)

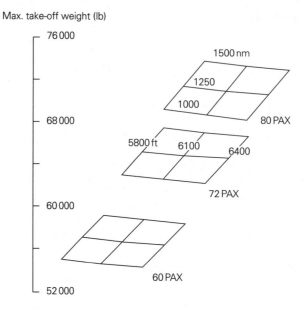

Fig. 2.18 Operational design study – take-off mass (example)

Seat mile cost (cents)

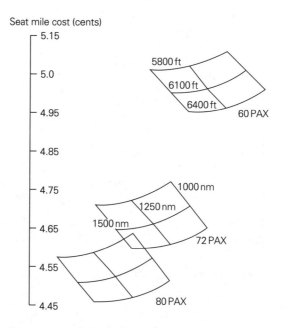

Fig. 2.19 Operational design study – seat mile cost (example)

2.10 Final baseline configuration

The scope and depth of the trade-off and parametric studies undertaken during the conceptual design phase will depend on the time and effort available. The final baseline configuration will benefit from such in-depth studies but often decisions are required before such work can be completed. It is important that all decisions on the configuration are made with enough time left to perform the final analysis on the design.

2.10.1 Additional technical considerations

As this stage in the initial design process represents the end of the technical work to be done, some extra details may be considered. For example:

- an appreciation of the structural framework for the aircraft,
- consideration of the inboard and sectional profiles through the aircraft,
- assessing the location and installation of the main systems and components (i.e. engine including intake and nozzle, cockpit layout, fuel tankage, weapons, payload, services).

If time permits a first-pass analysis of the aircraft stability and control should be made to ensure that the tail control surfaces are adequate.

2.10.2 Broader-based considerations

In contrast to the detailed technical analysis that has been the focus of much of the later stages of the design process, the final assessments should be concerned with broader-based aspects. In this respect each project will be different. The following list may help you in formulating the wider considerations for the project.

1. Manufacture
   - How and where
   - Required and available skills
   - Materials (availability and sizes)
   - Timescales
   - Developments
2. Flying issues
   - Pilot visibility and awareness
   - Handling and control
   - Training
   - Developments
3. Operational issues
   - Refuelling
   - Loading and unloading
   - Provisioning
   - Turn-round
4. Servicing
   - Engines (accessibility)
   - Stores
   - Systems
   - Regular inspections
   - Repairs
5. Environmental issues
   - Noise
   - Emissions
   - Recycling
   - Handling potentially dangerous systems and substances
6. Safety
   - Airworthiness regulations
   - Operational regulations
   - Manufacturing regulations
   - Certification procedures
   - Crashworthiness
   - Failure analysis
   - Reliability
7. Developments
   - Stretch
   - Multi-roles
   - Improvements
   - Technical developments
   - Flight testing

8. Programme management
   - Stretch
   - Cost
   - Facilities
   - Teambuilding
   - Availability
   - Risk management

9. Overall assessment
   - (S) Strengths
   - (W) Weakness
   - (O) Opportunities
   - (T) Threats

The prompts in the list above are not exhaustive. There may be other specific issues that apply to particular projects that are not mentioned above. Also, some of the topics mentioned might be irrelevant to your design but be careful not to dismiss any too hastily.

2.11   Type specification

At the completion of the initial design phase all details that are known about the aircraft are summarised in a report called the 'Aircraft Type Specification'. It is the project manager who is responsible for this report. He is accountable for the accuracy of the data and he will be expected to guarantee its validity. In a company, the sales and legal departments will use this document in contract negotiations. The technical specification therefore defines the guarantees the company will offer clients and thereby the liability it accepts in the contract to buy and use the aircraft and systems. Within this context the document is treated seriously in the design organisation. It will not contain speculative statements or unsupportable data. The report consists of textural descriptions, drawings, diagrams, numerical data, graphs and charts. As the design of the aircraft progresses through later phases of the design process the document will be systematically reviewed and updated to include the latest information.

For student work it is good practice to simulate this procedure. Project management should require the production of a document which defines the aircraft characteristics. As the project matures more details can be added to the report.

In Chapter 11 (section 11.4) detail recommendations for creating a student design project report are presented. While a student design team report may not always cover all of the items suggested in the following professional report example, the list provides suggestions for topics which could be considered for inclusion in the final team report.

2.11.1   Report format

The type of information that is included in the document will vary depending on the nature of the aircraft project. The following list is representative of the sections included in a professional document:

1. Introduction
2. General design requirements
3. Geometric characteristics
4. Aerodynamic and structural criteria
5. Weight and balance
6. Performance
7. Airframe
8. Landing gear

9. Powerplant (and systems)
10. Fuel system
11. Hydraulic and pneumatic systems
12. Electrical system
13. Avionics
14. Instruments and communication
15. Flight controls
16. Interior accommodation
17. Environmental control
18. Safety systems
19. Weapon systems (armament)
20. Servicing
21. Exceptions to regulations
22. Definition and abbreviations

2.11.2 Illustrations, drawings and diagrams

The Type Specification document contains several engineering drawings, schematic diagrams, system block diagrams, graphs, charts and general diagrams. The list below is not exclusive but provides a guide to the type of supporting illustrations to the text:

Aircraft three-view general arrangement	Fin structure
Inboard fuselage profile	Undercarriage (main and nose)
Fuselage sections	Nosewheel steering
Fuselage internal plan view (cabin arrangement)	Engine installation
	Engine power off-takes
Aircraft geometry	Engine controls
Mission profiles	Fuel system and tankage
Flight envelopes	Electrical system
Fatigue spectrum	Antenna and sensor locations
Undercarriage vertical velocity spectrum	Avionics
Runway loading	Hydraulics system
Weight and C.G. diagram	Pneumatic system
Fuselage structural framework	Flight control systems
Fuselage cross-section	Environmental control system
Floor loadings	Cabin pressurisation schedule
Cockpit view diagram	Instrumentation
Wing structural framework	Ejector seat installation
Wing/fuselage joint	Auxiliary power unit
Flap details	Access panels
Tailplane structure	Ground service

Obviously, some of the finer details contained in the above lists will not be known in the conceptual design phase. They have been included in the list to give a flavour of the type of detailed work that is still to be done on the aircraft design in subsequent phases.

References

1 Federal Aviation Administration (FAA-DOT) Airworthiness standards, FAR:
    Part 1 Definitions and abbreviations
    Parts 11, 13, 15 Procedural rules
    Parts 21 to 49 Aircraft regulations
  (details can be found on the FAA website www.faa.gov).
2 JAR (details can be found on the JAA website www.jaa.nl).
3 FAR part 36 – Noise standards: aircraft type and airworthiness certification.
4 Jenkinson, L. R., Simpkin, P. and Rhodes, D., *Civil Jet Aircraft Design*. AIAA Education Series and Butterworth-Heinemann, 1999, ISBN 1-56347-350-X and 0-340-74152-X.

5 Raymer, D. P., *Aircraft Design: A Conceptual Approach*. AIAA Education Series, 1999, ISBN 1-56347-281-0.
6 Brandt, S. A. *et al.*, *Introduction to Aeronautics: A Design Perspective*. AIAA Education Series, 1997, ISBN 1-56347-250-3.
7 *Aviation Week Source Book*, published annually in January.
8 Mattingly, J. D., *Aircraft Engine Design*. AIAA Education Series, 1987, ISBN 0-930403-23-1.
9 Eshelby, M. E., *Aircraft Performance – Theory and Practice*. Butterworth-Heinemann and AIAA Education Series, 2000, ISBN 1-56347-250-3 and 1-56347-398-4.

3

Introduction to the project studies

The design process has been described in detail in the previous chapters. All the steps that are necessary to successfully complete the preliminary design stages have been identified. The amount of effort and time spent in each stage depends on the overall schedule for the project. It is essential to complete the process with a feasible baseline design, therefore it is necessary to programme and manage the work in association with all other commitments. Although the design method has been shown as a sequential process, it is possible to run some of the steps in parallel. It is also possible to do some preparation work (e.g. develop estimating methods and spreadsheets) ahead of the later stages. This is particularly useful if the project is to be done by a group, or team, of people. In such cases, it would be essential to allocate all tasks and to set a rigid timetable for the completion of the work (see Chapter 11 for more details on team working).

Some of the case studies that follow are laid out in the standard format shown below. This format mirrors the sequence of the work to be done in the preliminary design of any aircraft.

Introduction to the project

1. Project brief
2. Problem definition
3. Design concepts
4. Initial sizing and layout
5. Initial estimates
6. Constraint analysis and trade-offs
7. Revised baseline layout
8. Further work
9. Study review

Some of the projects in the following chapters have been included to illustrate design investigations into specific operational environments and therefore do not strictly follow the sequence above.

The first three projects (Chapters 4, 5 and 6) have been shown in more detail than some of the subsequent studies. They are chosen to illustrate the design aspects of different parts of the aeronautical industry; namely, civil, military and general aviation respectively. Each of the later projects is selected because of an unusual operational or design aspect.

Chapter 4 considers the design of a new type of civil aircraft. The expected development of exclusive executive/business scheduled services that provide small capacity, long range operation is the stimulus for the project. The design of the aircraft is not difficult but as such types of aircraft have not been built before, there is no information to use as the starting point for the design. The example illustrates the iterative process that is essential in such cases. The conclusion of the study raises questions that could stimulate several other design studies in this area.

The second project (**Chapter 5**) relates to the design of a new military trainer. This project has been selected as it shows how a systems approach to the solution of a design problem can offer substantial benefits. The aeronautical design of the aircraft is relatively straightforward once the operational issues have been decided. The aircraft in the context of the training environment represents only one element of the total system. Other parts of the training environment include pre- and post-flight simulation experience, ground-based instructor stations and modern electronic communication and data links. Adapting technologies developed in other aeronautical applications (in this case, flight testing) allows more efficiency and flexibility to the aircraft design and the total training system. The project definition for the aircraft is shown to be influenced by issues relating to the development of the aircraft family. Single and twin seat versions are eventually shown to be desirable. This complicates the design process but such considerations are not unusual in actual project work. The design process for this aircraft has been shown in more detail than for other studies in the book as it combines many interacting constraints.

General aviation is the largest sector in the aviation business. **Chapter 6** shows how the design of a simple leisure aircraft can be combined with advanced technology developments. The project is set in the highly competitive field of air racing. Many racing aircraft have powerplants developed from automobile engines. Following this trend, this project postulates the introduction of electric propulsion to form a new type of racing formula. Developing and installing the current automotive fuel cell systems into an aircraft is investigated. Light aircraft design, from the Wright brothers onwards, has traditionally been used to test and develop new technologies. This project is chosen to simulate such situations.

The remaining four chapters each present a project that has unique operational requirements that significantly affect the basic layout of the aircraft. Such complications are often part of novel design specifications. The first of these (**Chapter 7**) deals with an aircraft concept that has long stretched the imagination of aircraft designers, namely the roadable aircraft. To combine the attributes of an automobile and a light aircraft would offer a highly desirable mode of transport. It would possess the convenience of the car for short journeys with the flexibility and time saving of an aircraft for long trips. The design problem concerns the matching of road and airworthiness regulations without compromising the operation in either transport mode. Although the completed preliminary design study lacks refinement in several technical areas, the student project won the NASA design prize for innovation in general aviation for the year 2000. A novel aspect of the work on this project was the integration of the design and analysis between student groups in UK and USA. This demonstrates that undergraduate design work does not have to be centred solely in one course, one department, one institution or even in one country. Dispersing the design team focused attention on the management and communication aspects of the design process, simulating modern industrial practice.

Many project studies arise from a request to consider an operational requirement outside existing experience. Such work can be classed as 'Feasibility Studies'. **Chapter 8** deals with such a study in the field of air offensive operations. This project formed the basis of the 2001/02 AIAA undergraduate design competition (see Chapter 8 for more details of this annual contest). Wars in the 1990s and since have demonstrated the need to gain air superiority over the war zone quickly. This leads to a requirement for aircraft to penetrate hostile territory in the early part of the offensive and 'neutralise' the air defensive capability of the enemy. Such initial strike aircraft are called interdictors. They must be stealthy and fast to avoid detection and have sufficient firepower to destroy heavily protected targets. This combination together with the long range required to attack deep inside an enemy country provides a challenging design problem.

The design study shown in **Chapter 9** also has origins in the international conflicts in Europe and Asia in the 1990s. These demonstrated the need for improved local surveillance over potentially hostile territory. To provide this safely in an unstable area of conflict calls for the design of a high-altitude, long-endurance uninhabited aircraft. This defines the mission requirement for the project. This study illustrates the difficulties to be encountered in designing an aircraft to fly outside the normal operational environment. To add to the unorthodox mission requirements, the study also investigates an unusual aircraft configuration (i.e. a high aspect ratio, swept-forward, braced wing layout).

The final project (**Chapter 10**) returns to the problems faced by the early aircraft designers, namely, operating aircraft from water. In the more remote parts of the world, light aircraft provide the most convenient form of transport. In such places level-ground landing surfaces may not be available. Stretches of water (lakes and sheltered bays) provide a suitable alternative. Amphibious aircraft combine both aerodynamic and hydrodynamic requirements that must be met to produce a successful design. This project shows how these criteria are combined to produce a feasible aircraft to operate from either land or water.

From the studies described in Chapters 4 to 10, it can be appreciated that each design task is unique. Projects can take several different forms of investigation. Each one requires a different form of study. This is illustrated in the variation of work described. One of the tasks for the project management team in the early stages of the design process is to identify the type of work that is necessary to successfully complete the project.

The selected projects have intentionally covered unusual and difficult design problems, set in civil, military and general aviation operating environments. The common theme in all the studies is the sequential nature of the preliminary design process. Working through these projects will provide an understanding of the stages to be followed in other design studies. Some helpful guidance on the best way to handle such projects in an educational environment is given in **Chapter 11**.

Project study: scheduled long-range business jet

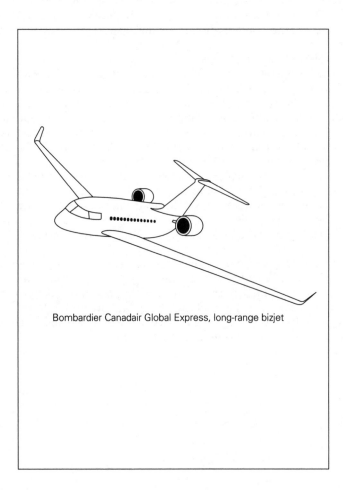

Bombardier Canadair Global Express, long-range bizjet

4.1 Introduction

Up to the events of 11 September 2001, all of the professional aeronautical industry market analysts predicted that scheduled airline business over the next 20 years was likely to increase at an annual growth rate of between 3.5 and 5 per cent. Such unexpected and tragic events illustrate the vulnerability of airline market projections to influences outside the control of the industry. However, it is expected that after a period of industrial recession the previous projections will be resumed. Although this is welcome news for the aircraft manufacturers and airlines, as more passenger movements equate to a growth in business, it also means that existing airports and associated infrastructures will become increasingly inadequate to satisfy this expansion. Already many of the world's international airports are working beyond capacity at peak operating periods. The expected doubling of demand over the next 15 to 20 years is generally incompatible with the planning approval and building timescales for airport expansion. The political, social and economic factors that accompany airport building projects lie outside the control of the aeronautical industries. In the past, planning enquires and environmental pressure groups have delayed many of the proposed airport development projects. There is no evidence that this situation will improve in the future.

Some of the problems at airports may improve when the new, supercapacity aircraft are introduced but even this development will not solve the passenger capacity problems at airports. Moving airline operations to larger aircraft is not new. Most airlines now use larger capacity aircraft on services that smaller types satisfied a few years earlier. This trend is likely to continue. This development allows an increase in passenger movements without increasing aircraft movements (i.e. increases passengers per flight 'slot'). However, this practice does not solve the problems of increased passenger demand on the airport terminal facilities. Handling larger aircraft and greater numbers of passengers requires an associated expansion of airport infrastructure.

Analysis shows that although the main airports are working at full capacity, over 70 per cent of all aircraft movements involve relatively small aircraft. These aircraft do not need the service provided at the large airports (i.e. runway length for take-off and landing, and terminal lounge capacity). Many of these flights are related to regional 'feeder' services that provide linking flights to international scheduled services. The mixture of small and large capacity services at airports leads to an inefficient use of the facilities available. This inefficiency is the source of many of the delays and disruption currently endemic at large airports.

Business surveys show that delays at airports will increase as demand on the services increases in the future. Delays and disruptions in the service affect all passengers. Airlines provide exclusive facilities at airports for their business travellers but this does not pacify a customer who misses an important meeting because of a flight delay. Such passengers demand more certainty in their travel arrangements than can be provided by the current and future operations. An expensive alternative to the current situation is for the business traveller to use a small, exclusive business jet for the journey but this may not be within the budget of most commercial travellers.

It has been suggested by researchers that the current problems at large airports could be eased if the feeder services were transferred to satellite airports. Such developments would potentially increase the capacity provision at the larger airport without the need to make changes to the present runway or terminal facility. However, as the traveller will need to transfer to and from the large airport there will be a requirement to provide or improve the ground transport provision between the two airports. This type of development is slowly taking place at the main 'hub' airports. The downside to this scheme is that the traveller is then subjected to extra potential delays from congestion at both airports and the ground interchange.

An essential element to any airline's success is the ability to attract the 'business' traveller. Business travellers pay significantly higher prices for travel than tourist-class passengers and represent a more dependable source of income than travellers who opt for first class. Capturing the loyalty of the business traveller is high on the agenda of every major airline. This is often accomplished on longer-range flights, where there is sufficient space, by creating a separate 'business-class' cabin. In this the seat widths, seat pitch and cabin amenities are set between those of the first-class and tourist-class sections. Business-class passengers are allowed early boarding and a wider choice of in-flight movies and passenger services, etc. Business travellers, of course, pay for these advantages but ticket-pricing schemes are devised in which the business-class ticket costs little more than the 'list' price of one in economy class. And, unless one wishes to purchase his or her ticket a couple of weeks in advance of the flight and is willing to stay at his or her destination over the weekend, the 'list' price is the best available. These special business-class amenities are usually provided only on longer-range flights since it is assumed that on trips of a couple of hours or less the benefits of such service are questionable.

There have been a few attempts to create specialised, all business-class airlines using aircraft such as the Boeing 727 or 737 fitted with only about half to two-thirds the usual number of seats. However, these and similar aircraft like the Airbus 320 class are usually limited in range and are unsuited to the long transcontinental or international flights where business-class amenities can really make a difference to the target group of travellers. So far, these attempts have failed to attract enough customers to make a profit. This may be due to insufficient perceived advantage in the wider seats and better service against the higher price on shorter-range flights. Alternatively, it may be due to the insufficient flight frequency of the special services compared to flight schedules of the existing airlines.

Is there a market for a business-class only aircraft or airline? For success, it must meet the preferences of the business traveller, which include the following amenities:

- Larger, more comfortable seats with more leg room than those in tourist class.
- Pemium in-flight service (better meals, free drinks, more selection of movies and a wider choice of entertainment options).
- Separation from tourist-class passengers in airport lounges during boarding, and on board the aircraft (for mixed-class operations).
- Faster flight check-in and post-flight luggage retrieval.
- Direct flights without delays at airports, especially on longer journeys.

The first two of the above points are currently available to business-class travellers on larger, longer-range flights with most airlines. The next two preferences can only be achieved with special 'business-class only' flights. In addition, the last of these may require a reduced dependence on hub airports.

Most airlines depend on a mix of passenger classes and fare levels to operate profitably. First-class passengers or their company pay dearly for their extra comfort and amenities. However, on many flights, the first-class seats are filled with business nor tourist-class passengers who have used accumulated frequent flyer miles to 'upgrade' their seats. At the other end of the airplane, the tourist-class passenger may have paid anything from a couple of hundred dollars to over two thousand dollars for a transoceanic or transcontinental flight. The 'list' price for tourist class is often very near that for the discounted business class. This serves as an inducement for passengers who cannot meet the requirements for discount tickets (typically 14 to 17 days' advance purchase and travel which includes a weekend stay) to purchase business-class tickets. Those who can meet the discount requirements can often fly for very low cost. Airlines use sophisticated seat management software to optimise the price of each ticket; seats in 'economy' are sold at a different price depending on availability and demand. The goal is to fill every seat, and having a 90 per cent discount passenger is better than an empty seat. Like soft fruits, scheduled airline seats are perishable goods that must be sold before the 'shop shuts' or aircraft departs!

There is a question as to whether, an 'all business class' airline can fill enough seats on enough flights to make a profit. If not, it must become like every other airline and offer either managed discounts or some other 'class' of seating with a scheme to fill these in order to make each flight at least break even in cost. Such an airline would also have to come close to matching the flight frequency of regular airlines and it may need to offer at least slightly lower ticket prices to induce business travellers away from their frequent flyer club loyalties. Success would probably also require a new class of airliner which could fly transcontinental and transocean ranges with passenger capacities similar to today's B-737s and A-320 class aircraft. Most existing airliners are designed as either a long-range/large-capacity, or a short-range/small-capacity operation, compatible with the current spoke and hub system of flight routing.

Meeting the last of the comments above will also require either a departure from the hub and spoke route model, or the use of aircraft designed to fly faster. The ideal airliner for this goal is probably a B-737, A-320 size aircraft with a range of 7000 nm and a cruise speed of Mach 0.9 or higher.

Any departure from the hub and spoke system may prove problematic given the saturated state of hub airports from which longer-range flights generally operate. A solution may be found in the use of other airports. Perhaps former metropolitan airports, which are currently used primarily for private and corporate aircraft operations, could be used. This would require the aircraft to be able to take off and land on shorter runways. It would also require the construction of high-speed ground transportation systems which could move business travellers between airports. This would further need the provision of gate-to-gate transfer of passenger and baggage without requiring additional baggage or security checks.

4.2 Project brief

From the analysis of existing and future air travel conditions above, it is possible to postulate a new type of airline service; one that is aimed at the profitable business travel market. This project study involves the development of a new scheduled business-exclusive international service from smaller airports.

An initial survey of the location of airports in the developed world was undertaken. This showed that within a radius of 50 miles around most current international airports, there is at least one regional airport that could be used for such a service. However, this may require the establishment of facilities to deal with international flights.

For existing airlines such a service would provide an improved and exclusive premium-class service and would allow an expansion of economy-class business at existing busy airports. New airlines may be set up to exploit the perceived market opportunity. Over the past decade, with the relaxed 'deregulatory' airline service, several new airlines have evolved to provide quick, easy and cheap (bus-type) alternative scheduled services. Some of these failed to achieve profitable operation but a significant number survived to compete with the older and larger established airlines. Such developments show that the airline business is dynamic enough to respond to novel market opportunities. A new aircraft type would create a unique, convenient, exclusive high-class business service that would compete with the current business-class sections in existing mixed-class scheduled services.

4.2.1 Project requirements

The following design requirements and research studies are set for the project:

- Design an aircraft that will transport 80 business-class passengers and their associated baggage over a design range of 7000 nm at a cruise speed equal or better than existing competitive services.
- To provide the passengers with equivalent, or preferably better, comfort and service levels to those currently provided for business travellers in mixed-class operations.
- To operate from regional airports.
- To use advanced technologies to reduce operating costs.
- To offer a unique and competitive service to existing scheduled operations.
- To investigate alternative roles for the aircraft.
- To assess the development potential in the primary role of the aircraft.
- To produce a commercial analysis of the aircraft project.

4.3   Project analysis

Project analysis will consider, in detail, each of the design requirements described in the previous section.

4.3.1 Payload/range

With only 80 seats, the aircraft is considered as a small aircraft in commercial transport operations. This size of aircraft is normally used only on short-range, regional routes. Table 4.1 (data from a *Flight International* survey of world airliners) shows the existing relationship between aircraft size (number of passengers (PAX)), design range and field capability.

By considering the range requirement of 7000 nm, the new aircraft falls into the large aircraft category but the passenger capacity of only 80 defines it as a small aircraft. This contradiction defines the unique performance of the new aircraft. The closest comparison to the specification is with the corporate business jet. However, this type of aircraft has a much smaller capacity (usually up to a maximum of 20 seats).

To investigate the significance of the 7000 nm range requirement, an analysis of the 50 busiest international airports was undertaken. This compared the great circle distances between airport pairs. It showed that very few scheduled routes exceeded 7000 nm. The list below shows the exceptions:

> US east coast – Sydney, Singapore, Thailand
> US central – Sydney, Thailand
> US west coast – Singapore
> Europe central – Sydney

All of the above routes could be flown with a refuelling stop at Honolulu for the US flights and Asia for the European flights. This analysis showed that 7000 nm was a

Table 4.1

	PAX	Range (nm)	Field length (m)
Small aircraft			
728Jet	70	1430	1676
CRJ 700	70	1690	1880
F 70	79	1870	1583
928Jet	95	1900	1950
RJ 100	112	2090	1275
F 100	107	1680	1715
B717-200	106	1460	1480
A318-100	107	2350	1400
Medium aircraft			
A310-300	218	5180	2300
B767-200	186	3000	2100
B767-300	245	3460	2550
A300-600R	266	4160	2280
Large aircraft			
A340-500	313	8504	3050
B777-200ER	305	7775	3020
MD11	285	6910	3110
B747-400	416	6177	3020
A380	555	7676	2900

reasonable initial assumption. This distance could be reduced if the design was shown, in subsequent trade-off studies, to be too sensitive to the range specification.

4.3.2 Passenger comfort

Long-range flights obviously equate to long duration. At an average speed of 500 kt, the 7000 nm journey will take 14 hours. Increasing the speed by only 5 per cent (e.g. from M0.84 to M0.88) will reduce this by 45 minutes. Anyone who has travelled on a long flight will agree that this reduction would be very welcome. Business travellers may accept a premium on the fare for such a saving in time and discomfort. As flight duration and comfort are interrelated, it is desirable to provide a high cruise speed for long-range operations. Reduced aircraft block time will also provide an advantage in the aircraft direct operating costs providing that extra fuel is not required for the flight.

As journey time relates directly to perceived comfort level, airlines have traditionally provided more space for business-class travellers. In the highly competitive air transport industry many other facilities and inducements have been used to attract this high value sector of the market. A new aircraft design will need to anticipate this practice and offer, at least, equivalent standards. This will impact directly on the design of the aircraft fuselage and in the provision of cabin services and associated systems.

4.3.3 Field requirements

The requirement to operate from regional airports effectively dictates the aircraft maximum take-off and landing performance (see Table 4.1). Operation from smaller airports will also affect the aircraft compatibility to the available airport facilities.

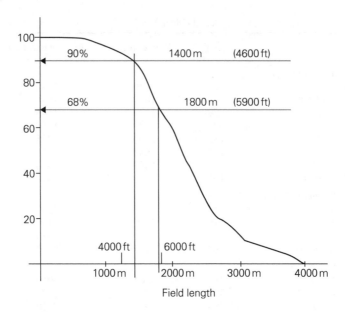

Fig. 4.1 Runway length survey

To understand this in more detail a survey is required to determine the available runway length of regional airports. Such a survey was undertaken in an aircraft design study[1] for a conventional feederliner. Figure 4.1 shows the results of this survey.

The frequency distribution of major European, regional airport, runway lengths (mostly UK, France and Germany) indicates that the 90 percentile equates to a minimum field length of 1400 m (4600 ft). Many of the aircraft operating within this field requirement are general aviation types. For an aircraft of 80 or more seats, this short distance may be regarded as too demanding on the aircraft design. It would force the wing to be too large or require a complex flap system. Both, or either, of these would increase drag in the cruise phase and thereby the aircraft direct operating costs. A sensitivity study on this aspect of the design could be conducted later in the design process when more details of the aircraft are available. Increasing the field length to 1800 m (5900 ft) will allow operation from 70 per cent of the airports surveyed. Comparing this choice to current, regional aircraft characteristics shows it is equivalent to the Avro RJ, Fokker and Boeing types. For this reason, the longer (1800 m) length will be specified for the design.

4.3.4 Technology assessments

The requirement to incorporate advanced technology into the design raises several questions relating to commercial risk, technical viability and economics. A design study that included a detailed assessment of new technologies applied to regional aircraft was presented by a Virginia Tech (VT) team at an AIAA meeting in 1995.[2] This considered some emerging technologies in propulsion, aerodynamics, materials and systems. In the final configuration of their aircraft, they selected ducted-direct-drive prop-fans as the powerplant. This showed substantial fuel saving over normal, high-bypass turbofans. They accepted the relatively slow cruise speed (M0.7) because their specification only called for a 3000 nm range. As much of the flight duration on short stage distances is spent in climb and decent, a reduced cruise speed is not too critical. For our design, such a slow cruise speed would not be acceptable, as it would significantly compromise the performance (flight duration) against existing scheduled services. For this reason, the prop-fan engine is not a suitable choice for our aircraft. A conventional high-bypass turbofan engine that is already certified and in use on other aircraft types will be our preferred choice. Although this will not show the fuel savings identified in the VT study, it will be comparable to the competitive aircraft. In addition, adopting a fully developed engine will reduce commercial risk and lower direct operating costs.

From an aerodynamic standpoint, the VT study proposed the incorporation of natural laminar flow aerofoil sections with boundary layer suction on the upper leading edge profile. Research results from NASA Langley were quoted to validate this approach. The hybrid laminar flow control system was shown to reduce aircraft drag and therefore fuel consumption. The study proposed the use of wing tip vortex turbines to power the boundary layer suction system. As such devices have not been developed in the time since the report was published, it is not considered wise to adopt this concept for our design. This will leave the wing tips clear for winglets to reduce induced drag in cruise. These are now well established on many long-range aircraft, therefore the technology is well understood. Boundary layer suction will need to be provided from bleeds from the engines. Later in the design process, a study will need to be undertaken to determine the effectiveness of the laminar flow system against the reduction in engine thrust in cruise caused by the demand from the air bleed system. On the turbulent flow parts of the aerofoil, it is proposed to incorporate the surface striation researched by Airbus and NASA in the late 1990s.

The use of new materials in the construction of civil aircraft is now becoming commonplace. To continue this trend composite materials will be used for wing skins, control surfaces, bulkheads and access panels. Advanced metallic materials will be used in high load areas (landing gear, flap mechanisms, engine and wing attachment structures). As proposed in the VT study, micro-perforated titanium, wing-leading-edge skins will be used for the boundary layer suction structure. A conventional, aluminium-alloy, fuselage pressure shell will be proposed as this is well proven and adds confidence to the aircraft structural framework. Filament wound composite structures may offer mass reductions for the pressure cabin but this technology is still unproven in airliner manufacture, so it will not be used on our aircraft.

Aircraft systems will follow current technology trends. This will include a modern flight deck arrangement. Aircraft system demand will increase due to the improvement in provision for the passenger services and comfort. This will include better air conditioning in the cabin to provide an increase in the percentage of fresh air feed into the system, more electronic in-flight passenger services and business (computing and communication) facilities. The aircraft will be neutrally stabilised to reduce trim drag in cruise and therefore require redundancy in flight control systems.

4.3.5 Marketing

Our aircraft type lies between the conventional mixed-class scheduled service and the exclusive corporate jet. The aircraft and operator will be offering a unique service. A comparison to the old 'Pullman'-class service operated by the railways at the beginning of the last century is appropriate. Avoiding major airports and the associated, and increasing, congestion and delays will be a significant feature of the service. Segmentation of the premium ticket passengers away from the low-cost travellers will be another positive marketing feature. Providing commercial/office facilities and a quieter environment during the flight will be another improvement over the existing mixed-class operations. All of these advantages will need to be set against the premium fare that the service will need to charge to offset the higher cost of operating the aircraft compared to existing services. In an analysis of the pricing policy of the new service it may be difficult to assess the elasticity of the ticket price because the service is new and untried. In the past, a sector of the travelling public has been attracted to the Concorde service. The reason that the extra ticket price was accepted is not clear. Either the time saving from supersonic flight or the exclusivity of the service, or both, may have been the feature that the customer was attracted to. It is felt that a premium above the existing business-class fare of 30 per cent is probably the limit of acceptance by the market sector. At this stage in the development of the project, this is only a 'guesstimate'. Market research would be necessary to identify the exact premium. A more in-depth market analysis will be needed before confidence in this figure is possible. There will always be a number of people who would use such a service. But as the ticket price rises, this number reduces. The number of passengers willing to pay the extra price must be seen to be greater than the number required to make the service commercially viable. The price at which companies regard the airfare as excessive must be determined.

4.3.6 Alternative roles

Developing an aircraft exclusively for a specialised role in civil aviation would be regarded as commercial madness. All aircraft projects should consider other roles the aircraft may fulfil. Our aircraft will have a fuselage size that is more spacious than normally associated with an 80-seat airliner. The long-range requirement will demand a high fuel load and this will make the aircraft maximum design weight heavier than normal for 80-seat aircraft. Both of these aspects suggest that the aircraft could be transformed into a conventional higher capacity, shorter-range airliner. A study will be required to investigate such variants. This type of investigation may result in recommendations to change the baseline aircraft geometry to make such developments easier to achieve. For example, increasing the fuselage diameter may allow a change from five to six abreast seating in the higher capacity aircraft to be made. Without such a change, six abreast seating may be unfeasible.

Other variants of the aircraft could be envisaged for military use. The long-range and small field features of the design are compatible with troop and light equipment transport operations. The ability to move military personnel without the need to refuel would avoid some diplomatic problems that have arisen in the past. The long endurance feature would make the aircraft suitable for maritime patrol, reconnaissance, surveillance and communication roles. The military variants should not be considered in the design of the baseline aircraft, as this would unduly complicate the conceptual design process. Such considerations should be left until the current design specification is better realised.

4.3.7 Aircraft developments

All aircraft projects must consider future development strategies to avoid complicated and expensive modifications in the development process. In modern civil aircraft design, it is common practice to consider the aircraft type as a 'family'. Airbus and Boeing use this approach successfully in their product lines. Stretching, and in some cases shrinking, the original design is now normal development practice. All new aircraft projects consider this in their definition of the initial design. It is essential to consider the

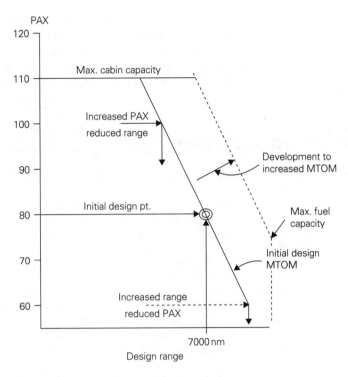

Fig. 4.2 Aircraft development (payload/range) options

consequences of this approach in the conceptual design phase. In this way, constraints to the development of the aircraft are reduced.

Apart from making geometrical changes around the initial, maximum design mass, it is common to expect a growth in this limit over the lifetime of the aircraft type.

Typically a 35 per cent growth in max. take-off mass may be expected over the lifetime of the type. Figure 4.2 shows how such developments are planned. The payload (PAX) – range (nm) diagram shows the initial design specification of the aircraft. The sloping maximum design mass line shows the initial layout options (trading passengers for range and vice versa). The dashed line represents a developed higher mass aircraft. This shows the growth (PAX and range) potential for an MTOM increase. Such investigations are required in the early conceptual design phase to guide the aircraft development path. It may be found necessary to slightly compromise the best layout of the initial aircraft to provide for such developments.

4.3.8 Commercial analysis

This last topic in the analysis of the aircraft project considers the commercial viability of the whole project. Although this cannot be assessed in detail at the start of the project due to a lack of technical data, it is possible to prepare for a commercial analysis later in the design process.

This preparation will identify the potential market for the aircraft, the potential customers for the aircraft, and the main competitors. The design team will need to know what are the principal commercial parameters that potential customers (airlines and passengers) will use to judge the attractiveness of the new service in the total market. One of the obvious issues to be considered is aircraft costs. This includes the purchase price and various direct operating cost (DOC) parameters.

Finally, assessment of the operating issues relating to the new service will need to be understood. This will include the customer service for both pre- and in-flight parts of the operation.

4.4 Information retrieval

As mentioned earlier in this chapter, this aircraft specification lies between long-range bizjets and regional feeder liners. The aircraft specified range is similar to the Gulfstream V but this bizjet only carries up to 15 passengers. The passenger capacity is similar to regional jets but they only fly about 1300 nm. To assess the design parameters that might be used in later sizing studies Table 4.2 has been compiled, which shows some of the details of these two different types of aircraft.

Table 4.2 shows that the thrust to weight ratios (T/W) for the two types are significantly different. The reasons for this lie in the requirements for higher climb/cruise performance and short field performance for the bizjets. These are parameters that our aircraft should have, so a thrust/weight ratio of 0.32 (the lower value for bizjets and the upper one for regionals) will initially be assumed for our aircraft.

Wing loading (W/S) is also seen from the data in the table to be statistically different between the two aircraft groups. There may be a variety of operational criteria for this division but for the same reason as above, a value lying between the two sets will be selected. A value of 450 kg/sq. m, being low for regional jets but high for bizjets, will be used. This decision may mean that 'high-performance' flaps will be required.

Mass ratios are always difficult to assess from published data as there are often conflicting variations in the definition of terms. For example, empty weight ratio will be higher for smaller aircraft and smaller for long-range aircraft. It should be relatively

Table 4.2

	PAX	Range (nm)	M_{TO} (kg, lb)	T/W	W/S (kg/sq.m, lb/sq.ft)	M_E/M_{TO}
Business jets						
Falcon 2000	19	3000	15 875, 35 000	0.327	323, 66.3	0.563
Gulfstream V	14	6500	40 370, 89 000	0.332	382, 78.3	0.526
Learjet 45	10	2200	8 845, 19 500	0.359	359, 73.6	0.600
Canadair RJER	50	2270	23 133, 48 800	–	478, 98.1	0.591
Beechcft 400A	8	1690	7 303, 16 100	0.360	326, 66.8	0.624
Hawker 100	10	3010	14 061, 31 000	0.340	404, 82.8	0.581
Citation	11	3300	15 650, 34 500	0.371	–	0.586
Commercial jets						
Fokker 100	107	1680	44 450, 98 000	0.308	475, 97.3	0.556
Romero 1-11	109	1480	47 400, 104 500	0.289	494, 101.2	0.500
RJ100	112	2090	44 000, 97 000	0.290	572, 117.2	0.573
B717-200	106	1460	49 895, 110 000	0.291	536, 110.0	0.614
A318-100	107	2350	64 500, 142 200	0.330	526, 107.8	0.627*

* A derivative of a larger aircraft.

easy to reassess the selected mass ratio following the first detailed mass estimations. Until this data is available it is necessary to make sensible 'guesstimates'. Values of 0.52 for the empty mass fraction and 0.35 for the fuel fraction seem reasonable, at this time.

For comparison, the values for these parameters for the VT study aircraft[2] are quoted as: 0.32, 535, 0.42, 0.32. Some of these differences can be explained by the larger size (165 PAX), shorter-range design specification of the VT study.

4.5 Design concepts

The previous section has shown that all of the potential competitors to the new design are of conventional configuration. They have trapezoidal, swept, low-mounted wings, with twin turbofan engines and tail control surfaces. Obviously, one of the concepts to consider is to follow this arrangement. The conservative airline industry may prefer such a choice. An alternative strategy is to adopt a novel/radical layout.

The 'new look' would set the aircraft apart from the competition and offer a marketing opportunity. In adopting such a design strategy, care must be taken to reduce technical risk and to show improved operational efficiency over the conventional layout.

Four design options are to be considered:

- Conventional layout
- Braced wing canard layout
- Three-surface layout
- Blended body layout

4.5.1 Conventional layout(s) (Figure 4.3)

This must be regarded as a strong candidate for our baseline aircraft configuration as it is a well-proven, low-risk option. The technical analysis is relatively straightforward and has a high confidence level in the accuracy of the results. Its main advantage is that it is similar to the competitor aircraft and thereby with airport existing facilities and operations.

There are some drawbacks to choosing this layout. These relate to the geometrical difficulties of mounting a high-bypass engine on a relatively small aircraft wing (relating mainly to ground clearance below the engine nacelle). This is illustrated in drawing A on Figure 4.3. There are two possible, alternative aircraft arrangements that could overcome this problem. Version B, shown on Figure 4.3, shows the engines mounted at the rear of the fuselage structure. This avoids the ground clearance problem but introduces other difficulties. Since a large component of aircraft mass is moved rearwards the aircraft centre of gravity also moves aft. This requires the wing to be moved back to balance the aircraft. The movement of the wing lift vector rearwards shortens the tail arm and consequently demands larger control surfaces. This increases profile drag and possibly trim drag in cruise. The second alternative layout is shown in version C on Figure 4.3. In this option the wing is moved to the top of the fuselage section (a high mounted wing). This lifts the engine away from the runway and provides adequate ground clearance. The high wing position, although used on some aircraft, is regarded as less crashworthy. The fuselage and therefore the passengers are not cushioned by the wing structure in the event of a forced landing. This is regarded as particularly significant in the case of ditching into water, as the fuselage would be below the floating wing structure. For an aircraft that is likely to spend long periods over water, airworthiness considerations may deter airlines from this type of layout.

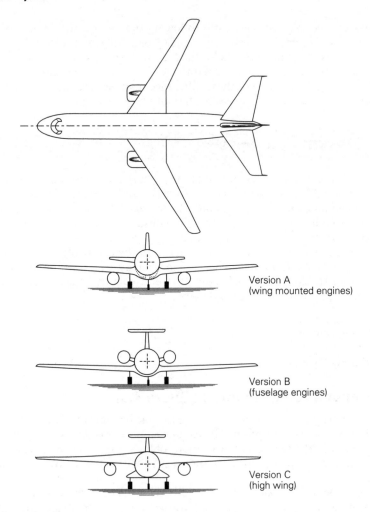

Fig. 4.3 Conventional layouts

A problem not necessarily restricted to the conventional layout is the potential lack of fuel tankage. A long-range aircraft will require substantial fuel storage and this may not be available in a conventional wing layout.

4.5.2   Braced wing/canard layout (Figure 4.4)

Although this configuration looks radical, it is technically straightforward with well-proven, and understood, analysis that provides technical confidence.

The canard and swept forward wings offer low cruise drag possibilities. The rearward positioning of the engines reduce cabin noise. The bracing structure should reduce wing loads and allow a thinner wing section to be used. This, in combination, may reduce wing structural mass and aircraft drag. The main weakness of the layout lies in the uncertainty of the positioning of the canard, wing and engine components, and the

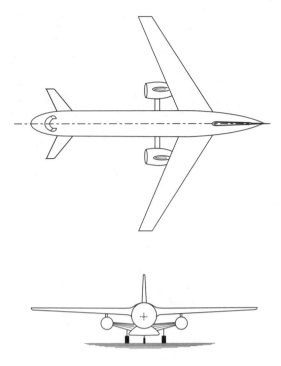

Fig. 4.4 Braced wing layout

interference effects of the airflow at the brace structure junctions. There is also some uncertainty about the effect of the brace on future stretch capability.

4.5.3 Three-surface layout (Figure 4.5)

This configuration has the advantage of low trim drag in cruise. The combination of forward and rear control and stability surfaces can be used to trim the aircraft in cruise with an upward (lift) force which will unload the wing. Two different wing layouts can be considered – swept forward or swept back. These options are shown in Figure 4.5.

It is anticipated that the swept forward configuration will be more suited to the development of laminar flow but may be heavy due to the need to avoid structural divergence. The bodyside wing chord will need to be sufficient to permit laminar flow systems to be installed. This is easier to arrange on the swept back layout. The increased internal wing volume created by the larger root chord will also provide increased wing fuel tankage. This together with the better flap efficiency of the swept back wing makes it the preferred choice of layout. The rear mounted engines will reduce cabin noise and visual intrusion although increase aircraft structural mass.

This layout is a strong contender for the preferred layout of our aircraft as the technical risks involved are low yet the configuration is distinctive. There may be a slight problem in positioning the forward passenger door due to the canard location but this should be solvable.

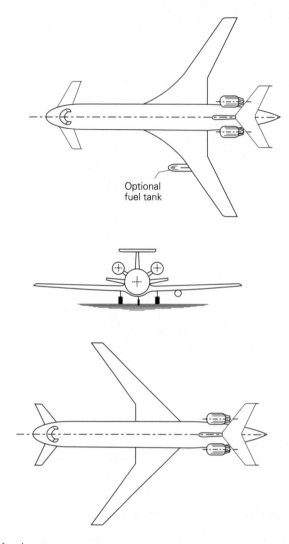

Optional
fuel tank

Fig. 4.5 Three-surface layout

4.5.4 Blended body layout (Figure 4.6)

There has been a lot of interest in this configuration for the new supercapacity (550–1000 seats) aircraft. It is not novel. Several previous aircraft designs (mostly military) have adopted the layout. Aerodynamically, this layout is very efficient and lends itself to the installation of laminar flow control systems. For a long-range aircraft this is a major advantage, as fuel consumption will be reduced. The large internal volume of the wing should provide sufficient fuel tankage.

The main disadvantages relate to the difficulty of providing cabin windows and ensuring passenger evacuation in the case of an accident. Some innovation will be necessary to overcome these problems and satisfy airworthiness authorities. Airlines may be cautious of making this 'step into the unknown' due to the uncertainty of passenger

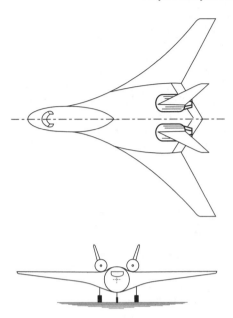

Fig. 4.6 Blended body layout

acceptance. A further problem, inherent in this layout, is the difficulty of stretching the integrated aircraft structure during programme development.

4.5.5 Configuration selection

From a narrow commercial viewpoint, the conventional layout should be chosen, as it is a low-risk, low-cost option. However, there are doubts regarding the adequate provision of fuel tankage and the lack of a new 'aesthetic' for the service. Of the conventional layouts, the best is version B (rear fuselage mounted engines). This makes the passenger cabin less noisy, which would be seen as an advantage for an executive-class aircraft operating long endurance flights.

Although the conventional design is the natural choice, we will select one of the more radical configurations. In this way, it should be possible to compare, in more detail, the strengths and weaknesses of the design relative to the alternative (competitive) strategy of using a modified version of an existing aircraft. This comparative study could form the conclusion to the study.

Of the three novel configurations described above, the most radical is the braced wing layout. This option presents a larger commercial risk therefore it will not be pursued. The blended body aircraft is potentially a strong contender as the integrated structure/aerodynamic concept provides a technically efficient layout. The generous internal volume of the aircraft will suit development potential and offer adequate fuel tankage. The main difficulty is the lack of understanding of the internal structural framework. The integration of the wing air loading and the fuselage pressurisation loads is not easy to envisage. For a larger aircraft, this problem is eased due to the internal space available. For our smaller aircraft, such separation of load paths may not be feasible. Even if the structural problems could be solved, the integration of structure and aerodynamic designs would make it difficult to stretch the design into a family of layouts (e.g. Airbus A318, 319, 320, 321).

The three-surface configuration has been successfully used on other aircraft and has been shown to offer performance advantage over the conventional layout. The saving in fuel during cruise will reduce the tankage requirement. As mentioned previously, the canard will make the front fuselage design more complicated but this difficulty can be overcome by detail design. If necessary, a shortage of fuel volume may be avoided if external tanks are added to the wing or if fuselage tankage is allowed.

Based on the arguments above the three-surface layout will be selected for our baseline configuration.

4.6   Initial sizing and layout

At this point in the design process, we can begin to realise the aircraft geometrical configuration. It is necessary first to make an estimate of the aircraft mass. Using this value, the engine and wing sizes can be determined. The fuselage shape is determined from the internal layout and tail requirements. With the main components individually defined, it is then possible to produce the first scale drawing of the aircraft. Crude estimates from similar aircraft types are necessary to complete the layout (e.g. tail and landing gear).

4.6.1   Mass estimation

From section 4.4 the following mass parameters were suggested:

$$\text{Empty mass fraction} = 0.52$$

$$\text{Fuel mass fraction} = 0.35$$

The mass estimations below are shown in kg only (conversion factor: 1 kg = 2.205 lb).

The payload is specified as 80 business-class passengers and their baggage. For this type of 'premium-ticket' operation the mass allowance per passenger (including baggage) will be larger than normal. We will allow 120 kg per passenger. The flight rules will dictate at least two pilots. Airlines will want to provide high-class service in the cabin so we will assume four cabin attendants are required. It is common practice[1] to allow 100 kg for each flight crew and 80 kg for each cabin attendant.

Hence, the payload is estimated to be:

$$M_{\text{pay}} = (80 \times 120) + (2 \times 100) + (4 \times 80) = 10\,120\,\text{kg}$$

To cover incidental flight services, allow an extra 5 kg per passenger. This adds 400 kg to the payload mass, making the aircraft 'useful mass' = 10 520 kg.

Using the equation described in Chapter 2 (section 2.5.1) with the values above gives:

$$\text{MTOM} = 10\,520/(1 - 0.52 - 0.35) = 80\,923\,\text{kg} \ (178\,435\,\text{lb})$$

The initial mass statement is:

	kg	% M_{TO}
Operational empty	42 080	52
Extra services	400	
Crew (2 + 4)	520	13
Passengers	9 600	
Fuel	28 323	35
$M_{TO}=$	80 923	100
	(178 435 lb)	

4.6.2   Engine size and selection

The literature survey (section 4.4) indicated a thrust to weight ratio of 0.32 was appropriate.
Hence:

Engine total take-off thrust $= 0.32 \times 80\,923 \times 9.81 = 254\,kN$ (57 100 lb)

With two engines this equates to 127 kN per engine (28 550 lb)

A choice of engines from different manufacturers is always the preferred commercial position for the airframe manufacturer. This ensures that the engine price and availability is more competitive. It also provides the potential airline customer with more bargaining power when selecting the aircraft/engine purchase.

There are several available engines that would suit our requirement. All of them are currently used on civil aircraft operations therefore considerable experience is available. The engines below are typical options:

- CFM56-5B as used on the A320
- CM56-5C as used on the A340
- IAE-V2533 as used on the MD90 family
- IAE-V2528 as used on the A321.

The details* below are representative of these engines:

Bypass ratio	5.5
Thrust ISA-sea-level static	31 000 to 34 000 lb (138 to 151 kN)
Typical cruise thrust (max.)	5840 to 6910 lb (26 to 31 kN)
Cruise specific fuel consumption	0.594 to 0.567
Length	102 in (2.6 m)
Diameter	68.3 to 72.3 in (1.7 to 1.8 m)
Engine dry weight (mass)	5250 lb (2381 kg)

*Note: engine manufacturers commonly quote values in Imperial units.

These details will be enough for initial performance and layout purposes but as the design progresses it will be necessary to periodically review the choice of engines to be used on the aircraft.

4.6.3   Wing geometry

The recommended wing loading is 450 kg/sq. m, hence:

Wing gross area $(S) = 80\,923/450 = 180$ sq. m (1935 sq. ft)

Selecting a high aspect ratio (AR) will lower induced drag in cruise and save fuel. A value of 10 is to be used. The choice of aspect ratio will need to be reviewed in a trade-off study later in the design process. Using the wing area and aspect ratio we can determine:

$$\text{Wing span } (b) = (\text{AR} \times S)^{0.5} = 42.4 \, \text{m} \, (135 \, \text{ft})$$

$$\text{Mean chord } (c_m) = (b/\text{AR}) = 4.24 \, \text{m} \, (13.5 \, \text{ft})$$

Selecting a taper ratio of 0.3 gives (approximately):

$$\text{Tip chord} = 2.0 \, \text{m} \, (6.6 \, \text{ft})$$

$$\text{Centre line chord} = 6.52 \, \text{m} \, (21.4 \, \text{ft})$$

Using published data on critical Mach number analysis, at our preferred cruise speed of M0.85, a sweepback angle of 30° will allow a maximum wing thickness of 15 per cent without incurring wave drag penalties. This wing thickness will allow space for the proposed laminar flow control system and offer extra fuel tankage. An unswept inboard trailing edge will make the flap more effective and provide space for the main undercarriage, retraction mechanism. This will add extra area so the wing chords will be changed from the values calculated above to 1.8 m and 6.0 m (5.9 and 19.7 ft) respectively to retain approximately 180 sq. m (1935 sq. ft) area.

The basic wing geometry is shown in Figure 4.7. This also shows the location of the wing fuel tanks and the position of the mean aerodynamic chord (MAC). As an initial assumption the longitudinal position of the wing on the fuselage will be arranged to line up the position of the MAC quarter-chord with the aircraft centre of gravity.

Fuel volume considerations

At this point in the design process it is necessary to determine the size of the required fuel volume estimated in the initial mass estimation and then to compare this to the available space in the wing. This involves transforming the fuel mass into a volume. Fuel volume/capacity is often quoted in terms of 'gallons'. This must be converted into linear units (cubic metres or cubic inches) to relate the size to the aircraft geometry. To do this conversion it is necessary to understand the various systems of units used. The calculation is shown in detail below because it is seldom to be found in other aircraft design textbooks although it is always a significant consideration.

In SI units – one litre is defined as the volume required to hold one kilogram of water. It is further defined as 1 litre = 1000 cubic centimetres (i.e. 0.001 cubic metres).

$$\text{Hence, } 1000 \, \text{kg of water occupies } 1.0 \, \text{m}^3$$

In USA – one US gallon equates to the volume required to hold 8.33 lb of water. For water at 62.43 lb/ft^3 a US gallon therefore corresponds to a volume of 231 cubic inches.

$$\text{Hence, } 1000 \, \text{lb of water occupies } 120 \, \text{US gallons } (=16.02 \, \text{ft}^3)$$

Warning: In the UK the definition of the gallon is different to the USA gallon!

In the UK – the Imperial gallon is used to measure liquid capacity. This is defined as the volume required to hold 10 lb of water. This makes the Imp gal. = 277.42 cubic inches.

$$\text{Hence } 1000 \, \text{lb of water occupies } 100 \, \text{Imp. gallons } (= 16.02 \, \text{ft}^3)$$

Note: the density of a liquid (water) does not change with the system of units!

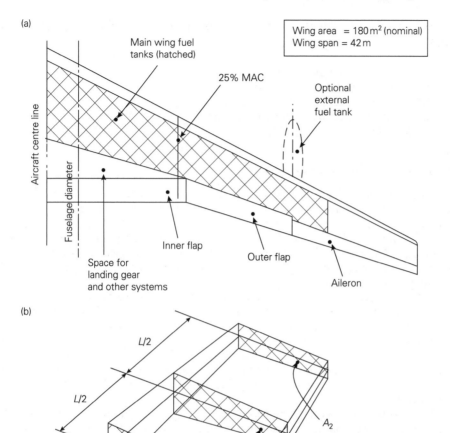

(a)

Main wing fuel
tanks (hatched)

25% MAC

Optional
external
fuel tank

Aircraft centre line

Fuselage diameter

Wing area = 180 m² (nominal)
Wing span = 42 m

Inner flap

Outer flap

Space for
landing gear
and other systems

Aileron

(b)

L/2

L/2

A₂

A_mid

Cross-sectional
area: A₁

Fuel tank volume = $\left(\frac{L}{6}\right)(A_1 + A_2 + 4A_{mid})$

Fig. 4.7 (a) Initial wing planform geometry (b) Fuel tank volume

Here are some useful conversion factors:

1 US gal = 0.833 Imp. gal   1 Imp. gal = 1.2 US gal
1 US gal = 3.79 litres   1 Imp. gal = 4.55 litres
1 cubic foot = 28.32 litres
1 cubic metre = 1000 litres   1 cubic metre = 35.3 cubic feet

Specific gravity is the unit that relates the density of a liquid to water. For aviation fuel, specific gravity varies with the type of fuel (e.g. JP1, 3, 4, 5, 6, or kerosene) between values in the range 0.82 to 0.76. For civil aircraft fuel, a value of 0.77 can be assumed, hence:

1000 kg of fuel occupies $1.3 \, m^3$
1000 kg of fuel occupies 1300 litres
1000 lb of fuel occupies 155.8 US gallons
1000 lb of fuel occupies 129.9 Imp. gallons
1000 lb of fuel occupies $20.8 \, ft^3$

Estimation of wing fuel volume

To determine the usable capacity of a fuel tank it is possible to calculate the external volume and then reduce this value to account for internal obstructions caused by structural and system components within the tank. Typically, reduce the available internal volume to 85 per cent for integral tanks and to 65 per cent for bladder or 'bag tanks'. Note: these factors do not account for landing gear or other significant intrusions into the available space. Such factors must be considered separately when deciding the overall location of fuel tanks.

For the aircraft in this project, we can consider the fuel to be held in wing tanks on each side of the aircraft fuselage as shown in Figure 4.7. Each tank occupies the space between the leading edge and trailing edge high-lift structure and associated mechanisms. The tanks will be of the integral type. The space ahead of the ailerons will not be used for fuel tankage, as the wing section here is too thin. Also, the space in front of the inboard flap is not used for fuel volume, as this is likely to be where the main landing gear will be stowed. The generalised geometry of the fuel tank is shown in Figure 4.7a.

The cross-section areas of the spanwise ends $(A_1$ and $A_2)$ and mid-span (A_{mid}) sections of each tank are determined by multiplying the average tank thickness (chordwise) (T) by the distance between the front and rear spars (W). The cross-sectional areas of each end of a tank are added and then multiplied by half the spanwise distance between the ends (L) to give the profile volume. This volume is then multiplied by the appropriate factor for the type of tank. Hence:

$$\text{Average thickness } T = k \cdot (t/c) \cdot c$$

where $k =$ factor to relate the average tank depth to the max. wing profile depth. The value depends on the shape of the section profile and the allowance made for structure. Typical values lie between 0.8 and 0.5.

$(t/c) =$ wing section profile thickness ratio (this will vary from thicker values near the bodyside to thinner values towards the tip).

$c =$ the local wing chord length

$$\text{Cross-sectional area } (A) = T \cdot W$$

where W is the width of the tank

$$\text{Tank profile volume} = (L/6)[A_1 + A_2 + 4A_{mid}]$$

where A_1 and A_2 are the cross-sectional areas of the ends of the tank
A_{mid} is the cross-sectional area at the mid-length position
L is the length of the tank

Table 4.3

Section	A/C centre	Mid.	Outer
Wing chord (c)	8.5	5.0	3.0
Wing thickness (t/c)	0.15	0.13	0.11
Thickness factor (k)	0.6	0.65	0.65
Thickness (T)	0.765	0.422	0.214
Widths (W)	4.2	2.6	1.7
Cross-sect. areas (A)	3.06	1.52	0.32
Tank length (L) = 16.0	Tank max. volume (profile) = 21.2 m^3		

The tank measurements (in metres) are quoted in Table 4.3. Total tank volume (both sides), including an 85 per cent factor for integral tankage:

$$\text{Available tank volume} = 2 \times 21.2 \times 0.85 = 36.1 \, \text{m}^3$$

$$\text{Required fuel mass (from section 4.6.1)} = 28\,323 \, \text{kg} \, (62\,450 \, \text{lb})$$

Using the volume conversion shown above, for typical aviation fuel:

$$\text{Required tank volume} = (28\,323/1000) \cdot 1.3 = 36.82 \, \text{m}^3 \, (590 \, \text{ft}^3)$$

Hence, within the accuracy of the calculation, the required fuel can be accommodated in the wing profile tanks. Extra fuel volume will be useful to extend the range of the aircraft for reduced payload operations. This could be provided by the optional external wing tanks but these would add extra drag.

4.6.4 Fuselage geometry

For most aircraft, the fuselage layout can be considered in isolation to the wing and other control surfaces. The internal space requirements, set by the aircraft specification, are used to fix the central section of the fuselage. For civil aircraft, this shape is governed by the passenger cabin layout.

The fuselage width is set by the number of seats abreast, the seat width and the aisle width. The depth is set to accommodate the cargo containers below the floor and the headroom above the aisle. A circular section is preferred for an efficient structural pressure shell. This requirement may impose constraints on the preferred width and depth sizes. Although this aircraft is designed principally as an executive aircraft, we must make sure that the size is suitable for any other variants that we may want to consider as part of the aircraft 'family'. For an aircraft of 80+ capacity, the conventional seating (mixed class) would be five abreast for economy and four abreast for business, with a single aisle. For our executive layout, four abreast would be sensible. As the aircraft mission is long range, it is necessary to provide a high comfort level. A typical maximum first-class seat is 0.7 m (27.5 in) wide. Providing a generous 0.6 m (24 in) aisle would make the cabin width 3.4 m (136 in). Adding 0.1 m (4 in) each side for structure makes the fuselage outside diameter 3.6 m (142 in). This width would allow five abreast 'tourist' seating with a seat width of 0.56 m (22 in). This is currently regarded as a very generous provision for this class. At six abreast the 'tourist/charter' seat width is 0.47 m (18.5 in). This is narrow for normal tourist provision but generous for the charter operation. The fuselage layout options are as shown in Table 4.4.

Table 4.4

Class	Seats abreast	Seat width	Cabin internal width
Executive	4	0.70 m	$(4 \times 0.7) + 0.6 = 3.4$ m
Tourist	5	0.56 m	$(5 \times 0.56) + 0.6 = 3.4$ m
Charter	6	0.47 m	$(6 \times 0.47) + 0.58 = 3.4$ m

Adding 0.2 m for the pressure cabin structure makes:

Total fuselage external diameter equal to 3.60 m (11.8 ft or 142 in)

The fuselage cross-section must also be considered in relation to the cargo pallet sizes to be accommodated below the cabin floor. This may require the fuselage profile to be altered to suit the geometry of standard containers. For example, the Boeing 757 fuselage section is 10 in deeper than the circular cabin shape. It is too early in the design process to consider such details but this aspect must be carefully studied later.

The length of the cabin is determined by the seat pitch. This varies as the class. Typical values are: executive class is 1.0 to 1.1 m (40 to 43 in), tourist is 0.8 to 0.9 m (31 to 35 in), and charter is 0.7 to 0.8 m (28 to 31 in). Using the longest executive seat pitch with four abreast seating requires a cabin length of 22 m (72 ft). With this length of cabin, the number of tourist passengers that can be accommodated is 140 and 120 for the short and longer pitches respectively. A similar calculation for the charter layout would provide 192 or 140 passengers respectively. It may not be possible for technical reasons (e.g. provision of emergency and other services) to accommodate the larger capacities calculated here. Nevertheless, the 22 m cabin length seems to offer a good starting point for the initial layout.

It is desirable to split the cabin into at least two separate sections. This makes the in-flight servicing easier and allows more options for the airline to segregate different classes. For the exclusive executive layout, this division will allow a quieter environment within the cabin. A service module (catering or toilets) is positioned at this location. External service doors and hatches are positioned here and these can act as emergency exits. The provision of service modules and the 'wasted' space adjacent to the doors will add about 4 metres (13 feet) to the cabin length.

The fuselage length is the sum of the cabin and the front and rear profile shaping. The front accommodates the flight deck and the rear provides attachment for the engines (in our case) and the tail surfaces. From an analysis of similar aircraft, the non-cabin length is about 15 metres.

Hence, the total fuselage length is $(22 + 4 + 15) = 41$ m (134 ft)

The resulting fuselage layout and geometry are shown in Figure 4.8.

4.6.5 Initial 'baseline aircraft' general arrangement drawing

With details of the engine, wing and fuselage available, it is now possible to produce the first drawing of the aircraft. The control surface sizes are estimated from area and tail volume coefficients of other similar aircraft. The aircraft general arrangement (GA) drawing is shown in Figure 4.9. With the sizes of the major components of the aircraft available from the GA, it is possible to make the initial technical assessments.

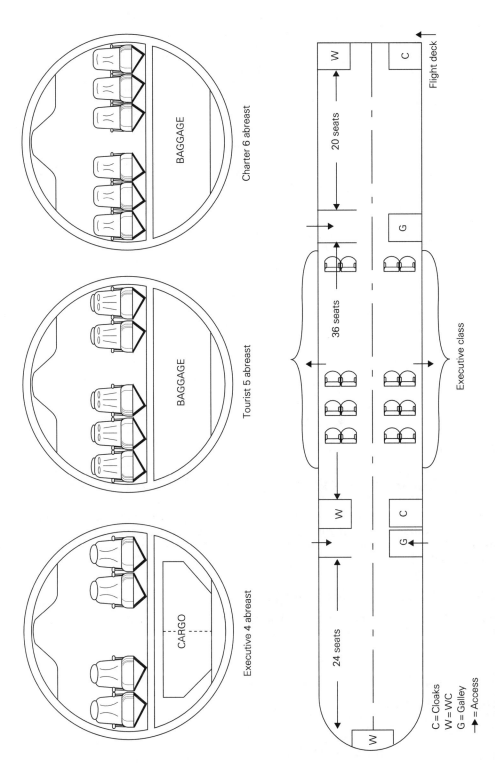

Executive 4 abreast

Tourist 5 abreast

Charter 6 abreast

CARGO

BAGGAGE

BAGGAGE

W

20 seats

G

36 seats

Executive class

24 seats

W

G

C

C

W

Flight deck

C = Cloaks
W = WC
G = Galley
➤ = Access

Fig. 4.8 Fuselage layout options

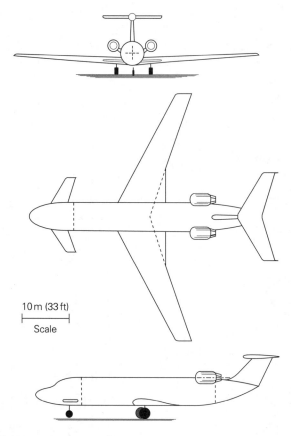

10 m (33 ft)

Scale

Fig. 4.9 Initial 'baseline' layout

4.7   Initial estimates

Before we can estimate aircraft performance, we must more accurately determine the aircraft mass, drag, lift and engine characteristics. Detailed calculations are not shown in the sections below as they follow conventional methods. Where appropriate reference is given to the methods used.

4.7.1   Mass and balance analysis

Each of the aircraft component masses will be estimated separately and then summed. For presentational convenience, the values below are quoted in kg only (conversion $1 \, \text{kg} = 2.205 \, \text{lb}$).

Wing structure

Using the estimation formula in reference 1, with:

$$\text{MTOM} = 81\,000 \, \text{kg} \qquad N_{\text{ult}} = 3.75$$

$$\text{Wing area} = 180 \, \text{sq. m} \qquad \text{Wing aspect ratio} = 10$$

Wing taper ratio $= 0.3$

Wing av. thickness $= 15\%$ Wing LE sweep $= 30°$

$R = (M_{wing} + M_{fuel}) = (8100 + 32\,400)\,\text{kg}$

Hence, the wing is calculated at: $M_{wing} = 11\,209\,\text{kg}$

This is 13.8 per cent of MTOM. This is uncharacteristically high for this type of aircraft. The formula is based on old and existing aircraft types. The average value of aspect ratio for the source aircraft is about 6. The much higher aspect ratio of our design (10) seems to have caused a large increase in wing mass. In addition, the formula was based on traditional metallic construction whereas our design will incorporate substantial composite structure. For these reasons, we will apply a reduction factor of 25 per cent to the estimated mass:

$$M_{wing} = 11\,209 \times 0.75 = 8407\,\text{kg}\ (10.4\%\ \text{MTOM})$$

As we have used the wing gross area, we will assume that this mass includes the flap weight.

Tail structure

With little knowledge of the tail design at this time, we will assume a representative percentage. We know that the extra control surface (canard) will add some weight so we will use a slightly higher percentage than normal for this type of aircraft (2.5 per cent). As we will be constructing these surfaces in composite materials, we will apply a technology reduction factor of 25 per cent.

$$M_{tail} = 0.025 \times 81\,000 \times 0.75 = 1519\,\text{kg}\ (1.9\%\ \text{MTOM})$$

Fuselage structure

The mass of the body will be estimated using a formula for body[1] with the parameter values shown below:

$$\text{MTOM} = 81\,000\,\text{kg};\ O/A\ \text{length} = 40.0\text{m};$$
$$\text{max. diameter} = 3.75\,\text{m};\ V_D = 255\,\text{m/s}$$

Gives:

$$M_{body} = 9232\,\text{kg}$$

Increasing by 8 per cent for pressurisation, 4 per cent for tail engine location and reducing by 10 per cent for modern materials and construction gives:

$$M_{body} = 9306\,\text{kg}\ (11.5\ \text{MTOM})$$

Nacelle structure

Based on an engine thrust of 254 kN:

$$M_{nacelles} = 1729\,\text{kg}\ (2.1\%\ \text{MTOM})$$

Landing gear

We will assume this to be 4.45 per cent MTOM:

$$M_{\text{landing gear}} = 3604\,\text{kg}$$

Surface controls

Using a typical value of $0.4 \times (\text{MTOM})^{0.684}$:

$$M_{\text{s/controls}} = 911\,\text{kg}$$

Aircraft structure

Summing the above components gives the aircraft structural mass:

$$M_{\text{structure}} = 25\,475\,\text{kg}$$

This is 31.5 per cent MTOM which is representative of this class of aircraft.

Propulsion system

Using the quoted engine dry weight of 5250 lb (each) and a system multiplying factor of 1.43, gives:

$$M_{\text{propulsion}} = 6810\,\text{kg}\ (8.4\%\ \text{MTOM})$$

Fixed equipment

A typical value for this type of aircraft is 8 per cent but as we will be providing more cabin services we will increase this to 10 per cent MTOM:

$$M_{\text{fix/equip}} = 8100\,\text{kg}$$

Aircraft empty (basic) mass

Summing the structure, propulsion and fixed equipment masses gives:

$$M_{\text{empty}} = 40\,385\,\text{kg}\ (49.9\%\ \text{MTOM})$$

Operational empty mass (OEM)

Adding the flight crew ($2 \times 100 = 200\,\text{kg}$), cabin crew ($4 \times 70 = 280\,\text{kg}$), cabin service and water (@21.5 kg/pass. $= 1724\,\text{kg}$) to the aircraft basic mass gives:

$$M_{\text{OEM}} = 42\,589\,\text{kg}\ (52.8\%\ \text{MTOM})$$

This is close to the assumed value from the literature search.

Aircraft zero-fuel mass (ZFM)

This is the OEM plus the passengers ($80 \times 80 = 6400\,\text{kg}$) and the passenger baggage ($40 \times 80 = 3200\,\text{kg}$), giving:

$$M_{\text{ZFM}} = 52\,189\,\text{kg}\ (64.4\%\ \text{MTOM})$$

Maximum take-off mass (MTOM)

In the analysis above, the MTOM has been assumed to be 81 000 kg.

Fuel mass

The aircraft zero-fuel mass (ZFM) and the assumed maximum take-off mass define the available fuel mass:

$$M_{fuel} = \text{MTOM} - \text{ZFM}$$

Hence,

$$M_{fuel} = 81\,000 - 52\,189 = 28\,811\,\text{kg (35.6\% MTOM)}$$

This is less than previously assumed so it will be necessary to recalculate the fuel mass ratio using a more detailed method. The Breguet range equation can be used if assumptions are made for the aircraft (L/D) ratio and the engine fuel consumption (c).

$$\text{Range} = (V/c)(L/D)\log_e(M_1/M_2)$$

where V = cruise speed = $M0.85^* = 255\,\text{m/s} = 485\,\text{kts}$
c = assumed engine fuel consumption = 0.55 N/N/hr
(L/D) assumed to be = 17 in cruise
M_1 = start mass = MTOM = 81 000 kg
M_2 = end mass = ZFM = 52 189 kg

*the cruise speed is set to avoid incurring significant drag rise. Typically, a 20 point drag count (one drag count = 0.0001) rise sets this speed.
With the speed in knots, this gives:

$$\text{Range} = 6589\,\text{nm}$$

Although this may seem close to the specified range of 7000 nm, it is necessary to account for the fuel allowances. Using the formula shown below,[1] the required design range can be used to calculate the equivalent still-air-range (ESAR). This includes the fuel reserves (diversion and hold) and other contingency fuel.

$$\text{ESAR} = 568 + 1.06\,\text{design range}$$

Hence, the required ESAR (for our specified design range of 7000 nm) = 7988 nm
Reversing this process with the 6589 nm range calculated above would only give a design range of 5927 nm. This shows that there is a substantial shortfall in the design range. The original assumption of 0.35 for the fuel fraction seems to be in error for our design. This is a major error as the aircraft is not viable at an MTOM of 81 000 kg. We can use the Breguet equation above to determine a viable fuel ratio for the 7988 nm ESAR.

$$7988 = (485/0.55)17(\log_e(M_1/M_2))$$
$$(M_1/M_2) = 1.704$$
$$M_2 = M_1 - M_{fuel}$$
$$(M_{fuel}/\text{MTOM}) = 1 - (1/1.704) = 0.413$$

This is a big change to the value used in the initial MTOM prediction. We will use the aircraft empty mass ratio of 0.495 determined in the component mass evaluation above, in a new estimation of MTOM:

$$\text{MTOM} = 110\,520/(1 - 0.495 - 0.413) = 114\,348\,\text{kg}\ (25\,214\,\text{lb})$$

Note: the denominator in the expression above is only 0.092. This makes the evaluation very unstable. For example, if the empty mass and fuel mass ratios are incorrect by only $+/-1$ per cent the MTOM would change to 146 111 and 93 928 kg respectively. These values are 28 per cent more and 18 per cent less than the predicted value. This illustrates the inappropriate use of the initial MTOM prediction method when the denominator is small. However, as we do not have another prediction, we will have to use the 114 348 value and, as quickly as possible, validate it with a detailed component mass prediction.

As we have still not evaluated the aircraft performance we will need to use the thrust and wing loading values (0.32 and 450) determined in the literature survey. The new value of MTOM will force a change in the engine thrust and wing area:

$$\text{Engine thrust (total)} = 359\,\text{kN}\ (80\,710\,\text{lb})$$

$$\text{Wing area (gross)} = 254\,\text{sq. m}\ (2730\,\text{sq. ft})$$

As other alterations are likely to follow, changes to the engine selection caused by the above will not be considered at this point in the design process.

The values above, together with the resulting heavier MTOM, will change the component mass predictions made earlier. Using the same methods, the aircraft mass statement (kg) is calculated as listed below:

Wing structure	= 13 224	(11.6% MTOM)
Tail structure	= 2859	(2.5% MTOM)
Body structure	= 10 278	(9.0% MTOM)
Nacelle structure	= 2441	(2.1% MTOM)
Landing gear	= 5088	(4.45% MTOM)
Surf. controls	= 1153	(1.0% MTOM)
STRUCTURE	= 35 043	(30.6% MTOM)
Propulsion	= 9625	
Fixed equip.	= 11 435	
A/C EMPTY	= 56 103	(49.1% MTOM)
Operational items	= 1724	
Crew	= 480	
OEM	= 58 307	(51.0% MTOM)
Passengers	= 6400	
Baggage	= 3200	
ZFM	= 67 907	(59.4% MTOM)
Fuel	= 46 441	(40.6% MTOM)
MTOM	= 114 348	(100% MTOM)
	(252 137 lb)	

Note: the fuel ratio is still slightly under the requirement. The calculation should be done again to obtain the correct ratio.

Applying the Breguet range equation with values determined above ($M_1 = 114\,348$ and $M_2 = 67\,068\,\text{kg}$) gives a range of 7812 nm. (A spreadsheet method with an iterative calculation function is very useful in this type of work.) As we have still made some gross assumptions in the calculations above (e.g. if the aircraft L/D ratio is 18 instead of 17 the range would increase to 8272 nm), we will continue the design process using the 114 348 MTOM value.

Before moving on to the aerodynamic calculations, it is necessary to redraw the aircraft with larger wings, control surfaces and engines. The fuselage shape will not change. The overall aircraft layout will be similar to that shown later in Figure 4.11. Assuming that the internal wing volume increases as the cube of the linear dimensions, the wing will be able to hold 52 668 kg (116 134 lb). This will be large enough to hold the extra fuel mass of the bigger aircraft.

4.7.2 Aerodynamic estimations

Conventional methods for the estimation of aircraft drag can be used at this stage in the design process. As it is assumed that, with careful detail design, the aircraft can fly at speeds below the critical Mach number, substantial additions due to wave drag can be ignored. Therefore, only zero-lift and induced drag estimations are required.

Parasitic drag is estimated for each of the main component parts of the aircraft and then summed to provide the 'whole aircraft' drag coefficient. The component drag areas are normalised to the aircraft reference area (normally the wing gross area).

$$\text{Component parasitic drag coefficient, } C_{Do} = C_f F Q [S_{wet}/S_{ref}]$$

where C_f = component skin friction coefficient. This is a function of local
               Reynolds number and Mach number
        F = component form (shape) factor which is a function of the geometry
        Q = a multiplying factor (between 1.0 and 1.3) to account for local
               interference effects caused by the component
        S_{wet} = component wetted area
        S_{ref} = aircraft drag coefficient reference area (normally the wing gross area)

Aircraft not in the 'clean' condition (e.g. with landing gear and/or flaps lowered, with external stores or fuel tanks) will also be affected by extra drag (ΔC_{Do}) from these items. The extra drag values will be estimated from past experience. Several textbooks (e.g. references 1 to 5) and reports provide data that can be used.

$$\text{Whole aircraft parasitic drag, } C_{Do} = \sum[\text{component } C_{Do}] + \sum[\Delta C_{Do}]$$

From the previous analysis the reference area will be 254 sq. m (2730 sq. ft).

Cruise (at 35 000 ft and M0.85)

The component drag estimations for the aircraft in this clean configuration are shown in Table 4.5.

From reference 1, induced drag coefficient,

$$C_{Di} = (C_1/C_2/\pi A)C_L^2 + (0.0004 + 0.15 C_{Do})C_L^2$$

where   (C_1 and C_2) are wing geometry factors (close to unity) and (A) is the wing aspect ratio.

For our design the equation above gives,

$$C_{Di} = 0.035 C_L^2$$

Table 4.5

Component	R. No.*	C_f	F	Q	Swet	(ΔC_{Do})
Wing	3.32	0.00234	1.50	1.0	432.0	0.00593
H controls	1.86	0.00255	1.31	1.2	59.7	0.00094
V control	2.99	0.00237	1.32	1.2	33.7	0.00050
Fuselage	2.65	0.00175	1.07	1.0	437.4	0.00321
Nacelles (2off)	3.42	0.00231	1.5	1.0	84.6	0.00116
Secondary items						0.00192
					$\sum$ (aircraft C_{Do})	0.01376

* *R. No.* = Reynolds number $(\times 10^{-7})$

4.7.3 Initial performance estimates

Cruise

Hence, at the start of cruise:

$$C_D = 0.0137 + 0.035C_L^2 \quad \text{and} \quad C_L = 0.339,$$

Making,

$$C_D = 0.01\,774$$

Therefore, at the start of cruise,

$$\text{Aircraft drag} = 54.3\,\text{kN}$$

Assuming, at this point, the aircraft mass is (0.98 MTOM), then L/D ratio = 19.1

Engine lapse rate to cruise altitude = 0.197 (based on published data[1])
Hence, available engine thrust = $0.197 \times 359 = 70.7\,\text{kN}$

This shows that the engine cruise setting could be 77 per cent of the take-off rating.

At the end of the cruise phase, assuming that aircraft mass is (0.65 MTOM) the aircraft C_L reduces to 0.225 if the cruise height remains constant. This reduces the aircraft L/D ratio to 14.5. This would increase fuel use. To avoid this penalty the aircraft could increase altitude progressively as fuel mass is reduced to increase C_L. This is called the 'cruise-climb' or 'drift-up' technique during which the aircraft is flown at constant lift coefficient. At the end of cruise, the aircraft would need to have progressively climbed up to a height of 43 600 ft. To reach such an altitude may not be feasible if the engine thrust has reduced (due to engine lapse rate) below that required to meet the cruise/climb drag.

Cruise/climb

At the initial cruise height, the aircraft must be able to climb up to the next flight level with a climb rate of at least 300 fpm (1.524 m/s).

This will require an extra thrust of 6758 N.

Adding this to the cruise drag gives 61.1 kN. This is still below the available thrust at this height (approximately 86 per cent of the equivalent take-off thrust rating).

Performing a reverse analysis shows that an aircraft (T/W) ratio of 0.276 would be adequate to meet the cruise/climb requirement.

Landing

The two-dimensional (sectional) maximum lift coefficient for the clean wing is calculated at 1.88. The finite wing geometry and sweep reduce this value to 1.46. Adding simple (cheap) trailing edge flaps ($\Delta C_{L\max} = 0.749$) and leading edge device ($\Delta C_{L\max} = 0.198$) produces a landing max. lift coefficient for the wing of 2.41.

At this stage in the design process, it is sufficient to estimate the landing distance using an empirical function. Howe[3] provides as simplified formula that can be used to estimate the FAR factored landing distance. The approach lift coefficient ($C_{L\mathrm{app}}$) is a function of the approach speed. This is defined in the airworthiness regulations as 1.3 times the stall speed in the landing configuration. Hence $C_{L\mathrm{app}}$ is $(2.4/1.69) = 1.42$. Assuming the landing mass is (0.8 MTOM), the approach speed is estimated as 64 m/s (124 kt). This equates to a landing distance of:

$$\text{FAR landing distance} = 1579\,\text{m (5177 ft)}$$

This is less than the design requirement of 1800 m.

Take-off

Reducing the flap angle for take-off decreases the max. lift coefficient to 2.11.

As for the landing calculation, it is acceptable at this stage to use an empirical function to determine take-off distance (TOD). For sea level ISA conditions, reference 3 gives a simplified formula for the FAR factored take-off distance. Assuming lift-off speed is 1.15 stall speed, the lift coefficient at lift-off will be $(2.11/1.15^2) = 1.59$, with $(T/W) = 0.32$ and $(W/S) = 450 \times 9.81 = 4414$ N/sq. m, the following values are calculated:

Ground run $= 1292.6$ m, Rotation distance $= 316.1$ m, Climb distance $= 81.6$ m,

FAR TOD $= 1690$ m (5541 ft)

This easily meets the previously specified 1800 m design requirement.

Second segment climb with one engine inoperative (OEI)

For the second segment calculation the drag estimation follows the same procedure as described above but in this case the Reynolds number and Mach number are smaller. The undercarriage is retracted and therefore does not add extra drag but the flaps are still in the take-off position and will need to be accounted for in the drag estimation. The failed engine will add windmilling drag and the side-slip (and/or bank angle) of the aircraft will also add extra drag.

Using published methods to determine flap drag[3] and other extra drag items[1]:

$$C_L = 1.59, \qquad C_{DO} = 0.0152, \qquad C_{DI} = 0.0376,$$

$$\Delta C_{D\mathrm{flaps}} = 0.015, \quad \Delta C_{D\mathrm{wdmill}} = 0.0033, \quad \Delta C_{D\mathrm{trim}} = 0.0008$$

These values determine an aircraft drag $= 116.1$ kN
Thrust available (one engine), at speed $V_2 = 161.5$ kN
This provides for a climb gradient (OEI) $= 0.0405$
This is better than the airworthiness requirement of 0.024
To achieve this requirement would demand only a thrust to weight ratio of 0.254

Later in the design process, it will be necessary to determine the aircraft balanced field length (i.e. with one engine failing during the take-off run).

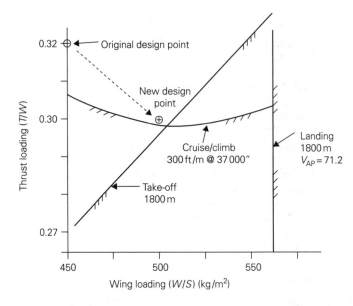

Fig. 4.10 Constraint diagram

4.7.4   Constraint analysis

The four performance estimates above have indicated that the original choice of aircraft design parameters (T/W, W/S) may not be well matched to the design requirements as each of the design constraints was easily exceeded. The assumed thrust and wing loadings were selected from data on existing aircraft in the literature survey. It seems that as our design specification is novel, this process is too crude for our aircraft. As we now have better knowledge of our aircraft geometry, it is possible to conduct a more sensitive constraint analysis. The methods described above will be used to determine the constraint boundaries on a T/W and W/S graph. The results are shown on Figure 4.10.

Moving the design point to the right and downwards makes the aircraft more efficient. The constraint graph shows that it would be possible to select a design point at T/W at 0.3 and W/S at 500 kg/sq. m (102.5 sq. ft). Recalculating the aircraft mass using the same method as above and with these new values gives:

Wing structure	= 11 387
Tail structures	= 2025
Body structure	= 10 050
Nacelle structure	= 2161
Landing gear	= 5088
Surface controls	= 1109

STRUCTURE MASS = 31 538 (29.2%)

Propulsion mass	= 8520
Fixed equipment	= 10 800

AIRCRAFT EMPTY MASS = 50 858 (47.1%)
OPERTN EMPTY MASS = 52 862 (49.0%)
ZERO FUEL MASS = 62 462 (57.8%)

Fuel mass = 45 538 kg (42%)
$$\text{MAX. MASS (MTOM)} = 108\,000\ (100\%)$$
$$(23\,814\ \text{lb})$$

Using this mass and our new thrust and wing loading ratios gives:

- Total engine thrust (static sea level) = 317.8 kN (71 450 lb)
- Gross wing area (reference area) = 216 sq. m (2322 sq. ft)

Assuming the wing tank dimensions are proportional to the wing linear size, the new wing area could accommodate 41 460 kg (91 400 lb) of fuel. This is less than predicted above (by 9 per cent). As we have made several assumptions and have not made a detailed analysis of the geometry and performance, we will delay the effect of this on the design of the wing until later in the design process.

4.7.5 Revised performance estimates

Range

With the cruise speed of 250 m/s (485 kt), assumed SFC of 0.55 force/force/hr, aircraft cruise L/D ratio of 17, initial mass (M_1) = MTOM (108 000 kg), and final mass (M_2) = ZFM (62 462 kg) gives:

$$\text{Range} = 8209\ \text{nm}$$

This is slightly longer than the previously estimated ESAR of 7988 nm but is within our calculation accuracy. The fuel ratio in the new design is 42.2 per cent whereas only 41.3 per cent is required therefore we have about 900 kg slack in the zero fuel estimation.

Cruise

With the new mass and geometry, the drag polar (start of cruise, 35 000 ft @ M0.85) is calculated as:
$$C_{\mathrm{D}} = 0.0148 + 0.0352 C_{\mathrm{L}}^2$$

At the start of cruise, the lift coefficient is 0.40, hence $C_{\mathrm{D}} = 0.0204$.
  This equates to a drag = 53.1 kN (11 938 lb), and hence a cruise $L/D = 19.5$
The engine lapse rate at cruise is 0.197. Therefore the available thrust at the cruise condition = 0.197 × 317.8 = 62.6 kN (14 073 lb)
This gives an engine setting in cruise of 85 per cent of the equivalent take-off rating

Cruise climb

Adding a climb rate of 300 fpm at the start of cruise makes the required thrust at the start of cruise = 59.4 kN (13 354 lb). This is 95 per cent of max. take-off thrust rating.

Landing

The approach speed is 64.5 m/s (125 kt)
This seems reasonable for regional airport operations
The landing distance is calculated as 1594 m (5225 ft)
This is well below the 1800 m design requirement

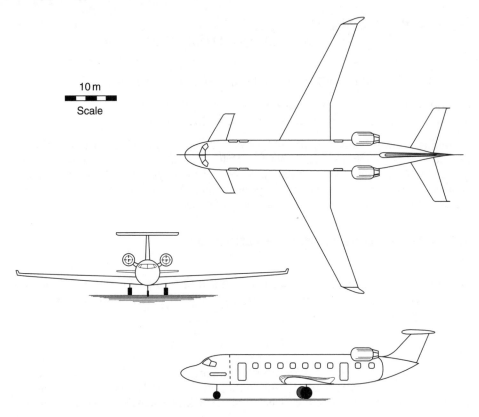

Fig. 4.11 Refined baseline layout

Take-off

The take-off distance is 1790 m (5869 ft)
The balanced field length is 1722 m (5647 ft)
These satisfy the design requirement of 1800 m
The second segment climb gradient (OEI) = 0.033
This satisfies the airworthiness requirement of 0.024

All of the design requirements have been achieved with the new aircraft geometry.
  It is now possible to draw the refined general arrangement of our aircraft, Figure 4.11.

4.7.6 Cost estimations

Using the methods described in reference 1:

For an aircraft OEM = 52 862 kg, the aircraft purchase price will be $42M (1995)
Assuming an inflation rate of 4 per cent per year
This brings the 2005 aircraft price = $62M

For engines of about 40 000 lb TO thrust, the price would be $4.0M (1995)

For two engines (2005 prices)   = $12M
Airframe cost = $62M – $12M   = $50M   (i.e. aircraft price less engines)
Assume 10% spares for airframe   = $5.0M
Assume 30% for engine spares   = $3.6M
                                   ──────
Total investment   = $70.6M

Assuming depreciation to 10 per cent over 20 years
Annual depreciation = $(0.9 \times 70.6)/20$ = $3.18 M
Assume interest on investment cost of 3.5% per yr   = $2.47 M
Assume insurance 0.5% per yr of investment   = $0.35 M
                                                      ──────
Total standing charges per year   = $6.00 M

For the cruise range of 7000 nm at 485 kt, the flight time will be 14.4 hr
Add 0.75 hours to account for airport ground operations = 15.15 hr

Total block time = 15.15 hr

Some cost methods use this time in the calculation of DOC. Others use the flight time only. We will use the flight time in the calculations below.
    Assume aircraft utilisation of 4200 hr per year (typical for long-range operations)

Standing charges per flying hour = $1429

Crew costs (1995) per hr = 2×360 for flight crew $+ 4 \times 90$ for cabin crew

= $1080 = $1594 per hr (2005)

Landing and navigation charges per flight = 1.5 cents/kg MTOM = $1620 per flight
Ground handling charge = $3220 per flight
Total airport charges = $4840 per flight = $336 per flight hr
From the mass and range calculations: fuel used for ESAR (8209 nm) = 45 538 kg
Estimated fuel used for the 7000 nm design range = 40 300 kg
Assuming that little fuel is burnt in the ground,
Fuel used per flight hour = 40 300/14.4 = 2798 kg
Fuel volume = 2798/800 = 3.5 sq. m = 3500 litres = 3500/3.785 = 924 US gal
Assuming the price of fuel is 90 cents per gal,

Fuel cost = $832 per flight hour

As maintenance costs are too difficult to assess at this time in the design process, we will assume them to account for 15 per cent of the total operating cost.

Total operating cost $ per flight hour

Standing charges	= 1429
Crew cost	= 1594
Airport charges	= 336
Fuel cost	= 832
Maintenance costs	= (15%) (739)
	──────
	= $4930 per flight hour

Hence DOC,   Total stage cost = 4616×14.4 = $70 996
                Aircraft mile cost = 66 477/7000   = $10.14
                Seat mile cost (100% load factor)   = 12.68 cents

Operators who lease the aircraft use 'cash DOC' to determine flight cost. They add the lease charges to their indirect costs as they are committed to this expense regardless of the aircraft utilisation. Cash DOC is calculated in the same way as above but without the standing charges. As aircraft maintenance is unaffected by the 'accountancy' method used to determine DOC, the cost is assumed to be the same as used above.

Cash DOC, Operational cost = $3504 per hr
           Total stage cost = $50 451
           Aircraft mile cost = $7.21
           Seat mile cost   = 9.01 cents

Assuming that the ticket price (LHR–Tokyo) is $4500 for the executive-class fare:
Revenue per flight (assuming 65 per cent load factor) = $234 000
This compares favourably with the direct operating stage cost of $70 996
Even allowing for a 100 per cent indirect operating cost (IOC) factor added to DOC, the operation would be viable

The seat mile costs calculated above are substantially larger than those quoted for high-capacity mixed-class services in which about 75 per cent of the seats are assigned to economy-class travellers. The revenue from such customers is significantly lower than from the executive class as they will be charged only about 20 per cent of the higher price fare. Without a detailed breakdown of the financial and accounting practices of an airline, it is impossible to determine the earning potential of the new service compared with the existing operation. However, the revenue assessment shown above is encouraging enough to continue with the project.

4.8   Trade-off studies

There are many different types of trade-off studies that could be undertaken at this stage in the design process. These range from simple sensitivity studies on the effect of a single parameter or design assumption, to extensive multi-variable optimisation methods. The studies shown below include trade-off plots that are used to determine the best choice of aircraft geometry. Wing loading and wing aspect ratio are chosen as the main trade-off parameters. These are regarded as the most significant design parameters for the short field, long-range requirements of the aircraft operation.

The studies shown in this section are presented as typical examples of the type of work appropriate at this stage of aircraft development. Many other combinations of aircraft parameters could have been selected and in a full project analysis would have been performed.

4.8.1   Alternative roles and layout

As mentioned in section 4.6.4 (fuselage layout), for all aircraft design studies it is necessary to consider the suitability of the aircraft to meet other operational roles. Although the principal objective of the project is to produce an efficient large business exclusive aircraft, we must also consider other mixed-class variants. In this way, a family of aircraft can be envisaged. This will increase the number of aircraft produced and reduce the design and development overhead per aircraft. Recognising this requirement, the fuselage diameter was designed to be suitable not only for the four abreast executive class seating but also five and six abreast layouts of higher capacity options. The cabin length of 22 metres plus 4 metres for services and egress space is a fixed parameter and

Table 4.6

		Rear cabin	Centre cabin	Front cabin	Total seats
A	Executive	24	32	24	80
B	Mixed*	35 econ.	50 econ.	24 exec.	85/24 = 109
C	Economy	35	50	35	120
D	Charter**	48	72	48	(168) = 150

* This provides 22 per cent business occupancy.
** The maximum capacity is reduced by about 10 per cent to account for extra
  spacing at emergency exits.

will control the layout and capacity of alternative roles. Within this length, various combinations of passenger layouts can be arranged. The position of doors and service modules (toilets, cupboards and galleys) is fixed but these can be used to provide natural dividers between classes.

From the previous fuselage layout drawing (Figure 4.8), the rear cabin is 6.5 metres, centre cabin 9.0 metres and front cabin 6.5 metres long. Using seat pitches of 1.1, 0.85, and 0.75 metres for executive/business, economy and charter classes respectively results in the layouts shown in Table 4.6.

For civil aircraft, it is common practice to stretch the fuselage in a later development phase. Typically, this may increase the payload by 35 per cent. Using this value (approximately), the single-class, economy version would grow to 160 passengers. At the 0.85 metre seat pitch, this would equate to a lengthening of the fuselage by 6.8 metres. To maintain aircraft balance a 2.8 m plug would be placed in the rear cabin and a 4.0 m plug forward of the wing joint. In this version the capacity of the aircraft would increase to the values shown below:

A Executive (single class) = 104 seats
B Mixed class   = 141 seats (105 econ. and 36 exec.)
C Economy (single class) = 160 seats
D Charter (single class)   = 204 seats

The extra capacity would require more passenger service modules and extra emergency exits to be arranged in the new cabin. This would reduce the space available for seating and slightly reduce the capacities shown above. Alternatively, the fuselage stretch would need to be increased by about a further 1.0 to 1.5 metres (40 to 60 in). Figure 4.12 shows some of the layout options described above.

Non-civil (military) versions of the aircraft could also be envisaged. With only 0.7 seat pitch for troop carrying a total of 186 soldiers could be carried in the original aircraft and 246 in the stretched version. The large volume cabin (for a small aircraft), the long endurance and the short field capabilities would be suitable for reconnaissance and electronic surveillance roles. In such operations, the reduced payload mass could allow extra fuselage fuel tanks to be carried to extend the aircraft duration. The high-speed, long-range performance could be useful for military transport command. Using this aircraft would avoid diplomatic complications caused by the need to refuel in foreign countries in conflict scenarios.

Many other versions of the aircraft may be envisaged (e.g. freighter/cargo, corporate jet, and communication platform) but these would not significantly affect the design of the current aircraft configuration.

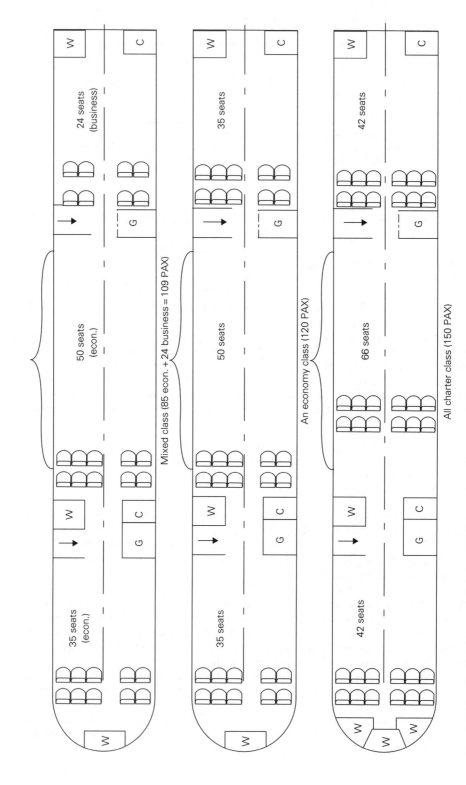

Fig. 4.12 Fuselage development options (see also Figure 4.8 for all executive (baseline) layout)

4.8.2   Payload/range studies

For any aircraft design, it is uncommon to consider just the capability of the aircraft at the design point. Trading fuel for payload with the aircraft kept at the max. design mass results in a payload range diagram (shown in Figure 4.13).

Point A (the design point) shows the aircraft capable of flying 80 executive-class passengers over a 7000 nm range. Points B and C relate to the alternative cabin layouts described in the section above. At point B the payload is 11 380 kg. This means that 1780 kg of fuel is sacrificed for payload. At point C, 2400 kg of fuel is lost. Assuming the aerodynamic and engine efficiencies remain unchanged, the available range in these two cases reduces to 6811 and 6675 nm respectively. The reduction of about 530 nm in range for a 50 per cent increase in passenger number is a result of the low value of $(M_{\text{pay}}/M_{\text{TO}})$ on this design. Stretching the design to accommodate 160 passengers as described above would therefore be relatively straightforward on this design. This development is also shown on the payload range diagram. Even after allowing for an increase in structure and system mass of 1000 kg, the range in this configuration only reduces to 5625 nm.

As the aircraft is seen to be relatively insensitive to changes in payload, it is of interest to determine the effect of passenger load factor. Commercial aircraft do not always operate at the full payload condition. For this type of operation, an average load factor of 70 per cent is common. With less payload, the aircraft could increase fuel load (providing that space is available to accept the extra fuel volume). At 70 per cent passenger load factor with extra fuel, the Breguet equation gives an increase in range of 668 nm. If extra space is not available, the aircraft at 70 per cent load factor and with normal fuel load would be able to fly a stage length of about 7500 nm. The sensitivity

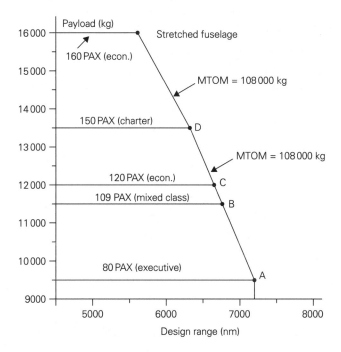

Fig. 4.13 Payload/range diagram (developments)

of the range calculation to passenger load factor raises the question of the choice of the realistic design payload. Designing for the 0.7 × 80 passenger load would significantly reduce the aircraft max. take-off mass and fuel load. This would considerably reduce the stage cost and aircraft price.

The payload/range study has shown that the loading conditions around the design point must be carefully considered. As the effect of range and associated fuel is uncharacteristically sensitive on this aircraft it is important to reconsider the original design specification to account for this aspect.

4.8.3 Field performance studies

The evaluation of field performance calculated earlier (section 4.7.3) was concerned with the aircraft at the design condition only. The predictions can be recalculated for variations in the aircraft parameters. In the earlier work, the take-off performance was seen to be well matched to the 1800 m requirement. However, it is also necessary to understand the sensitivity of the calculation to changes in the main, aircraft design parameters (e.g. thrust and wing loadings). A carpet plot can be constructed to show these effects (Figure 4.14).

Note: none of the study points achieves the 1400 m field length originally considered (section 4.3.3). The aircraft thrust loading is shown most influential in reducing take-off distance. To investigate this further, a second trade-off study has been conducted. Keeping the wing loading constant, the wing max. lift coefficient and thrust loading have been varied. The results are shown in Figure 4.15.

The carpet plot shows that increasing the thrust loading to 0.32 would allow a reduction in take-off lift coefficient to 2.05. This would reduce wing structure complexity and thereby wing mass. Obviously, an increase in engine thrust would also involve a corresponding increase in propulsion group mass. A more detailed study would need to be done later in the design process to draw a firm conclusion to this trade-off.

In the revised aircraft layout, the landing performance was shown to be well within the 1800 m design constraint. This suggests that changes could be considered to the aircraft. For a fixed wing area the two parameters that have an effect on the landing

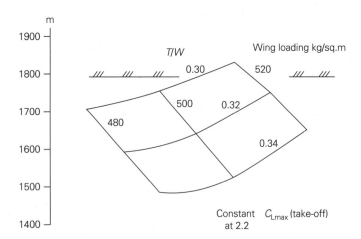

Fig. 4.14 Take-off distance study (T/W and W/S)

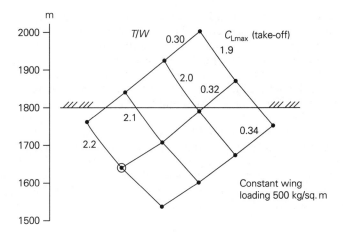

Fig. 4.15 Take-off distance study (T/W and C_{Lmax})

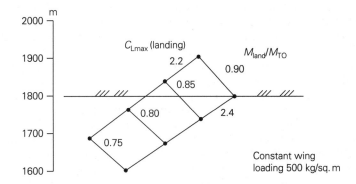

Fig. 4.16 Landing distance study

performance are aircraft max. lift coefficient (in the landing configuration) and the landing mass ratio ($M_{landing}/M_{TO}$). The trade-off study results are shown in Figure 4.16.

This shows that the aircraft is capable of landing in the 1800 m field at 90 per cent MTOM. The max. landing lift coefficient could be reduced to 2.2 and still allow a 82 per cent MTOM landing mass. This lift coefficient is the same as previously used for the take-off condition therefore a further set of trade-off studies should be done to select the best combination of lift coefficients for take-off and landing. This would fix the flap type and deflection angles to give the optimum design combination. To do this, more detailed aerodynamic analysis is required than is available at this stage in the development of the aircraft.

4.8.4 Wing geometry studies

To conduct a full and accurate analysis of the wing parameters (e.g. area and aspect ratio) would involve a full, multivariate optimisation method. As most of the design parameters are interconnected this would be a complex process. At this early stage in the design process and with limited resource and time available, it is not possible to undertake such a comprehensive study. Some simplifying assumptions are necessary to enable a sensitivity study to be done. For example, we may assume that the engine parameters are kept constant and that the less sensitive mass components (e.g. surface controls, systems, etc.) are held constant or considered to be directly proportional to MTOM. With such assumptions, the results of the study can only be used to indicate the sensitivity of variations to the design and the direction of possible changes to the existing configuration.

A study has been completed around the current aircraft configuration to investigate the effect of changes to wing area and aspect ratio on some of the aerodynamic and mass parameters. A series of carpet plots (Figures 4.18 to 4.23) illustrate the results of the study. To allow comparison to the earlier work, the study used wing loading to represent area variations. The resulting wing area values are plotted in the carpet plots. Wing loading (kg/sq. m) and aspect ratio values selected for the study are shown below:

Wing loading   400, 450, 500, 550
Aspect ratio   8, 10, 12, 14

To appreciate the geometrical implications of these changes, the extreme layouts (400/14 and 550/8) together with the existing baseline configuration (500/10) are illustrated in Figure 4.17. (Note: the drawings of these aircraft are illustrative only as they have not been balanced.)

As expected, the study shows that increasing the size of the wing (i.e. reducing wing loading) and/or increasing aspect ratio increases the wing mass. Figure 4.18 illustrates these effects clearly and provides quantitative data of the mass changes around the design point (500/10). The increasing slope of the aspect ratio lines shows the progressive mass penalty, especially for the larger area wings.

Wing mass, although important, represents only a component of aircraft mass. The combined effect on aircraft empty mass is illustrated in Figure 4.19. Although a similar pattern is seen on this plot, the changes represent a smaller proportion (about a quarter of the previous percentage values). For example, moving from the design point to point 550/8 is shown to reduce wing mass by about 35 per cent but the empty mass is reduced 8 per cent. Note: the wing loading of 550 was shown to violate the original take-off constraint (Figure 4.10). Making this move would require the take-off and possibly the climb performance to be reconsidered.

Making the wing smaller and increasing aspect ratio has a significant effect on both parasitic and induced drag. Both will be reduced. Figure 4.20, which plots aircraft lift/drag ratio, shows how the aerodynamic efficiency of the aircraft is improved. Note that the design point shows a value higher than that originally assumed (i.e. $L/D = 17$). Over the range of geometrical changes investigated the L/D ratio varies between 16 and 21. This is a significant variation that shows the sensitivity of choice of wing geometry.

The aircraft L/D ratio, and max. take-off mass (discussed below) are important parameters in the calculation of the required fuel mass to fly the 7000 nm stage length. Assuming that the cruise speed and engine specific fuel consumption remain unchanged from their previous values, the resulting fuel mass calculations are shown in Figure 4.21. At each of the wing loading lines the 'optimum' aspect ratio value moves progressively from about 9 for the large wing to 14 for the smallest wing. At the design wing loading of 500, there appears to be a small advantage to increasing aspect ratio from the design value of 10 to 12. Extending to 14 is not seen to be worthwhile.

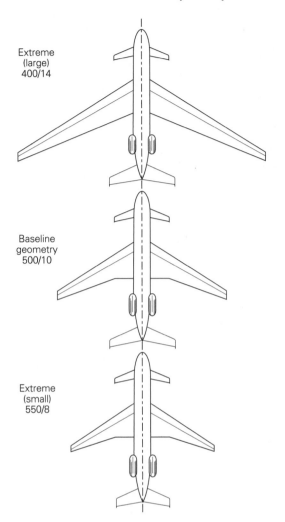

Extreme
(large)
400/14

Baseline
geometry
500/10

Extreme
(small)
550/8

Fig. 4.17 Geometrical variations

Aircraft take-off mass (MTOM) is dependent on both structure mass and fuel mass. The studies above have shown that the wing geometrical changes may increase structure mass but then reduce fuel mass. The combined effect is shown in the MTOM carpet plots (Figure 4.22). Due to the reduced fuel mass, the significance of the structure mass changes is eroded but the overall pattern remains similar to the empty mass plots discussed earlier. A move from the design point, to a wing loading of $550 \, \text{kg/m}^2 \, (112.6 \, \text{lb/ft}^2)$ and aspect ratio of 8 (i.e. 550/8) would reduce MTOM by about 5 per cent. This is a significant reduction and is worth investigating further if the economic studies described below confirm this advantage.

Wing area is a function of wing loading and MTOM. To show the dimensional effects of the changes in these parameters the absolute values for wing area have been plotted (Figure 4.23). Note the significance of aspect ratio on the larger wings and the relatively low sensitivity for small wings.

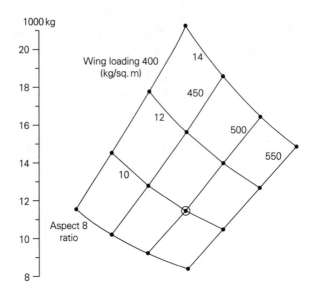

Fig. 4.18 Trade-off study: wing mass

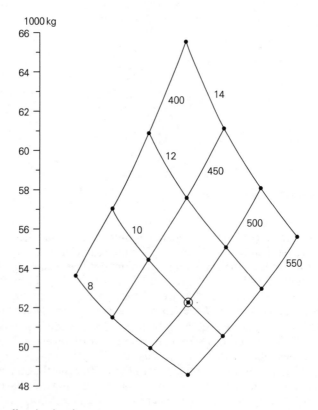

Fig. 4.19 Trade-off study: aircraft empty mass

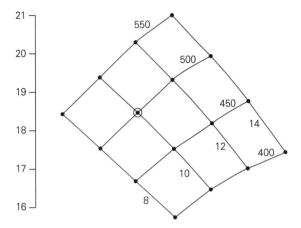

Fig. 4.20 Trade-off study: cruise L/D ratio

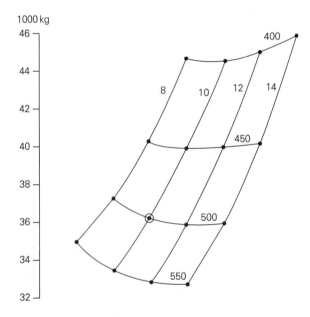

Fig. 4.21 Trade-off study: stage fuel mass

4.8.5 Economic analysis

The results from the studies above can be used, together with operational data, to assess the economic viability and sensitivity to the aircraft geometrical changes. The aircraft price is related to the aircraft empty mass and engine size. The cost of fuel is proportional to fuel mass. Other operational costs are related to aircraft take-off mass. Hence, changes to the aircraft configuration will affect both aircraft selling price and operating costs. For civil aircraft designs, these two cost parameters are often selected as the principal design drivers (optimising criteria). Although the aircraft configuration

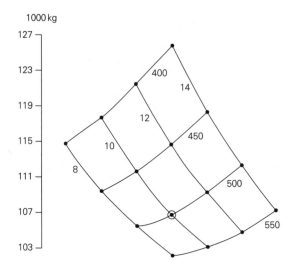

Fig. 4.22 Trade-off study: aircraft max. TO mass

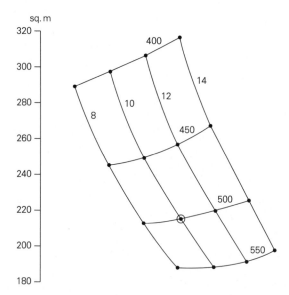

Fig. 4.23 Trade-off study: wing area

may not be selected at the optimum configuration for these parameters, the design team will need to know what penalty they are incurring for designs of different configuration.

All of the cost calculations have been normalised to year 2005 dollars by applying an inflation index based on consumer prices. Several separate cost studies have been performed as described below.

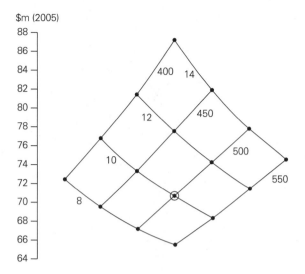

Fig. 4.24 Trade-off study: aircraft price

Aircraft price

Aircraft price is one component in the evaluation of total investment. This includes the cost of airframe and engine spares. For this aircraft, the total investment is about 12 per cent higher than the aircraft price.

Figure 4.24 shows the variation of aircraft price for the geometrical changes considered previously. At the design point the price is estimated to be $70.5 m. The carpet plot shows that this price would fall by about 5 per cent if the aspect ratio was reduced to 8, and by about 9 per cent if the configuration was moved to point 550/8. The effect of reducing wing loading progressively increases aircraft price (e.g. reducing wing loading to 400 kg/sq. m increases the price by 9 per cent). Similarly, increasing wing aspect ratio increases price (e.g. moving from 10 to 14 increases price by over 10 per cent). Without consideration of other operating costs, the main conclusion of this study is to move the design point to lower values of both wing loading and aspect ratio.

Direct operating cost (DOC) per flight

There are two fundamentally different methods of estimating aircraft DOC. The traditional method includes the depreciation costs of owning the aircraft. On this aircraft, this would be about 33 per cent of the total DOC. If the aircraft operator leases the aircraft, the annual cost of the aircraft is regarded as a capital expenditure. This would be considered as an indirect aircraft operating cost. In this case, the aircraft standing charges (depreciation, interest and insurance) are not included in the calculation and the resulting cost parameter is termed 'Cash DOC'. It is important to calculate both of the DOC methods. The results of the DOC calculations are shown in Figures 4.25 and 4.26.

The DOC per flight at the design point (500/10) is $72 740. This figure would be reduced by 3 per cent if the design was moved to point 550/8 and still satisfy the technical design requirements. Increasing wing area and/or aspect ratio from the design point is not seen to be advantageous. At the design point the Cash DOC is estimated to be

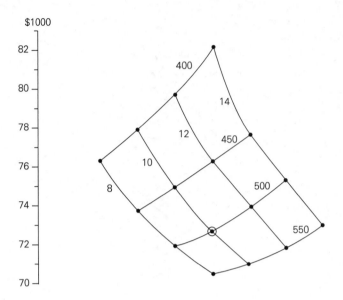

Fig. 4.25 Trade-off study: DOC per flight

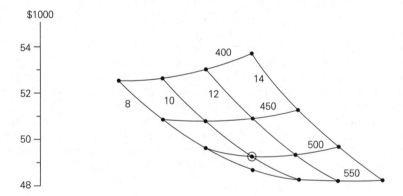

Fig. 4.26 Trade-off study: cash DOC per flight

$49 470. In this case, curves are seen to be flatter than for the full DOC values. This results in optimum points for aspect ratio. At the design point, the existing value of aspect ratio is seen to be optimum. Moving to the higher wing loading (550), if feasible, would reduce Cash DOC by about 2 per cent.

It is of interest to note that the design conclusions from the two DOC methods are different. This implies that the design strategy to be adopted is conditional on the accounting practices used by the operator. This is a good example of the need for the designers to understand the total operating and business environment in order to select the best aircraft configuration.

Seat mile cost

The cost of flying the specified stage (design range) is dependent on the payload. In the DOC calculations above, the aircraft has been assumed to be operating at full payload. This is conventional practice as it allows the maximum seats to be used in the evaluation of seat mile costs. Flying at max MTOM, the DOC per flight does not vary with passenger numbers. The seat mile cost (SMC) shown in Figures 4.27 and 4.28 (for DOC and Cash DOC respectively) are evaluated for the 80-seat executive version of the aircraft.

Other versions have been evaluated at the design point and are listed in Table 4.7.

Note the powerful effect of passenger numbers in reducing SMC and the substantial reduction in the Cash SMC method. When using values from other aircraft it is important to know the basis on which cost data has been calculated.

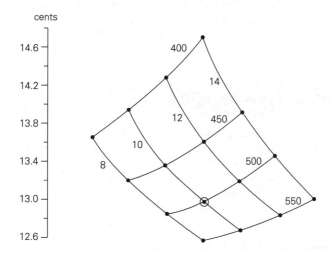

Fig. 4.27 Trade-off study: seat-mile cost (SMC)

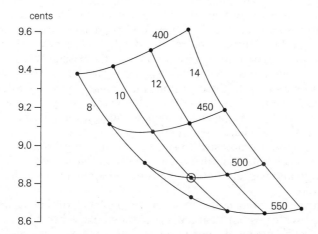

Fig. 4.28 Trade-off study: cash SMC

Table 4.7

Layout	Passengers	SMC	Cash SMC
Executive only	80	12.99	8.83
Mixed class	107	9.74	6.62
Economy only	120	8.70	5.91
Charter only	150	6.85	4.68

The SMC and Cash SMC carpet plots show a similar pattern to the DOC figures. The SMC curves show an advantage to reducing aspect ratio for all values of wing loading but this advantage reduces with higher wing loadings. For the design point, a reduction of aspect ratio to 8 would be recommended. For the Cash SMC calculation, the best aspect ratio varies from 8 at low loading to 12 at the high loading. For the design point, the value of 10 is seen to be about 'optimum'.

4.9   Initial 'type specification'

At the end of the initial concept stage it is important to record all of the known details of the current design. This document forms the initial draft of the aircraft type specification. As the design evolves over subsequent stages, this document will be amended and enlarged until it forms the definitive description of the final configuration. The initial draft will form the input data for the next stage of the design process. The sections below are typical of a professional aircraft specification.

4.9.1   General aircraft description

This aircraft is designed for exclusive, business/executive, long-range routes from regional airports. Although apparently conventional in configuration, it incorporates several advanced technology features. These include natural laminar flow control, composite material and construction, enhanced passenger cabin services and comfort standards, and three-surface control and stability. The single aisle cabin layout is arranged to accommodate four abreast seating for the baseline executive configuration. In other configurations it will provide five abreast economy class seating and six abreast charter operations. In these versions the increased passenger numbers reduce the range capability of the aircraft. In the baseline executive layout, space for 80 sleeperettes at 1.1 m (44 in) pitch is available. For the other layouts, 120 economy seats at 0.8 m (32 in) pitch or 150 charter seats at 0.7 m (28 in) pitch are feasible. Toilet, galley and wardrobe provision is adjusted to suit the layout using fixed service facilities in the cabin. Emergency exits and other safety provisions meet FAR/JAR requirements. Underfloor cargo and baggage holds are positioned fore and aft of the wing/fuselage junction structure.

To reduce engine operating noise intrusion into the cabin, during the long endurance flights, the engines are positioned at the rear of the fuselage, behind the cabin pressure bulkhead. Several existing and some proposed new engine developments are suitable to power the aircraft. This provides commercial competitiveness and flexibility to the potential airline customers. All the engines are modern, medium-bypass (typically 6.0) turbofans offering efficient fuel economy.

The modern, aerodynamically efficient, high aspect ratio wing layout provides good cruise efficiency. A lift to drag ratio in cruise of 19 is partly achieved due to the aerodynamic section profiling and the provision of natural laminar flow. Leading and trailing edge, high-lift devices provide the short field performance required for operation from regional airfields.

The three-surface (canard, main wing and tail) layout offers a reduction to trim drag in cruise and improved ride comfort. Integrated flight control systems with fly-by-wire actuation to multi-redundant electric/hydraulic controllers provide high levels of reliability and safety.

Aircraft manufacture combines established high-strength metallic materials with new composite construction techniques. The combination of conventional and novel structural and manufacturing practices offers reduced structural weight with confidence.

4.9.2 Aircraft geometry

Principal dimensions

Overall length	= 43.0 m,	141 ft
Overall height	= 13.0 m,	42.6 ft
Wing span (total)	= 48.0 m,	157 ft

Main wing

Gross (ref.) area	= 216 sq. m,	2322 sq. ft
Aspect ratio	= 10	
Sweepback (LE)	= 22°	
Mean chord	= 4.65 m,	15.25 ft
Taper ratio	= 0.3	
Thickness (mean)	= 11%	

Control surfaces

Horizontal tail area	= 20.0 sq. m,	215 sq. ft
Vertical (fin) area	= 15.5 sq. m,	167 sq. ft
Canard area	= 7.0 sq. m,	75 sq. ft

Fuselage/cabin

Fuselage length	= 40.0 m,	131 ft
Cabin outside dia.	= 3.6 m,	142 in
Pass. cabin length	= 22.0 m,	72 ft

Landing gear

Wheelbase	= 18.0 m,	59 ft
Track	= 8.25 m,	27 ft

Engines (two)
Various types, static SL thrust (each) = 160 kN, 35700 lb

4.9.3 Mass (weight) and performance statements

Mass statement

Aircraft empty mass	50 858 kg,	112 142 lb
Aircraft operational mass	52 862 kg,	116 560 lb
Aircraft max. (design) mass	108 000 kg,	238 140 lb

Baseline (executive) version (80 PAX)
Zero fuel mass	62 462 kg,	137 729 lb
Payload	9600 kg,	21 168 lb
Fuel load	45 538 kg,	10 0411 lb
Still air range	7200 nm	

Mixed-class version(107 PAX)
Payload	11 380 kg,	25 093 lb
Fuel	43 758 kg,	96 486 lb
Still air range	6810 nm	

All economy version (120 PAX)
Payload	12 160 kg,	26 812 lb
Fuel	44 138 kg,	97 324 lb
Still air range	6675 nm	

Charter version (150 PAX)
Payload	13 450 kg,	29 657 lb
Fuel	41 688 kg,	91 922 lb
Still air range	6350 nm	

Stretched version (160 economy PAX)
Payload	16 000 kg,	35 280 lb
Still air range	5600 nm	

Performance statement

(baseline version with 80 PAX and fuel):
Cruise speed	= M0.85
Cruise altitude	= 36 000 ft
Initial cruise altitude climb rate	= 300 fpm
Take-off distance	= 1790 m (5870 ft)
Balanced field length	= 1720 m (5670 ft)
Second segment climb gradient	= 0.033
Approach speed	= 125 kt
Landing distance	= 1594 m (5225 ft)

4.9.4 Economic and operational issues

Cost statement

(baseline aircraft, 2005 US dollars)
Aircraft price	= $62.0 M
Total investmt/aircraft	= $70.6 M
Standing charges/yr	= $6.0 M
Standing charges/flt hr	= $1430
DOC/fl hr	= $4930
Stage cost	= $71 000
Aircraft mile cost	= $10.14
Seat mile cost (100% PAX)	= 12.7 cents
Cash DOC/flt hr	= $3500
Total stage cash cost	= $50 450
Aircraft cash mile cost	= $7.20
Cash seat mile cost (100%)	= 9.0 cents

Operational statement

The aircraft is capable of stretching to accommodate up to 204 charter seats. In a military role the baseline aircraft can seat 186 soldiers and in the stretched version 246 troops. In each of these versions it would be possible to fly 7000 nm (unrefuelled).

Other roles for the aircraft could include:

- Civil corporate jet
- Freighter
- Military refuelling tanker
- Communication platform
- Military surveillance aircraft
- Military supply aircraft

These versions of the aircraft have not been considered in the overall geometrical layout of the aircraft in the initial design process. A short study would be appropriate when the initial baseline study has been completed to identify any small changes to the aircraft layout to accommodate any of the above roles.

4.10 Study review

This aircraft project has shown how, for a relatively simple aircraft, the design process is taken from the initial consideration of the operational requirements to the end of the concept design phase. The intervening stages have shown how the aircraft design evolves during this process. This showed that the initial configurational assumptions for thrust and wing loadings, based on data from existing aircraft, were found to be in error because of the unique operational performance of the aircraft. A more efficient aircraft layout was identified. Even the revised configuration was shown capable of improvement by the trade studies. For most aircraft projects, this iterative process is commonplace.

The economic assessment of the aircraft indicated that the project was viable and therefore worth taking into the next stage of development.

Due to time and resource restrictions in the conceptual stage, several technical aspects of the design have not been fully analysed. These include:

- The stability and control analysis of the aircraft including the assessment of the effect of the three-surface control layout.
- The aerodynamic analysis of the laminar flow control system and the associated structural and system requirements.
- The aircraft structural analysis and the realisation of the combined conventional and composite structural framework.
- The aircraft systems definition and the associated requirements for the new executive-class communication and computing facilities.
- The special requirements for aircraft servicing and handling at regional airports.
- The detailed trade-off studies applied to the field requirements (e.g. the definition of aerodynamic (flap design and deflection), propulsion (T/W), structures, systems and costs).
- The assessment of the overall market feasibility of the project.

Each of the topics in the list above involves work that is either comparable with, or exceeds, the work that has already been done on the project. In industry, progressing to the next stage of aircraft development would involve a 20- to 50-fold increase in technical manpower. To commit the company to this expenditure is a significant investment. A decision to proceed would only be taken after discussions with potential airline customers.

If the type of operation envisaged by this project is seen to be attractive, it will stimulate competition. This may come from airframe manufacturers who could modify existing aircraft to meet the specification. It is essential that the design team of the new aircraft anticipate this threat. They will need to conduct their own studies on the modifications to the aircraft that may be used as competitors. These are studies that require substantial effort, but in completing them, the advantages of the new design can be identified. This information will be useful to the technical sales team of the new aircraft and used to counteract the threat from the 'old-technology', 'modified' existing types.

References

1 Jenkinson, L. R., Simpkin, P. and Rhodes, D., *Civil Jet Aircraft Design*, 1999, AIAA Education Series, ISBN 1-56347-350-X.
2 Mason, W. H. *et al.*, *Low-cost Commercial Transport – Undergraduate Team Aircraft Design Competition*, 1995, Virginia Tech. AIAA 95–3917.
3 Howe, D., *Aircraft Conceptual Design Synthesis*, 2000, Prof. Eng. Pub. Ltd, ISBN 1-86058-301-6.
4 Fielding, J. P., *Introduction to Aircraft Design*, 2000, Cambridge University Press, ISBN 0-502-65722-9.
5 Hoerner, S. F., *Fluid Dynamic Drag*, published by the author, Bricktown, NJ, 1965.

5

Project study: military training system

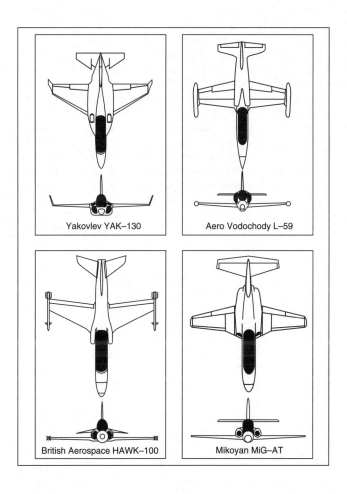

Yakovlev YAK–130

Aero Vodochody L–59

British Aerospace HAWK–100

Mikoyan MiG–AT

5.1   Introduction

A project similar to the one described below was the subject of a EuroAVIA design workshop sponsored by British Aerospace. Undergraduate students from ten European countries worked for three weeks in separate teams to produce specifications for new training systems. The study below represents a combination of the results from this workshop and some subsequent design work done on aeronautical courses in two English universities. Acknowledgement is given to all the students who worked on these projects for their effort and enthusiasm which contributed to the study described.

In the following analysis general references are made to aircraft design textbooks.[1–5] To avoid confusions in the text, a list of current popular textbooks, useful for this project, is included in the reference section at the end of this chapter. A fuller list of information sources can be found in Appendix B towards the end of this book.

5.2   Project brief

All countries with a national airforce need an associated programme for their pilot selection and training; therefore the commercial market for military training aircraft and systems is large. Designing training aircraft is relatively straightforward as the technologies to be incorporated into the design are generally well established. Many countries have produced indigenous aircraft for training as a means of starting their own aircraft design and manufacturing industry. This has generated many different types of training aircraft in the world. For many different reasons only a few of these designs have been commercially successful in the international market. The British Aerospace Hawk (Figure 5.1) family of aircraft has become one of the best selling types in the world with over 700 aircraft sold. It is a tribute to the original designers that this aircraft, which was conceived over 25 years ago, is still in demand. The maturity of the Hawk design is not untypical of most of the other successful trainers. Only recently have new aircraft been produced (mainly in East European countries) but these are still unproven designs and not yet competitive with the older established products.

Since the early 1970s when the Hawk and other European trainers were designed, front-line combat aircraft operation has changed significantly. The introduction of higher speed, more agile manoeuvring, stealth, together with significant developments

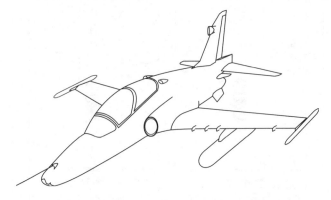

Fig. 5.1 Hawk aircraft

in aircraft and weapon systems generated a requirement for a new training system. As airframe and system development is expensive it is essential that an overall systems approach is adopted to this project.

The project brief for a new training system covers pilot training and selection from the *ab-initio* phase (assuming cadets have had 50 hours' flight training on a light propeller aircraft) to the start of the operational (lead-in) training on twin-seat variants of combat aircraft. This period covers the existing basic and advanced training phases covered by Hawk type aircraft. To represent modern fighter capabilities the new training system should also include higher flight performance and weapon system training which is not feasible on current (older) training aircraft.

The concepts to be considered are those associated with an integrated training system. This must account for the various levels of capability from the aircraft, synthetic training systems (including simulators) and other ground-based facilities. It will be necessary to define the nature of the training experiences assigned to each component of the overall training system.

The minimum design requirements for the aircraft are set out in the aircraft requirements section below but consideration should be given to the development of the training programme to include flight profiles with transonic/supersonic performance. Also, as all commercially successful training aircraft have been developed into combat derivatives, this aspect must be examined. To reduce the overall cost of the project to individual nations discussion must be given to the possibility of multinational co-operative programmes. All the issues above will be influential in the choice of design requirements for the aircraft.

5.2.1 Aircraft requirements

Performance

General   Atmosphere   max. ISA $+ 20°$C to 11 km (36 065 5 ft)
                                 min. ISA $- 20°$C to 1.5 km (4920 ft)
           Flight missions – see separate tables

           Max. operating speed, $V_{mo} = 450$ kt @ SL (clean)
                                    $V_{mo} = 180$ kt @ SL (u/c and flaps down)

Turning   Max. sustained g @ SL $= 4.0$
           Max. sustained g @ FL250 $= 2.0$
           Max. sustained turn rate @ SL $= 14°$/s
           Max. instantaneous turn rate @ SL $= 18°$/s

Field   Approach speed $= 100$ kt (SL/ISA)
           TO and landing ground runs $= 610$ m (2000 ft)
           Cross-wind capability $= 25$ kt (30 kt desirable)
           Canopy open to 40 kt
           Nose wheel steering

Miscellaneous Service ceiling > 12.2 km (40 000 ft)
               Climb – 7 min SL to FL250
                       (note: one flight level, FL $= 100$ ft)
               Descent – 5 min FL250 to FL20 (15° max. nose down)
               Ferry range $= 1000$ nm (2000 nm (with ext. tanks))
               Inverted flight $= 60$ s

Structural

- Flight envelope $n_1 = +7$, $n_3 = -3$
- Max. design speed M0.8
- $V_D > 500$ kt CAS
- Utilisation $= 500$ h/year
- Fatigue life $= 30$ yr

Operational

- Hard points $= 2$ @ 500 lb (227 kg) plus 2 @ 1000 lb (453 kg), all wet
- Consideration for fully armed derivatives
- Consideration for gun pod installation
- Provision for air-to-air refuelling

Cockpit

- Aircrew size – max. male 95 per cent, min. female 50 per cent
- Ejection – zero/zero
- All weather plus night operations
- Cockpit temperature, 15–25°C
- Oxygen system

Systems

- Avionics to match current/near future standards
- Consideration given to fly-by-wire FCS
- Consideration given to digital engine control
- Glass cockpit
- Compatibility to third and fourth generation fast jet systems where feasible

5.2.2   Mission profiles

Mission profiles used in the design of the aircraft are to be defined by the design team but they must not have less capability than described below:

1. Basic   This is to represent early stages of the flight training. Two sorties are to be flown without intermediate refuelling or other servicing.

Phase	Description	Height	Time (min)
1	Start, taxi	SL	4
2	Take-off	FL20	1
3	Max. climb	FL250	7
4	Cruise to training area	FL250	6
5	High-speed decent	FL20	5
6	General handling (Buffet control, etc.)	FL20	10
7	Max. climb	FL250	6
8	Manoeuvres (Turns, spin, etc.)	FL250	4
9	Cruise to base	FL250	5

Phase	Description	Height	Time (min)
10	Descent	FL20	5
11	Recover to base*	FL20	3
	– 100 nm fuel + 5% reserve or		
	– 5 circuits + 10% reserves		
12	Landing, taxi, shutdown	SL	4
		Mission elapsed time	**60**

(*reserve fuel is only applicable to the second sortie)

2. Advanced   This mission is typical of fighter handling at the advanced training stage.

Phase	Description	Height	Time (min)
1	Start, taxi	SL	4
2	Take-off	FL20	1
3	Max. climb	FL250	7
4	Cruise to training area	FL250	6
5	Weapon training	FL250	10
6	Aerobatics and high g	FL50	10
7	Low-level flying	250 ft	10
8	Climb to cruise	FL250	7
9	Cruise to base	FL250	6
10	High-speed descent	FL20	5
11	Recover to base*	FL20	6
	– 100 nm fuel + 5% reserve or		
	– 5 circuits + 10% reserves		
12	Landing, taxi, shutdown	SL	4
		Mission elapsed time	**76**

(*reserve fuel is only applicable to the second sortie)

Note: the times quoted in the above profiles are approximate and do not define aircraft performance requirements. (FL = flight level, 1FL = 100 ft.)

3. Ferry   This mission is required to position aircraft at alternative bases. The ferry ranges are specified in section 5.2.1. The ferry cruise segment may be flown at best economic speed and height. Reserves at the end of the ferry mission should be equivalent to that for the basic mission profile.

5.3   Problem definition

The main difficulty with this project lies is the broad spectrum of training activities that are expected to be addressed by the system. To cover all flight training from post- *ab-initio* to pre-lead-in will include the basic, intermediate and advanced training phases (Figure 5.2). In most air forces this involves the use of at least two different types of aircraft (e.g. a basic trainer like the Tucano and an advanced trainer like the Hawk). There will be about 90 hours of training in the selection and elementary phases. To reduce flight costs most of this will be done on modified light aircraft with a single piston/propeller engine and semi-aerobatic capability (e.g. Bulldog, Firefly). Such aircraft have a limited top speed of about 130 kt. The next phase (basic training) lasts for about 120 hours, using faster turboprop or light turbojet trainers (e.g. Tucano, L39). This includes visual flying experience (climbs, descents, turns, stall and spin) together with some aerobatics navigation training, instrument flying and formation

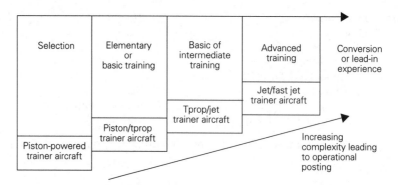

Fig. 5.2 Airforce flight training phases

flying. The advanced training phase is about 100 hours' duration and takes the pilot up to the point of transfer to an operational conversion unit (OCU). This phase will involve using an advanced turbojet trainer (e.g. Hawk) to provide experience at higher speeds (530 kt) and higher 'g' manoeuvres. The programme will include air warfare, manoeuvrability, ground attack, weapon training and flight control integration. The operational conversion unit will use two-seat derivatives of fast jets and provide the experience for lead-in to operational type flying.

To devise a training system for both basic and advanced phases based on a single aircraft type will present commercial opportunities to the manufacturer together with overall cost and operational advantages to the airforce. If innovation can be harnessed to produce a system to meet all the through-training requirements it would offer substantial advantages over all existing training aircraft and current projects which offer less capability. This is obviously a difficult task but the key to the successful solution to this problem lies in the careful exploitation of new technologies that have been used in other aeronautical applications.

Designing a new training system that introduces, develops and relies on innovation carries a commercial risk associated with the unpredictability of the technology. Although, as engineers we may have complete faith in new concepts, perhaps the principal drawback in using a novel, high-tech system lies in the conservative nature of our proposed customers (i.e. training organisations). Any new system must possess the ability to gradually evolve new features even if this means a temporary partial degrading of the overall concept in the early stages.

With the above considerations in mind we (the designers) are required to produce a technically advanced system to meet the defined training requirements yet exhibit sufficient capability to avoid initial scepticism from established customers. The system must show technical and economic advantages over existing equipment and possess the possibility to develop alternative combat aircraft variants based on the trainer airframe, engine and systems.

5.4 Information retrieval

Researching trade journals (e.g. the annual military aircraft reviews in aviation magazines, like *Flight International* and *Aviation Week*) provides data on existing and recently proposed training aircraft. Clearly the market is saturated with training aircraft of various types. The list below shows aircraft that are available to potential customers.

Aeromacchi (MB339/S211A/S260)	Italy
Aeromacchi/Alenia/Embraer/Aerospatial (AMXT)	International
Aero Vodachody (L39/L59/L139/L159B)	Czechoslovakia
AIDC (AT-TC-3A/B)	Taiwan
Avionne (JARA/G-4M)	Romania
BAE Hawk	UK
Boeing (T2/T28/T43)	USA
Boeing/McD (T45 Goshawk)	USA
Bombardier/Shorts (Tucano)	UK
CASA (C101DD)	Spain
Cessna (T37)	USA
Daewoo (KTX-1)	South Korea
Dassault-Breguet/Dornier, Alpha Jet	International
Denel (MB326M)	South Africa
Fuji (T3/T5)	Japan
HAI (Kiran 1A/MK2)	India
Israel Aircraft (TC2/TC7)	Israel
Kawasaki (T4)	Japan
Lockheed Aircraft (IA63)	Argentina
MAPO (MIG-AT)	Russia
Mitsubishi (T2)	Japan
NAMC (K8)	China
Northrop-Grumman (T38)	USA
PAC	Pakistan
Polskie (WSK PZL M-93V, I-22)	Poland
Raytheon-Beechcroft (T1/JPATS)	USA
Rhein-Flugg. (Fantrainer)	Germany
SAAB (SK 60W)	Sweden
Samsung/Lockheed (KTX-2)	South Korea
Socata (TB30/TB31)	France
UTVA (Soko G4)	Serbia
Yakovlev (Yak 130)	Russia
(Yak/Aeromacci) (Y130)	International

The list above is a 'mixed-bag' of aircraft including propeller types, derivatives of existing non-training aircraft, and some purely national projects. It is necessary to review the collection to select aircraft that we feel are more appropriate to this project. The following aircraft are regarded as significant:

1. B.Ae. Hawk (Mk60/100): this is one of the most successful training aircraft in the world with more than 700 produced and sold internationally.
2. L139/159: are 'westernised' versions of the very successful earlier Czech training aircraft (L39/59) which were used by airforces throughout the old Eastern Bloc. When fully developed it may present a serious competitor in future trainer markets.
3. MB339: is a derivative of the very successful Italian trainer (MB326). It has been extensively modernised with upgraded avionics and a modern cockpit.
4. MiG-AT: compared with the above aircraft this is a completely new design by the highly competent Russian manufacturer. It is in competition with other aircraft for the expected 1000+ order for the Russian airforce and their allies. It presents a serious competitor to this project.

5. Yak/AEM 130: this is a new subsonic trainer from a Russian/Italian consortium. It will compete with the MiG-AT for the Russian airforce order and could be a considerable challenge to the Hawk in future years.
6. KTX-2: is a new supersonic (M1.4) trainer from a South Korean manufacturer (in association with Lockheed Martin). It is expected to be sold in direct competition with all new trainer developments and with other light combat aircraft.
7. AMX-T: this is a trainer development of the original AMX attack aircraft. It is produced by an international consortium and will be a strong contender in future advanced trainer aircraft markets.

5.4.1 Technical analysis

Details of the aircraft in the list above have been used in the graphs described below to identify a suitable starting point for the design. Decisions on selected values to be used in the project are influenced by this data. To reduce format confusion the graphs are plotted in SI units only.

Empty mass data (conversion: 1 kg = 2.205 lb)

Figure 5.3 shows the empty mass plotted against maximum take-off mass for jet trainers. The graph also shows the constant 'empty mass ratio' radials. These radials can be seen to bracket 0.75 to 0.45. Our **selected value of 0.6** lies between the higher values for the Russian aircraft and the Italian MB338 but above those for the L159, Hawk and Alpha Jet.

Wing loading (conversion: 1 kg/sq. m = 0.205 lb/sq. ft)

Figure 5.4 is a graph of the maximum take-off mass versus wing reference area for existing aircraft. The wing loading radials bracket 500 to 200 kg/m^2. Our **selected value is 350 kg/m²**. Most of the specimen aircraft have higher wing loading but our specified low approach speed requirement will dictate a lower wing loading.

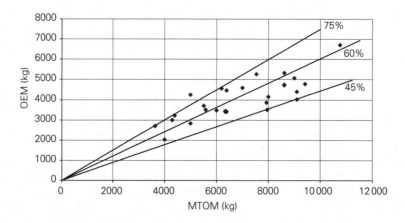

Fig. 5.3 Survey of empty mass ratio

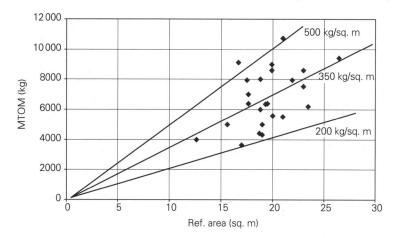

Fig. 5.4 Survey of wing loading (kg/m^2)

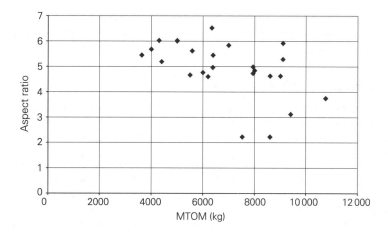

Fig. 5.5 Survey of wing aspect ratio

Aspect ratio

Figure 5.5 plots the wing aspect ratios for the trainer aircraft. Most seem to lie in the region of 5 to 6. A value of **5 will be used** as an initial guide to the wing planform geometry. In subsequent phases of the design process, it will be necessary to conduct detailed 'trade-off studies' to establish the technical 'best' choice of wing aspect ratio. At this stage in the development of the aircraft it is impossible to do such studies as sufficient details of the aircraft are unknown.

Thrust loading (conversion: 1 N = 0.225 lb)

Figure 5.6 shows the installed thrust versus maximum aircraft take-off mass. The radials show that modern aircraft lie along the 40 per cent SLS thrust line. As might be expected the manoeuvre and performance of these aircraft are similar. The new supersonic aircraft are above this line and older aircraft substantially below. This reflects the

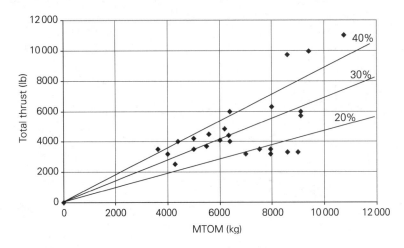

Fig. 5.6 Survey of thrust/weight ratio (lb/kg)

requirement for improved performance for newer aircraft. We **will select a 40 per cent based on the SSL thrust rating**.

5.4.2   Aircraft configurations

Looking in detail at the configuration of aircraft in the candidate list confirms the impression that most of the existing trainers are conventional in layout. They all have twin, tandem cockpits with ejector seats and large bubble canopies. Apart from the latest Russian designs they are single-engined with fuselage side intakes. The slower aircraft have thick (12 per cent) relatively straight wings. Some of the later designs have thinner swept wings to match the faster (supersonic) top speeds. The wing/fuselage position is mostly low set but with some at mid-fuselage. The Alpha Jet has a shoulder wing position. Tail position for all aircraft except the MiG is conventional with the tailplane set on the aft fuselage with the fin slightly ahead to give protection for post-stall control. The MiG originally had a 'T' tail but this was later changed to a mid-fin location.

5.4.3   Engine data

Engines suitable for trainer aircraft lie in the 9 to 29 kN (2000 to 6500 lb) thrust range. The engines shown in Table 5.1 are available.

Figure 5.7 shows the engine weight (mass) versus SSL thrust data.

5.5   Design concepts

To provide a stimulus for the design of the aircraft it has been decided that a radical (novel) solution to the problem should be investigated. This consists of specifying a total training system to cover all the required phases. It comprises an advanced simulator, a single-seat aircraft (see later comment), ground-based instructor console(s) and a modern communication and data linking facility (Figure 5.8). Removing the instructor

Table 5.1

Engine (Manufacturer)	Used on	Thrust SLS (kN/lb)	SFC (@ SLS) (–/hr)	Eng. mass (kg/lb)
FJ44-2A (Williams/RR)	–	10.20/2300	0.45	203/447
JT15D (P&W Can)	Citation	13.54/3045	0.55	284/627
Larzac 04-C20 (TM/Snec.)	MiG-AT	14.12/3175	0.74	302/666
J85-21 (Gen. Elec.)	F5/T38	15.57/3500	1.00	310/684
Viper 680 (Rolls Royce)	MB339	19.30/4339	0.98	–
PW 545A (P&W Canada)	Citation	19.79/4450	0.44	347/765
DV-25 (PS/Russian)	Yak 130	21.58/4852	0.60	450/992
TFE 731-60 (Allied-Sig.)	Citation	24.86/5590	0.42	421/929
CFE 738 (GE/ASE)	Falcon 2000	24.90/5600	0.37	601/1325
PW 306A (P&W Canada)	DO 328	25.35/5700	0.39	473/1043
Adour 871 (Rolls Royce)	Hawk/T45	26.81/6028	0.78	602/1328
F124-100 (Allied-Signal)	–	28.02/6300	0.81	499/1100
AE 3007C (AEC)	UAV	28.89/6495	0.33	717/1581

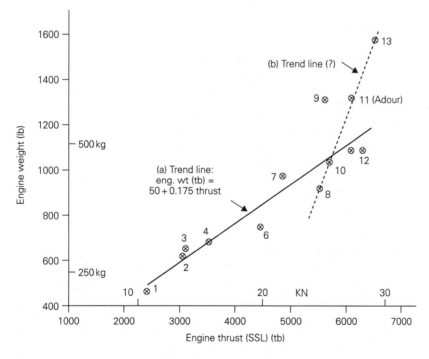

Fig. 5.7 Survey of engine weight versus SSL thrust

from the training aircraft is regarded as feasible with the adoption of new technologies that have been proven in other applications. Experience from flight test data links and recording gives assurance that technically the systems are available and feasible on which to develop a remote instructor system. Without a second seat the aircraft will be

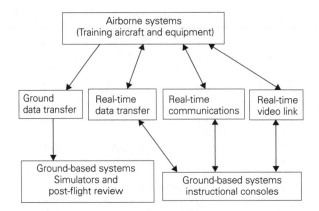

Fig. 5.8 Proposed total training system diagram

simpler, lighter and cheaper and flying solo the pilot will be in a more realistic operational environment. A further advantage lies in the development of the aircraft into a combat derivative. It will obviously be essential to carefully design the communications link and the instructor module to ensure reliable and safe operation. Modern electronic and video equipment should be capable of providing the necessary confidence. This decision was later reviewed.

Development of a two-seat simplified version of the aircraft will be possible by sacrificing some of the payload and performance capability. This may provide a means of avoiding some of the apprehension centred on the use of the system in the basic and early parts of the intermediate training phases. The two-seat version represents a relatively straightforward development of the aircraft.

The single-seat aircraft strategy makes it possible to set the design point for the aircraft at the upper end of the advanced training spectrum. This will guide the definition of the critical performance, payload and systems specification. As previously mentioned setting this specification will also provide a better baseline for the development of the combat aircraft derivative.

5.6 Initial sizing

Using the assumed values for empty mass ratio (0.6), wing loading (350 kg/m^2 (72 lb/sq. ft)), aspect ratio (5), thrust loading (0.40) and the specified useful load (pilot + operational equipment + 3000 lb weapon load, totalling an assumed 3308 lb (1500 kg)), it is possible to make estimates of the initial mass and sizes for the aircraft.

Using the equation from Chapter 2, section 2.5.1:

$$M_{TO} = \frac{M_{UL}}{1 - (M_E/M_{TO}) - (M_F/M_{TO})^*}$$

*Using a value of 0.15 for the fuel fraction (from Hawk data, this will need to be verified later) and substituting the known and assumed values gives:

$$\frac{1500}{1 - 0.6 - 0.15} = 6000\,kg\ (13\,230\,lb)$$

With this aircraft mass the assumed wing loading gives:

Wing reference area $(S) = (6000/350) = 17.14\,\text{m}^2$ (184 sq. ft)

Using an aspect ratio of 5 sets of wing span $(b) = (5 \times 17.14)^{-0.5} = 9.26\,\text{m}$ (30.4 ft)

This sets the mean chord $(c_{\text{mean}}) = (9.26/5) = 1.85\,\text{m}$ (5.9 ft)

Assuming a wing taper ratio of 0.25 sets the approximate values for the centre line chord of 3.0 m (9.8 ft) and tip chord of 0.75 m (2.5 ft)

In drawing the aircraft we will round off the measurements to give a span of 9.0 m (29.5 ft). This results in the slightly larger wing area of $16.88\,\text{m}^2$ (181 sq. ft)

The selected thrust loading of 0.4 in association with the estimated aircraft mass gives a required static sea level thrust of $(0.4 \times 6000 \times 9.81) = 23.54\,\text{kN}$ (5300 lb).

The choice of engines to provide this thrust rests between:

- the old and slightly overpowered Adour engine used in the Hawk,
- a more modern higher bypass engine from Allied Signal/P&W Canada/GE (used on business jets),
- a slightly underpowered Russian engine as used on the Yak 130, or two TM/Snecma engines as specified for the MiG-AT.

This presents a somewhat difficult choice as:

- the Hawk engine is thirsty,
- the higher bypass engines are larger diameter and lose thrust at altitude (a major disadvantage for the proposed combat variant),
- the Russian manufactured engines may not appeal to established Western customers,
- installing two engines will complicate the systems and cockpit (but would be representative of modern fighter configurations).

After careful consideration it has been decided to use the Adour engine as it is well respected by established customers, is reliable and will add confidence to our novel training system. The Adour 861 provides 5700 lb of thrust so it would be possible to derate the engine for the main specification (this would extend engine life). This strategy would make extra thrust available for the faster trainer and combat variants. With maximum thrust available, the thrust loading would be increased to 45 per cent.

The initial sizing above has provided sufficient data to consider in more detail the initial aircraft layout.

5.6.1 Initial baseline layout

With our understanding of the configurational options used on existing trainers together with representative sizes of components from our initial sizing, it is possible to consider the detailed layout of the aircraft (Figure 5.9).

The following decisions on the aircraft geometry have been taken:

- The aircraft will be of conventional layout with rear fuselage-mounted tail surfaces.
- A single RR Adour 861 engine will be mounted in the aft fuselage, length 2.0 m (80 in), width 0.76 m (30 in), height 1.04 m (42 in), intake diameter 0.7 m (28 in), nozzle diameter 0.6 m (24 in).
- Single seat cockpit with ejector seat and enclosed canopy/windscreen providing the required vision capability.

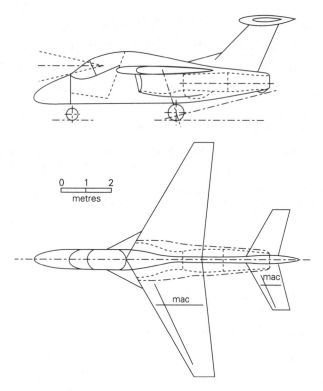

Fig. 5.9 Initial aircraft layout

- The fuselage aft of the cockpit will be suitably configured to permit an easy modification to accommodate a twin-tandem layout for the basic trainer variant.
- The wing will be mid-to-high fuselage (shoulder) mounted to provide generous ground clearance for underwing stores. It will also be manufactured as a single piece (tip to tip) structure which will be mounted above the upper fuselage longerons. This will avoid complicated wing-to-fuselage structural joints.
- The wing will be trapezoidal in planform with at least 30° leading edge sweepback and a thin (10 per cent) section (to provide for future higher-speed variants).
- Tail area ratios will match existing aircraft data (from a review of the aircraft data file these values seem appropriate; $S_H/S = 0.255$ giving $S_H = 4.3\,m^2$ (46 sq. ft) and $S_V/S = 0.185$ giving $S_V = 3.1\,m^2$ (33 sq. ft)).

(Note: an initial layout drawing of the aircraft showed that the short fuselage length makes a conventional fuselage-mounted tailplane suffer from a shortage of tail arm. Therefore a 'T-tail' arrangement has been adopted. This configuration improves both the horizontal and vertical tail effectiveness which allows a reduction in tailplane area to 22 per cent S ($=3.7\,m^2$ (40 sq. ft)). A reduction in fin areas could also be anticipated but this was not adopted, as the proposed two-seat variant will require more fin to balance the increased fuselage nose length. The T-tail arrangement will enable the provision of a communication/video pad to be installed at the top of the fin (at the tailplane junction). This will be useful to accommodate some of the extra equipment necessary for the remote instructor facility.)

- To assist in re-energising the airflow over the rear fuselage and fin in high-alpha manoeuvres, small wing leading edge extensions will be added to the planform.
- Underwing, fuselage-side intakes will be positioned below the leading edge extensions to ensure clean airflow into the engines in high-alpha manoeuvres.
- Conventional tricycle landing gear with wheel sizes representative of existing aircraft (from the existing aircraft data file, the main and nose-wheel diameters are 0.6/0.45 m (24/18 in) respectively).

5.7 Initial estimates

With a scale drawing of the baseline aircraft configuration and an understanding of the engines and systems to be used it is now possible to conduct a series of detailed calculations to estimate the aircraft mass, aerodynamic characteristics and performance.

5.7.1 Mass estimates

The mass of each component of the aircraft can be calculated using methods described in aircraft design textbooks. These are generally based on geometrical and aircraft load data and are often derived from analysis of existing aircraft configurations. Suitable adjustments need to be made in those areas where the proposed design is significantly different from past designs. In our case there are two such considerations:

- much more composite material will be used than in the predominately aluminium alloy aircraft built previously and,
- for this training system more sophisticated and extensive flight control and communication systems will be installed (allowance will need to be made for the reduced mass and volume of new electronic/computer systems).

A design take-off mass of 6000 kg will be assumed for determination of the aircraft structural components. Although this mass is likely to be higher than that estimated for the maximum take-off mass of the aircraft it will provide an insurance against future mass increase. If it is necessary to determine the minimum take-off mass for the aircraft, the estimation would need to be done iteratively using the calculated take-off mass as the design mass input for components mass estimations.

Detailed calculations for mass estimations have not been shown below but the input data on which the calculation was based is given (for reference). To simplify presentation, the data below is shown in SI units only. The completed mass statement is shown in dual units.

Wing: Area $(S) = 16.88\,\mathrm{m}^2$ Aspect ratio $= 5$ Wing thickness (average) $= 10\%$
Sweepback $(c/4) = 25°$ Taper ratio $= 0.25$ Control surface areas $= 15\%\,(S)$

Conventional mass estimation $= 462\,\mathrm{kg}$, assuming 15 per cent reduction for composites $= \mathbf{392\,kg}$

Fuselage: Length $= 9.5\,\mathrm{m}$ Depth $= 2.0\,\mathrm{m}$ Width $= 0.75\,\mathrm{m}$

Conventional mass estimation $= 576\,\mathrm{kg}$, assuming 10 per cent reduction for composites $= \mathbf{518\,kg}$

Horizontal tail: Area $= 22\%\,(S)$ Tail span $= 4.0\,\mathrm{m}$ Fuselage width $= 0.75\,\mathrm{m}$

Conventional mass estimation $= 76\,\text{kg}$, assume 15 per cent reduction for composites
$$= \mathbf{65\,kg}$$

Fin: Area $= 18\%\,(S)$,   T-tail structure

Conventional mass estimation $= 39\,\text{kg}$, assume 15 per cent reduction for composites
$$= \mathbf{34\,kg}$$

Undercarriage: $M_{\text{LAND}} = 90\%\,M_{\text{TO}}$ $n_{\text{ULT}} = 3 \times 1.5 = 4.5$
Main: Length $= 0.75\,\text{m} \to$ Mass $= 185\,\text{kg}$
Nose: Length $= 0.60\,\text{m} \to$ Mass $= 52\,\text{kg}$ Total $= \mathbf{237\,kg}$
No reduction on this mass for new materials

Engine: Dry $= \mathbf{577\,kg}$ (from manufacturers)

Engine and fuel systems: Assume 50 per cent engine dry mass $= \mathbf{288\,kg}$

Aircraft equipment: To account for new aircraft system requirements assume 20 per cent of aircraft design mass $\to \mathbf{1200\,kg}$

Fuel: This will be checked by the performance estimates, for now we will still assume 15 per cent of aircraft design mass $\to \mathbf{900\,kg}$

Crew: Assume one pilot with operational equipment $\to 300\,\text{lb} = \mathbf{136\,kg}$

Weapon load: Specified at $3000\,\text{lb} \to \mathbf{1360\,kg}$

Hence the initial mass statement for the aircraft can be compiled:

Component		kg/lb	% M_{TO}
Wing		392/664	6.9
Fuselage		518/1142	9.1
Horizontal tail		65/143)	
Fin		34/75) $\to$	1.7
Undercarriage		237/523	4.2
	Total structure	**1246/2747**	**21.8**
Engine (dry)		577/1272	
Engine and fuel systems		288/635	
	Total propulsion	**865/1907**	**15.2**
Equipment (total)		1200/2646	21.0
	Total aircraft empty	**3311/7300**	**58.0**
Fuel		900/1984	15.8
Crew		136/300	
Weapon load		1360/3000	
	Total Useful Load	**2396/5283**	42.0
	<u>**MAXIMUM TAKE-OFF MASS**</u>*	**5707/12584**	100.0
	Note: aircraft design take-off mass	6000/13230	

(*This compares with 9000 to 7500 kg (20 000 to 16 000 lb) for older trainers and is competitive with the newer designs.)

At this stage all the input values used to predict component masses are somewhat tentative, therefore the use of more extensive methods are inappropriate.

The aircraft aerodynamic and performance calculations that follow will use the above mass statement to determine aircraft masses for various stages in the flight profiles:

- At take-off, the aircraft can be loaded up to a variety of different conditions depending on the operation to be performed. The most critical take-off condition will be at maximum take-off mass. This will be used in the estimation of take-off distance.
- At landing, the aircraft could again be at different mass. The most critical flight case would be to land immediately after taking off with full fuel and payload. This would be an emergency condition and therefore it is not necessary to use this as a design case (in an emergency the pilot may jettison some of the weapons and/or part of the fuel load to reduce the landing mass). The minimum landing mass would relate to zero weapon load and only a small percentage of fuel remaining (e.g. 10 per cent). This may also be a rare condition and therefore not applicable to design calculations. The landing calculations are often performed at a landing weight of 90 per cent of the take-off weight. For the maximum take-off mass condition this relates to $5707 \times 0.9 = 5136$ kg (11 325 lb). Another take-off condition that may be considered is a take-off with zero weapon load (i.e. a basic training mission). Using 90 per cent of the take-off weight in this condition gives $(5707 - 1360) 0.9 = 3912$ kg (8626 lb). These two cases will be used in the landing distance and speed estimations.
- For the manoeuvring calculations (turns and climb) a mean flying weight will be assumed. This relates to half weapon and half fuel load (= 4577 kg/10 092 lb).
- For the ferry case the aircraft take-off condition will be without weapon load but may have carry external wing tanks.

5.7.2 Aerodynamic estimates

Using the geometrical data from the initial baseline layout and masses from the above section, it is possible to make initial estimates for the aircraft lift and drag in various flight conditions.

Aircraft drag coefficients

Using standard equations from aeronautical textbooks and the data below, it is possible to estimate the drag coefficients for the aircraft in different flight conditions.

In SI units:

Take-off mass (M_{TO})	5707 kg	Wing aspect ratio (A)	5
Wing ref. area (S)	16.88 m^2	Wing LE sweep (Λ)	30°
Wing loading	3212 kN/m^2	(337 kg/m^2)	

The following equations and parameters are appropriate:

- Induced drag factor $K = (1/\pi e A)$
- Planform factor $e = 4.61((1 - 0.045A^{0.68}) (\cos \Lambda)^{0.15}) - 3.1$
- Assumed skin friction coefficient $C_f = 0.0038$
- Profile drag coefficient $C_{DO} = C_f(S_w/S)$
  (assumed aircraft wetted area – estimated from the initial layout drawing (total) $S_w = 78$ m^2)
- Dynamic pressure $q = 1/2\rho.V^2$
  (ρ and V are the air density and aircraft speed at the flight condition under investigation)
- Aircraft lift $= C_L q S$

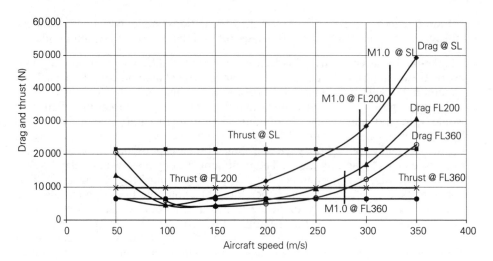

Fig. 5.10 Aircraft drag polar

- Aircraft drag $= C_D q S$
- Aircraft lift/drag ratio $= L/D = C_L/C_D$

The above data was input to a spreadsheet program to determine the lift and drag parameters for a range of aircraft speed (V), operating altitude (h) and load factor (n) for the aircraft in the 'clean' (no weapons, u/c and flaps retracted) flight condition. The range of values used in the calculations is shown below:

$$V \text{ (m/s)} \quad 50/100/150/200/250/300/350$$

$$h \text{ (ft)} \quad 0 \text{ (SL)}/25\,000 \text{ (FL250)}/36\,000 \text{ (FL360)}$$

$$n \text{ (g)} \quad 1/3/6$$

The spreadsheet results for sea level at $n = 1$ are shown in Figure 5.10. The thrust displayed in the graph is for the Adour engine (no variation of thrust with aircraft forward speed is typical for low bypass ratio engines).

The drag analysis described above assumes subsonic flow conditions but the upper value for aircraft speed is seen to be in excess of M1.0 at each altitude. Obviously aircraft drag will increase rapidly as supersonic flow is developed over the aircraft. Some allowance will need to be taken for the wave drag at higher speeds. Also, since it is intended to investigate the potential for increasing the aircraft top speed into the transonic range (e.g. M1.2) it is necessary to make suitable corrections.

Textbooks quote the Sears–Haack wave drag coefficient as:

$$C_{Dwave} = (9\pi/2)(A_{max}/L)^2(1/S)$$

where (A_{max}) is the maximum cross-sectional area of the aircraft
(L) is the distance from the nose to the position where the area is maximum
(S) is the reference wing area for the aircraft

From the aircraft drawing $A_{max} = 2.0\,\text{m}^2$ (21.5 sq. ft) and $1 = 4.5\,\text{m}$ (14.75 ft) are reasonable estimates. This data gives $C_{DW} = 0.14$. This value could be reduced to account for the swept wing but would need to be increased due to the poor area ruling of the design (due mainly to the effect of the canopy). At this early stage in the analysis of the aircraft it is assumed that these effects cancel leaving $C_{DO} = 0.14$ at M1.05.

Note: the increase due to wave drag at M1.05 approximately doubles the subsonic drag previously calculated. The wave drag will be progressively felt from the drag divergent Mach number, which for this wing sweep will be about M0.75. At this stage in the aircraft evaluation it will be sufficient to sketch a blend line from the subsonic drag polar to account for the values above. A suggestion from one textbook assumes half the wave drag increase to have been achieved at M1.0, therefore the blend will be from M0.75 and through the M1.0 and M1.05 points.

Aircraft lift coefficients

From an analysis of stall and approach speeds from existing aircraft it is acceptable to assume maximum lift coefficients for our aircraft of:

$$\text{Landing } C_{lmax} = 2.10$$

$$\text{Take-off } C_{lmax} = 1.70$$

5.7.3   Performance estimates

The calculation of the aerodynamic coefficients described above are relatively crude and do not take into account many of the detailed factors that are known to be significant. However, they do provide 'ballpark' values which can be used to initially assess aircraft performance. From these calculations it will be possible to identify the changes that are necessary to the design to meet the specified design criteria. Later analysis can account for the more detailed aspects of the predictions.

Maximum speed

Applying the wave drag increase to the subsonic drag polars at the three altitudes (SL, FL250, FL360), shows (Figure 5.10) that a top speed of 320 m/s (621 kt) is possible at SL-ISA with the full rating of the Adour engine. If the engine was derated to 85 per cent of maximum thrust the speed would reduce to 305 m/s (592 kt). Both of these are well in excess of the 450 kt specified in the design brief. However, these calculations have been done with the aircraft in the 'clean' condition. If an allowance for weapon drag ($\Delta C_D = 0.01$) is made, the speeds are reduced to 280 m/s (544 kt) and 260 m/s (505 kt) respectively. These are still faster than the specified speeds.

The conclusion to this part of the performance study is that the aircraft will easily meet the specified top speed requirement.

Turn performance

Aircraft performance textbooks quote the general equations as:

- turn rate $= \chi = g(n^2 - 1)^{0.5}/V$
  where $n =$ normal acceleration factor
          $V =$ aircraft speed
- radius of turn $R = V/\chi$
- aircraft bank angle $\phi = \cos^{-1}(1/n)$

These equations have been solved, using a spreadsheet, for values of:

$$V = (50/100/150/200/250/300/350 \, \text{m/s}) \quad \text{and}$$

$$n = (1/2/3/4/5/6/7/8)$$

(A reminder: 1 m/s is approx. 2 kt (more precisely 1.94).)

Note that the turn equation above is independent of aircraft parameters. The resulting curves are shown in Figure 5.11. This graph describes the overall manoeuvring design space for the aircraft.

For our aircraft the boundaries to the manoeuvring space are:

- aircraft maximum speed (assumed to be M0.8),
- stall speed (clean at mean flying weight with $C_{Lmax} = 1.10$),
- structural load limit of $n = 7$.

These provide the limiting lines shown on the sea-level manoeuvring diagram (Figure 5.12). The corner speed gives the maximum instantaneous turn rate. A value of 24°/s is predicted for our aircraft.

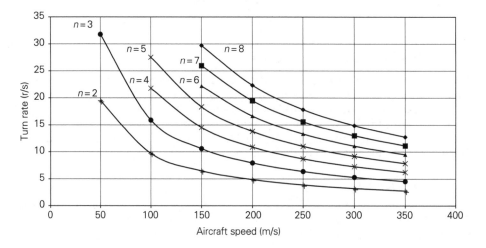

Fig. 5.11 Manoeuvring design space

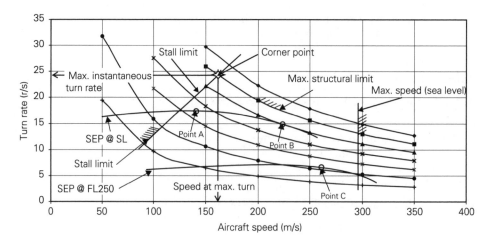

Fig. 5.12 Aircraft turn performance

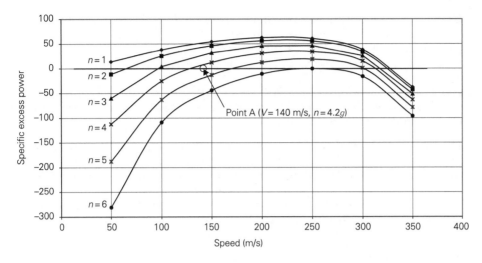

Fig. 5.13 Specific excess power at SL

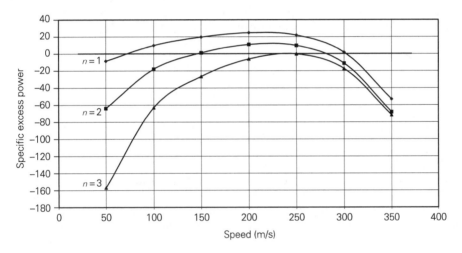

Fig. 5.14 Specific excess power at FL250

Determination of the sustained turn requires an analysis of the available specific excess power (SEP) for the aircraft. The zero SEP boundary provides the value for sustained turn rate. From textbooks the equation for specific excess power is:

$$SEP = V(T - D)/W$$

SEP will therefore vary according to the aircraft speed, engine thrust, aircraft drag and weight (i.e. M.n.g). Drag (and thrust) will vary with aircraft altitude and speed. The values determined for SEP for a range of aircraft speeds (50 to 350 m/s) and load factors (1 to 6), for both sea level and FL250, have been calculated and are plotted in Figures 5.13 and 5.14. Cross plotting the aircraft speeds at zero SEP from these graphs onto the manoeuvring diagram and joining with a smooth line provides the boundary for sustained turn rate.

- From Figure 5.12: maximum instantaneous turn rate = 24°/s at a flight speed of 165 m/s (the specified requirement is 18°/s).
- From Figure 5.13 with the results transferred to Figure 5.12 (point A): sustained turn rate = 17°/s at sea level at a flight speed of 140 m/s (the specified requirement is 14°/s).
- From Figure 5.12 with the zero SEP lines transferred from Figures 5.13 and 5.14: sustained turn g of 6 at sea level (point B) and just about 3 at (point C) and FL250 (the specified requirements are 4 and 2g).

The above calculations were done at aircraft maximum take-off mass with a clean configuration. Similar calculations were done with weapon drag added ($\Delta C_D = 0.01$) at mean flying mass (4577 kg/10 092 lb) but these were shown to be less critical than the clean-heavy case quoted above.

Note: the manoeuvre graph shows that an extra 5°/s instantaneous turn rate would be achieved if the aircraft structural limit was raised from the specified 7g to 8g.

Field performance

For take-off and landing calculations it is necessary to add the extra drag due to the extended undercarriage and flaps. The flap deflection for take-off is assumed to be 20° and for landing 40°. The following drag contributions are regarded as appropriate:

$$\Delta C_D \quad \text{undercarriage} = 0.0075$$

$$\Delta C_D \quad 20° \text{ flap} = 0.015$$

$$\Delta C_D \quad 40° \text{ flap} = 0.030$$

Lift co-efficients are as quoted earlier (i.e. take-off 1.7 and landing 2.1).

Take-off estimation
In the early part of the conceptual design process simplified take-off calculations can be done using Nicolai's (reference 4) simplified take-off parameter (this includes W/S, T/W, C_{lmax}). As we require only the ground run to be estimated, the take-off distance attributed to the climb segment can be ignored. The calculation for total take-off distance, assuming the aircraft at maximum take-off mass (5707 kg/12 584 lb) and with maximum sea level static thrust (5700 lb), gives:

$$\text{Total distance to 50 ft} = 2408 \text{ ft } (734 \text{ m})$$

Removing the distance covered in the climb gives the take-off ground run = 1856 ft (566 m) (the specified maximum take-off ground run is 2000 ft (610 m)).

Using a derated (85 per cent) engine thrust increases the ground run to 2184 ft (666 m) which is more than the required distance. In this case, to get off from a 2000 ft/610 m runway, the maximum aircraft mass would need to be limited to 5300 kg (11 685 lb).

Landing estimations
For landing, we will assume a landing mass of 90 per cent max. take-off mass. Therefore, $M_{LAND} = 5135$ kg/11 323 lb which gives an aircraft stall speed of 46.7 m/s (90 kts). The approach speed for military aircraft is set at $1.2V_{stall}$ giving 108 kt which means that the specified value of 100 kt is not achieved. In the design brief the appropriate aircraft landing weight for the approach speed requirement is not specified. If we assume the low approach speed is applicable to the basic training role we can assume the weapon

load to be zero. Assuming the landing weight is to be set at 90 per cent of the take-off weight gives a landing mass for the basic training role of:

$$M_{LAND} = 0.9\,(5707 - 1360) = 3912\,kg\,(8626\,lb)$$

At this mass the stall speed is 40.8 m/s (79 kt) making the approach speed $= 1.2 \times 79 = 95$ kt. This achieves the requirement but begs the question of the relationship of the approach speed requirement to the aircraft role.

Assuming a constant deceleration of 7 ft/s^2 (as assumed in the simplified Nicolai[4] estimation) and a touch-down speed of $1.15\,V_{\text{stall}}$. The landing distance is determined from:

$$\text{Landing ground run} = V_{TD}^2/(2 \times 7)$$
$$(V_{TD} = 1.15 \times 46.7 = 53.7\,m/s\,(176\,ft/s)$$
$$\therefore \text{Landing ground run} = 2212\,ft\,(675\,m)$$

This is also in excess of the specified distance of 2000 ft but the calculation was done at maximum landing mass (5136 kg/11 325 lb). Performing the same calculation at the lighter landing mass assumed above (3912 kg/8626 lb), using the same aircraft assumptions gives:

$$\text{Landing ground distance} = 1690\,ft\,(515\,m)$$

This easily achieves the specified distance of 200 ft (610 m) but again begs the question of the definition of landing weight.

The heavy and light landing mass assumptions used in the landing calculations provide a range of maximum/minimum values for the aircraft. Further discussion with the project customers would be necessary.

Mission analysis

The mission analysis allows us to estimate the fuel requirements. The project brief specifies three different mission profiles. At this stage, it is not obvious which one of these will be most critical, therefore each will be analysed to determine the required fuel. The calculations use the weight fractions suggested in Raymer's book[1] for the less significant segments of the missions.

Using the aerodynamic analysis described earlier it is possible to determine the lift/drag ratio variation with aircraft speed for the aircraft cruising at 25 000 ft. From this data a representative value of 9.0 will be assumed in the calculations. This decision will be verified later.

The specific fuel consumption of the Adour engine is 0.78 –/hr at the SSL thrust rating. This value will increase with aircraft altitude, speed and engine setting (e.g. cruise). Determining the exact extent of this rise requires engine performance data. At this stage in the design process such data is not available, a value of 0.95 has been assumed in the calculations below.

Basic training profile (two sorties)
The training profile is shown diagrammatically in Figure 5.15.

Analysis of the change in aircraft mass in each segment is shown in Table 5.2.

As this is the basic training mission it is done clean (no weapon drag) and at a lower take-off mass (no weapon load). The first of the two sorties will require the

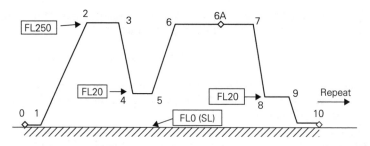

Fig. 5.15 Mission flight profile (basic training)

Table 5.2

Segment	Description	Parameters	$(M_{(n+1)}/M_n)^*$
0 to 1	Take-off		0.970
1 to 2	Climb		0.985
2 to 3	Cruise (out)	6 min	0.990
3 to 4	Descent		0.995
4 to 5	General handling	10 min	0.980
5 to 6	Climb		0.985
6 to 6A	Manoeuvres	4 min	
6A to 7	Descent	5 min	0.995
7 to 8	Recover		0.995
8 to 9	Land		0.995
9 to 10	Cruise (return)		0.980
			$\sum 0.877$

Fuel fraction for the above sortie $M_{fuel}/M_{TO} = (1 - 0.877) = 0.123$.
$^*(M_{(n+1)}/M_n)$ is the ratio of the aircraft mass at the end of the segment relative to that at the start.

following fuel:

$$\therefore M_{TO1} = 5707 - 1360 = 4347 \, kg \text{ (to avoid duplication remember, 1 kg = 2.205 lb)}$$
$$\therefore M_{fuel1} = 0.123 \times 4347 = 535 \, kg$$

As there is no intermediate refuelling the second sortie will be flown at a lower take-off mass than the first:

$$M_{TO2} = 4347 - 535 + 3812 \, kg$$

The same mission will be flown, therefore the same fuel fraction will apply:

$$\therefore M_{fuel2} = 0.123 \times 3812 = 469 \, kg$$
$$\text{Total fuel required} = M_{fuel1} + M_{fuel2} = 535 + 469 = 1004 \, kg/2234 \, lb$$

This is slightly more than the fuel load of 900 kg assumed in the mass calculations.

Advanced training profile (single sortie)
As this profile is flown with full weapon load the aircraft take-off mass will be 5707 kg. To account for the weapon drag the lift/drag ratio will be reduced to 7.5.

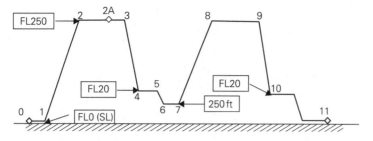

Fig. 5.16 Mission flight profile (advanced training)

Table 5.3

Segment	Description	Parameters	$(M_{(n+1)}/M_n)^*$
0 to 1	Take-off		0.970
1 to 2	Climb		0.985
2 to 3	Cruise (out)	6 min	
	Weapon training	10 min	0.967
3 to 4	Descent		0.995
4 to 5	Aerobatics	10 min	0.970
5 to 6	Descent		
6 to 7	Low level	10 min	0.970
7 to 8	Climb		0.985
8 to 9	Cruise (return)	6 min	0.990
9 to 10	Descent		0.995
10 to 11	Land		0.995
			$\sum 0.835$

$^*(M_{(n+1)}/M_n)$ is the ratio of the aircraft mass at the end of the segment relative to that at the start.

The profile is shown diagrammatically in Figure 5.16 and the segment analysis is shown in Table 5.3.

$$\text{Fuel fraction for the total mission} = (1 - 0.835) = 0.165$$

$$\therefore \text{Fuel required } M_{\text{fuel}} = 0.165 \times 5707 = 942\,\text{kg}/2077\,\text{lb}$$

This is also slightly higher than originally assumed but less than that estimated for the basic training profile above.

Ferry
The quoted ferry range is 1000 nm. This will be flown from a maximum take-off mass of 5707 kg and at optimum speed. It is assumed from Figure 5.17 that V_{cruise} is 150 m/s (291 kt). At this speed the aircraft lift/drag ratio is 9 for best lift/drag ratio (clean) as shown on Figure 5.18. The engine sfc (c) at this condition will be assumed to be 1.05.
To estimate the fuel fraction we will use the transformed Breguet range equation:

$$(M_1/M_2) = \exp[-(R.c)/(V.(L/D))]$$
$$= 0.72$$

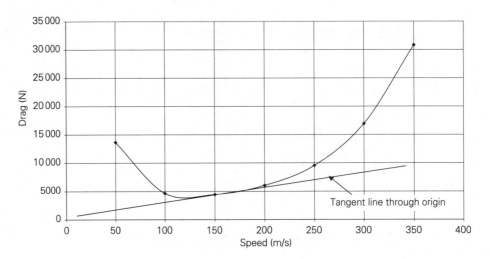

Fig. 5.17 Ferry mission drag polar

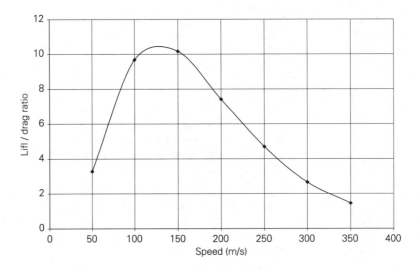

Fig. 5.18 Aircraft cruise lift/drag ratio

Multiplying the take-off, climb and landing mass fractions as used in the previous mission analysis (to account for the fuel used at the start and end of the ferry mission) to this gives the overall fuel mass fraction:

$$M_{start}/M_{end} = 0.97 \times 0.72 \times 0.995 = 0.696$$

$$\therefore \text{ Fuel required} = 5707(1 - 0.696) = 1733\,\text{kg (3821 lb)}$$

This is higher than the fuel that can be carried internally (assumed to be 900 kg), therefore external tankage will be necessary. This would reduce the (L/D) ratio used above due to the increased drag from tank profile. A rough estimate of this effect shows that about another 140 kg of fuel would be required.

Reversing the analysis above and converting the entire payload (weapon) to fuel shows that:

$$\text{Maximum fuel mass} = 1360 + 900 = 2260 \text{ kg } (4983 \text{ lb})$$

Subtracting the take-off, climb and landing fuel fractions reduces this to 2181 kg.

Assuming 5 per cent of the external fuel mass is required for the external fuel tank structure (i.e. $1360 \times 0.05 = 68$ kg) means that only $2181 - 68 = 2113$ kg is available for fuel.

From the Breguet range equation:

$$\text{Ferry range} = [V \times (L/D)/c]\log e(M_0/M_1)$$
$$= (291 \times 10/0.95)\log e(5560/3447) = 1464 \text{ nm}$$

(Note: this range estimation ignores any requirement for reserve fuel at the end of the flight.)

From the above calculation it appears that the requirement for 2000 nm ferry range is too difficult to meet. Keeping this requirement could seriously compromise the basic design of the aircraft. Within the accuracy of the calculations at this stage, it would be reasonable to request a reduction to this requirement down to a value of 1500 nm.

Climb and ceiling estimations

The rate of climb equation is:

$$(R \text{ of } C) = V(T - D)/W$$

Note: this is equivalent to the specific excess power expression.

Using the aerodynamic analysis from the earlier work it is possible to plot the rate of climb against aircraft speed for flight at SL, FL250 and FL360. Figure 5.19 shows the rate of climb at the mean mass condition with an allowance for an increase in drag for weapons ($\Delta C_D = 0.01$). The locus of maximum climb rate allows the data to be cross-plotted (Figure 5.20). This shows the service ceiling (i.e. the height at which the rate of climb falls to 100 fpm (0.5 m/s)) to be about 46 000 ft (against a specified requirement of 40 000 ft).

Figure 5.20 shows that the average rate of climb from sea level to FL250 is 34 m/s. This is used to predict that the time to climb to FL250 (7625 m) is 190 seconds (=3.2 min). This easily meets the specified requirement of 7 min. The same calculations were done to check the climb rate and time for maximum take-off mass with full weapon drag ($\Delta C_D = 0.017$). This showed an average rate of climb of 30 m/s, which leads to a time to climb to 25 000 ft of 4.2 min. At a similar condition but with engine derated to 85 per cent thrust gives a time of 5.8 min.

All the calculations above show that the climb and ceiling requirements are not critical.

Summary of initial performance analysis

- The initial estimate of take-off mass of 5707 kg will need to be revised to 5814 kg to account for the increase in fuel required for the basic training flight profile.

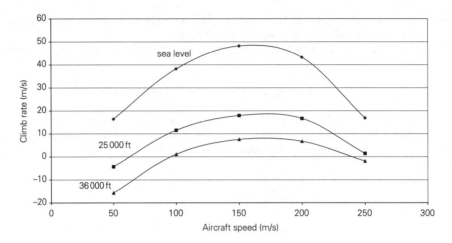

Fig. 5.19 Aircraft rate of climb

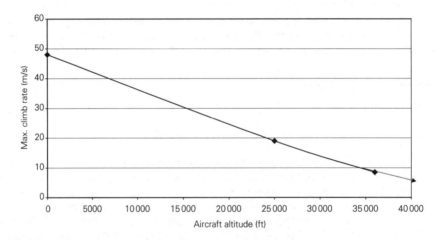

Fig. 5.20 Aircraft climb and ceiling evaluation

- The maximum speed even at 85 per cent thrust is 505 kt. This easily exceeds the specified requirement of 450 kt.
- All the turn performance criteria are easily met.
- Take-off ground run at 1856 ft is below the specified 2000 ft but with a derated engine of 85 per cent thrust, this increases to 2184 ft.
- The approach speed requirement of 100 kts cannot be met except by a lighter aircraft (no weapon load). In this case an approach speed of 95 kt is achieved.
- Landing ground run at 2215 ft also exceeds the specification of 2000 ft. Only with aircraft at lighter landing mass can the specification be met.
- The ferry mission of 1000 nm cannot be met with internal fuel but can be achieved if 833 kg of fuel is carried externally. The maximum range that could be flown is estimated at 1464 nm. This is substantially less than the 2000 nm specified. It is

suggested that this requirement be reviewed as in its present form it would seriously compromise the overall aircraft design.
- Climb and ceiling requirements are easily achieved.

5.8 Constraint analysis

From the project brief there are six separate constraints to be considered in this analysis:

1. Take-off distance less than 2000 ft.
2. Approach speed no greater than 100 kt.
3. Landing distance less than 2000 ft.
4. Combat turn, at least $4g$ at sea level.
5. Combat turn, at least $2g$ at 25 000 ft.
6. Climb rate to provide for 7 min climb to 25 000 ft.

5.8.1 Take-off distance

The equations to be used to determine the effect of the take-off criterion can be found in most textbooks (e.g. reference 4) as shown below:

$$(T/W) = \text{(constant)} \, (W/S)/(S_{\text{take-off}} \cdot C_{\text{Ltake-off}})$$

Obviously this represents a straight line on the (T/W) versus (W/S) graph. For our aircraft the lift coefficient in the take-off configuration ($C_{\text{Ltake-off}}$) is assumed to be 1.7. The value $S_{\text{take-off}}$ represents the total take-off distance (i.e. ground roll plus climb distance to 50 ft). Assuming a climb gradient from zero to 50 ft of 5° gives a ground distance covered of 571 ft. Adding this to the specified ground roll of 2000 ft gives $S_{\text{take-off}} = 2571$ ft (784 m).

The constant in the above equation is assessed from Nicholi's book[4] as 1.27 (in SI units with wing loading in kg/m^2), so

$$(T/W) = 1.27/(784 \times 1.7)(W/S) = 0.00095(W/S)$$

5.8.2 Approach speed

Assuming the approach speed $V_A = 1.2 V_{\text{STALL}}$ then:

$$(W/S)_{\text{landing}} = \beta(W/S) = 0.5 \times \rho(V_A/1.2)^2 \times C_{\text{Llanding}}/g$$

V_A is specified at 100 kts (52 m/s). β is the ratio of landing mass to take-off mass. At a maximum landing weight $\beta = 0.9$. At minimum landing weight (i.e. empty aircraft plus pilot plus 10 per cent fuel $= 3311 + 136 + 90 = 3537$ kg) $\beta = 0.62$.

Assuming the lift coefficient in the landing configuration ($C_{\text{Llanding}}) = 2.1$

$$(W/S) = (0.5 \times 1.225 \times 52 \times 52 \times 2.1)/(1.2 \times 1.2 \times 9.81) = 273.6 \ @ \ \beta = 0.9$$

$$= 397.1 \ @ \ \beta = 0.62$$

Note: these constraints are constant (vertical) lines on the (T/W) versus (W/S) graphs.

5.8.3   Landing distance

The approximate equation to determine ground run in landing can be rewritten as shown below:

$$(W/S) = (S_{\text{landing run}} \times C_{\text{Llanding}})/(\text{constant} \times \beta)$$

The landing ground run $S_{\text{landing run}}$ is specified as 2000 ft (610 m). The lift coefficient in the landing configuration (C_{Llanding}) is assumed to be 2.1 (as above). The expression will be evaluated for the two landing mass fractions used above (i.e. $\beta = 0.9$ and 0.62). The (constant) in the expression above (in SI units with W/S in kg/m^2) is 5.0.

$$(W/S) = (610 \times 2.1)/(5.0 \times \beta) = 284.7 \ @ \ \beta = 0.9 \text{ and } 413.2 \ @ \ \beta = 0.62$$

Note: these are also constant vertical lines on the constraint diagram.

5.8.4   Fundamental flight analysis

The fundamental equation used in the flight cases can be found in most textbooks. In terms of sea level, take-off thrust loading the equation is:

$$(T/W)_{\text{TO}} = (\beta/\alpha)[(q/\beta)\{C_{\text{DO}}/(W/S)_{\text{TO}} + k_1(n\beta/q)^2(W/S)_{\text{TO}}\}$$
$$+ (1/V)(dh/dt) + (1/g)(dV/dt)]$$

where $(T/W)_{\text{TO}}$ is the take-off thrust loading
$\alpha_1 = T/T_{\text{SLS}}$
T_{SLS} = sea level static thrust (all engines)
$\beta = W/W_{\text{TO}}$
C_{DO} and k_1 are coefficients in the aircraft drag equation, see below
$D = qS(C_{\text{DO}} + k_1C_{\text{L}}^2)$
$(W/S)_{\text{TO}}$ is the take-off wing loading (N/m^2)
n is the normal acceleration factor $= L/W$
g = gravitational acceleration
V is the aircraft forward speed
q is the dynamic pressure $= 0.5\rho V^2$
(dh/dt) = rate of climb
(dV/dt) = longitudinal acceleration

5.8.5   Combat turns at SL

In this flight condition the aircraft is in 'sustained' flight with no change in height and no increase in speed therefore the last two terms in the fundamental equation are both zero.

At sea level $\alpha = 1$

Assume that the turn requirement is appropriate to the mean combat mass (i.e. aircraft empty + pilot + half fuel + half weapon load = 3311 + 136 + 450 + 680 = 4577 kg/10 092 lb)

Hence $\beta = 4577/5707 = 0.8$.

From previous analysis (in SI units) the best speed for turning at SL is about 150 m/s.

$$\therefore q = 0.5 \times 1.225 \times 150^2 = 13\,781$$

From the drag analysis done earlier (at 4577 kg with an increase in drag coefficient to represent the stores on the wing) at a speed of 150 m/s, $C_D = 0.03 + 0.017C_L^2$.

As specified, the aircraft is subjected to a normal acceleration $n = 4$ in the turn.

$$T/W = 13\,781\{(0.03/(W/S) + 0.017 \times [4/13\,781]^2 \times (W/S)\}$$

5.8.6 Combat turn at 25 000 ft

This is similar to the analysis above but with $\alpha = 0.557/1.225 = 0.455$.

At 25 000 ft the best speed for excess power is 200 m/s (in SI units)

$$\therefore q = 0.5 \times 0.557 \times 200^2 = 11\,140$$

With β and C_D values the same but with load factor $n = 2$ gives:

$$T/W = (0.8/0.445)[(11\,140/0.8)\{(0.03/(W/S) + 0.017 \times [(2 \times 0.8)/11\,140]^2 \times (W/S)\}$$

5.8.7 Climb rate

This criterion assumes a non-accelerating climb, so the last term in the fundamental equation is zero but the penultimate term assumes the value relating to the specified rate of climb.

We will use an average value of climb rate of 18.15 m/s (i.e. 25 000 ft in 7 min) and make the calculation at the average altitude of 12 500 ft, at a best aircraft speed of 150 m/s.

$$\text{At } 12\,500\,\text{ft} \quad \alpha = 0.841/1.225 = 0.686$$
$$\text{At } 150\,\text{m/s} \quad q = 0.5 \times 0.841 \times 150^2 = 9461$$

Using the standard values for β at mean combat mass, and the drag coefficients (C_{DO} and K) previously specified, we get:

$$T/W = (0.8/0.686)[(9461/0.8)\{(0.03/(W/S) + 0.017 \times [(1 \times 0.8)/9461]^2 \times (W/S)\}$$
$$+ 18.15(1/150)$$

5.8.8 Constraint diagram

The above equations have been evaluated for a range of wing loading values (150 to 550 kg/m^2). The resulting curves are shown in Figure 5.21.

The constraint diagram shows that the landing constraints (approach speed and ground run) present severe limits on wing loading.

To identify the validity of the constraints relative to other aircraft, values appropriate to specimen (competitor) aircraft that were identified earlier in the study have been plotted on the same constraint diagram Figure 5.21. Some interesting conclusions can be drawn from this diagram:

- The S212, T45, MiG, L159 and, to a lesser extent, the Hawk aircraft appear to fit closely to the climb constraint line. This validates this requirement.

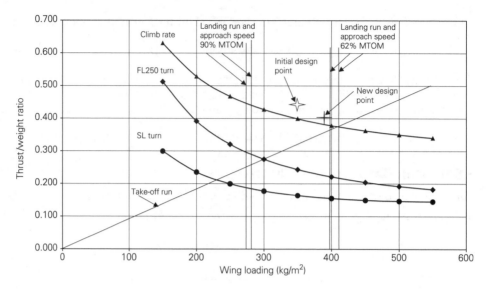

Fig. 5.21 Aircraft constraint diagram

- None of the existing aircraft satisfy the landing conditions at $M_{\text{LAND}} = 0.9M_{\text{TO}}$. This suggests that this requirement is too tight.
- The turn requirements do not present critical design conditions for any of the aircraft. The 25 000 ft turn criteria is seen to be the most severe. Some further detailed analysis suggests that the aircraft is capable of a $3g$ turn rate at this altitude.

Warning: The constraint analysis described above is a very approximate analytical tool as it does not take into account some of the finer detail of the design (e.g. detailed changes in engine performance with speed). It can only be used in the form presented in the initial design phase. Later in the development of the layout more detailed analysis of the performance will enable the effect of the various constraints on the aircraft design to be better appreciated. However, with this consideration in mind it is possible to use the constraint diagram to direct changes to the original baseline layout as discussed below.

5.9 Revised baseline layout

The main conclusion from the constraint analysis and aircraft performance estimations is that the aircraft landing requirements are too tight and should be renegotiated with the customers. To provide evidence on the effects of the landing constraints, the revised baseline layout will ignore them. The new design can be analysed to show what landing characteristics are feasible.

With the above strategy in mind the design point for the aircraft will be moved closer to the intersection of the take-off and climb constraint lines, i.e.:

$$(T/W) = 0.38 \quad \text{and} \quad (W/S) = 390 \,\text{kg/m}^2 \,(80 \,\text{lb/sq. ft})$$

Anticipating the need to increase aircraft mass to allow more fuel to be carried, the maximum take-off mass is increased to 5850 kg (and the structural design mass increased to 6100 kg). Using the new values for (T/W) and (W/S) the new thrust and wing area become:

$$T = 0.38 \times 5850 = 4900\,\text{lb (SSL)}$$

$$S = 5850/400 = 14.65\,\text{m}^2(136\,\text{sq. ft})$$

For an aspect ratio (AR) of 5, the new area gives a wing span $(b) = 8.56\,\text{m}$ and a mean chord $= 1.71\,\text{m}$. For an aspect ratio of 4.5 the wing geometry becomes $b = 8.12\,\text{m}$ and mean chord $= 1.80\,\text{m}$. Rounding these figures for convenience of the layout drawing gives:

$$c_{\text{mean}} = 1.75\,\text{m (5.75 ft)}\quad\text{and}\quad b = 8.5\,\text{m (28 ft)}$$
$$\therefore\text{ gives, } AR = 4.86\quad\text{and}\quad S = 14.87\,\text{sq. m/160 sq. ft}$$

This geometry will be used in the new layout.

Also, since the tip chord on the previous layout seemed small, the taper ratio will be increased to 0.33.

$$\text{Hence } C_{\text{mean}} = (C_{\text{tip}} + C_{\text{root}})/2 = 1.75\,\text{m (assumed)}$$
$$\text{With, } (C_{\text{tip}}/C_{\text{root}}) = 0.33$$
$$\text{This gives } C_{\text{root}} = 2.63\,\text{m/8.6 ft}, C_{\text{tip}} = 0.87\,\text{m/2.8 ft}$$

5.9.1   Wing fuel volume

It is now possible to check on the internal fuel volume of the new wing geometry. Assume 15 per cent chord is occupied by trailing edge devices and 33 per cent span is taken by ailerons (assume no fuel in the wing tips ahead of the ailerons).

Although previously the wing thickness was assumed to be 10 per cent, it has now become clear that the aircraft will require substantial internal volume for fuel storage. To anticipate this, the wing thickness will be increased to 15 per cent in the expectation that supercritical wing profiles can be designed to assist in the transonic flow conditions particularly for the high-speed development aircraft.

With the above geometry (see Figure 5.22) and assuming 66 per cent of the enclosed volume is available for fuel, gives an internal wing fuel capacity of $0.5\,\text{m}^3$. A total fuel load of 1050 kg equates to a volume of 305 Imp. gal. This requires a volume of $1.385\,\text{m}^3$. It is therefore necessary to house some fuel in the aircraft fuselage (namely $1.385 - 0.5 = 0.885\,\text{m}^3$). This is not uncommon on this type of aircraft. The preferred place to keep the fuel is in the space behind the cockpit and between the engine air intakes. This is close to the aircraft centre of gravity, therefore fuel use will not cause a large centre of gravity movement. For our layout it would be preferable to keep the fuel tank below the wing structural platform to make the wing/fuselage joint simpler. From the original aircraft layout this fuselage space would provide a tank volume of about $1 \times 2 \times 0.5 = 1\,\text{m}^3$. This is satisfactory to meet the internal fuel requirement. Using all of this space for fuel may present a problem for the installation of aircraft systems. To anticipate the need for extra space in the fuselage to house the electronic and communication systems an extra 0.5 m will be added to the length of the fuselage. Moving the engine and intakes back to rebalance the aircraft will also provide a cleaner installation of the intake/wing junction (i.e. moving the intake behind the wing leading edge).

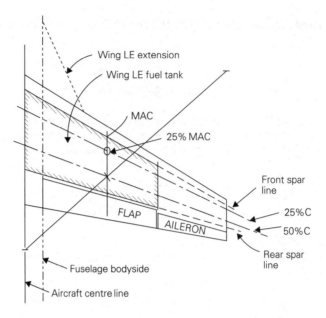

Fig. 5.22 Revised aircraft wing planform

Lengthening the fuselage has the effect of increasing the tail effectiveness. This may permit either a traditional low tailplane/fin arrangement, or more likely, a twin fin/tail butterfly layout. Subsequent wind tunnel tests and CFD modelling would be necessary to define the best tail arrangement. In the revised layout a butterfly tail will be shown to illustrate this option.

It is now possible to redraw the baseline layout to account for the above changes. At the same time it is possible to add more details to the geometry (Figure 5.23).

5.10   Further work

With the new baseline aircraft drawing available and increased confidence in the aircraft layout it is possible to start a more detailed analyses of the aircraft.

We start this next stage by estimating the mass of each component using the new aircraft geometry as input data for detailed mass predictions. Such equations can be found in most aircraft design textbooks. These formulae have, in general, been derived from data of existing (therefore older) aircraft. As our aircraft will be built using materials and manufacturing methods that have been shown to provide weight savings it will be necessary to apply technology factors to reduce the mass predicted by these older aircraft related methods. The factors that are applied must correspond to the expected degree of mass reduction. Different structural components will require individual factors depending on their layout. For example, the wing structure is more likely to benefit from a change to composite material than the fuselage. The fuselage has many more structural cut-outs and detachable access panels than the wing which makes it less suitable. The mass reduction factors for composite materials may vary between 95 and 75 per cent. The lower value relates to an all-composite structure (e.g. as used for control surfaces and fin structure).

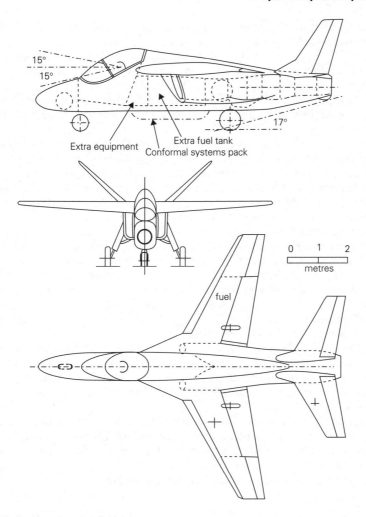

15°
15°
17°

Extra equipment
Extra fuel tank
Conformal systems pack

fuel

0   1   2
metres

Fig. 5.23 Revised baseline aircraft layout

Aspects other than the choice of structural material may also influence the estimation of component mass. Such features may include the requirement for more sophistication in aircraft systems to accommodate the remote instructor concept, the requirements related to the proposal for variability in the flight control and handling qualities of the aircraft to suit basic and advanced training, and the adoption of advanced technology weapon management systems. All such issues and many more will eventually need to be carefully considered when finalising the mass of aircraft components.

When all the component mass estimations have been completed it will be possible to produce a detailed list in the form of an aircraft mass statement. Apart from identifying various aircraft load states, the list can be used to determine aircraft centre of gravity positions. As the aircraft will be used in different training scenarios (e.g. basic aircraft handling experience to full weapon training) it is necessary to determine the aircraft centre of gravity range for different overall loading conditions. With this information it will be possible to balance the aircraft (see Chapter 2, section 2.6.2) and to accurately position the wing longitudinally along the fuselage. Up to this point in the design process the wing has been positioned by eye (i.e. guessed).

With the wing position suitably adjusted and a knowledge of the aircraft masses and centre of gravity positions, it is now possible to check the effectiveness of the tail surfaces in providing adequate stability and control forces. Until now the tail sizes have been based on the area ratio and tail volume coefficient values derived from existing aircraft. It is now possible to analyse the control surfaces in more detail to see if they are suitably sized.

The previously crude methods used to determine the aircraft drag coefficients can now be replaced by more detailed procedures. Using the geometry and layout shown in Figure 5.23 it is possible to use component drag build-up techniques or panel methods to determine more accurate drag coefficients for the aircraft in different configurations (flap, undercarriage and weapon deployments). Aircraft design textbooks adequately describe how such methods can be used. Likewise, more accurate predictions can now be made for the aircraft lift coefficient at various flap settings.

Before attempting to reassess aircraft performance it is necessary to produce a more accurate prediction of engine performance. If an existing engine is to be used it may be possible to obtain such data from the engine manufacturer. If this is not feasible it will be necessary to devise data from textbooks and other reference material. It may be possible to adapt data available for a known engine of similar type (e.g. equivalent bypass and pressure ratios) by scaling the performance and sizes. Design textbooks suggest suitable relationships to allow such scaling.

More detailed aircraft performance estimations will be centred on point performance. The results will be compared to the values specified in the project brief and subsequent considerations. The crude method used previously will be replaced by flight dynamic calculations (e.g. the take-off and landing estimations will be made using step-by-step time methods).

It is also possible at this stage to use the drag and engine performance estimations to conduct parametric and trade-off studies. These will be useful to confirm or adjust the values used in the layout of the aircraft geometry (for example, the selection of wing aspect ratio, taper, sweepback and thickness).

Further detailed work on the aircraft layout will include:

- The identification and specification of the aircraft structural framework.
- The installation of various aircraft system components. This will require some additional data on the size and mass of each component in the system (e.g. APU).
- A more detailed understanding of the engine installation. This will include the mounting arrangement and access requirements. It will also be necessary to consider the intake and nozzle geometry in more detail.
- Investigate the landing gear mountings and the required retraction geometry.
- Make a more accurate evaluation of the internal fuel tank volumes (wing and fuselage tanks).
- Detailed considerations of the layout requirements for wing control surfaces including flap geometry.

It is obvious that the above list of topics requires a great deal of extra work. All of this is necessary in order to draw the final baseline layout. It would be wasteful to do all of this work without first reviewing the project and considering the overall objectives against the predicted design. The following section outlines the nature of such a review process.

5.11 Study review

There are several different ways in which a design review can be conducted. At the higher level a technique known as a SWOT (strengths, weaknesses, opportunities, threats) analysis can be used. At a lower (more detailed) level an analysis similar to that described in section 2.10.2 could be followed. In this study we will adopt the SWOT method as this will illustrate the use of this technique in a design context. It must be emphasised that the low- and high-level methods of review are not mutually exclusive and that in some projects it is advisable to use both.

Before starting the review it must be mentioned that the descriptions below do not constitute a complete analysis. A project of this complexity has many facets and it would be too extensive to cover all of them here. The intention is to provide a guide to the main issues that have arisen in the preceding work.

5.11.1 Strengths

The most obvious advantage of this project lies in the overall life cycle cost (LCC) savings that are expected from introducing a new advanced technology, training system, approach. If such savings cannot be shown it will be difficult to 'sell' the new system to established air forces. The savings will accrue from the lighter modern aircraft. The use of composites will increase the purchase cost of the aircraft based on the price per unit weight. This would also require extra stringency in inspection of the structure. More elaborate systems will also increase the aircraft first cost. However, the new concept would avoid duplicity of aircraft types in the basic to advanced phase and this will reduce life cycle costs. In addition, the aircrew will have received a higher standard of training from the advanced training system, a consequential reduction of OCR training cost.

The second most powerful advantage for the new concept lies in the ability of the aircraft to more closely match modern fast-jet performance than is currently possible with training aircraft that were originally conceived and designed in the 1970s.

Another strength of the new system is the total integration of modern flight and ground-based systems into a total system design approach. Upgraded older aircraft types are not capable of achieving this aspect of the training system.

Many more advantages could be listed for the system. How many can you identify?

5.11.2 Weaknesses

There are three principal weaknesses to the project as currently envisaged. To reduce these deficiencies, if at all possible, it will be necessary to devise strategies or modifications to our design.

The main and intrinsic difficulty lies in the conservative nature of all flight training organisations. This is a natural trait as they take responsibility of human life and national security. As such they will be highly sceptical of the potential advantages of conducting advanced training in a single seat aircraft with a remote instructor. For our concept, as we currently envisage it, this difficulty is insurmountable. Therefore a change of design strategy must be considered to save the credibility of the project. It will be necessary to extend the design concept to encompass a two-seat trainer throughout the full (basic to advanced) training programme. The remote instructor concept can be developed as a separate part of the aircraft/system development programme (i.e. flight testing the aircraft without the instructor present as a proof of concept).

This would allow the design and validation of the ground-based instructor system and associated communication and data linking without jeopardising the success of the traditional design. As we had already accepted that the basic training role would require the development of a two-seat variant, the new strategy will only involve an upgrade to the design to allow the full payload to be carried in this version. Initial calculations suggest that the new aircraft will be about 500 kg (1100 lb) heavier than the existing design (i.e. approximately 10 per cent increase in MTOM). At this point in the development of the project it is obvious that significant changes to the baseline aircraft would be required. Therefore, the work on the present design must be delayed until a revised baseline layout is produced.

The second weakness is associated with the risk involved in the development of new technologies on which the whole system is reliant. If the changes described above are accepted this risk to the project will be avoided. The remaining technologies used in the design can be assured by their current adoption in new aircraft projects (e.g. Eurofighter, F22 and JSF).

The third area relates to the selection of engine for the existing design. From the previous work there are two aspects that require further consideration. First, the Adour engine is shown to be too powerful for our design. The original suggestion (to derate the engine) would only seem to be sensible if the full-rated engine was to be used in future aircraft variants. For the existing trainer aircraft, incorporating an engine larger than necessary effectively adds about 100 kg to the aircraft empty mass. A second propulsion issue relates to fuel usage. Previous calculations showed that the required ferry range was not feasible without seriously penalising the aircraft MTOM. Even to accommodate the fuel required to fly the training sorties it was shown necessary to extend the fuselage to house a larger fuel tank behind the cockpit. For each of the three missions investigated it was found necessary to increase the fuel load that had been previously assumed. As the fuel requirements are directly related to the engine fuel consumption, and thereby to operational cost, it would be advantageous to use a more fuel efficient engine.

Selecting a modern higher-bypass engine with slightly less static sea-level thrust would offer a better design option than using the Adour. Although the engine will be of larger diameter and therefore increase the size of the rear fuselage, it will be lighter and use less fuel. Overall, the change will lead to a lighter and potentially cheaper aircraft.

From the engine data collected earlier (section 5.4.3) there are three possible engines from which to choose (specific fuel consumption (sfc) in lb/lb/hr or N/N/hr):

1. TFE 731-60 manufactured by Allied Signal and used on the Citation and Falcon business jets (SSL thrust = 5590 lb, sfc = 0.42, L = 1.83 m, dia. = 0.83 m, depth = 1.04 m, dry mass = 448 kg).
2. CFE 738 (General Electric/ASE) used on the Falcon business jet (5725 lb, sfc = 0.38 SSL, sfc = 0.64 cruise, L = 2.5 m, W = 1.09 m, depth = 1.2 m, mass = 601 kg).
3. PW 306A (Pratt & Whitney of Canada) used on the Dormier 328 regional jet (5700 lb, sfc = 0.39, L = 2.07 m, W = 0.93 m, depth = 1.15 m, mass = 473 kg).

Aircraft manufacturers prefer to have a choice of available engines as this adds competition on price and delivery. The three engines above are all used on civil aircraft and this may further provide a cost advantage as engine manufacturers will identify an additional market for their product. This should result in a competitive commercial advantage. Approval for military applications will require some extra certification work but this extra cost will be negligible compared to that required to design and develop a completely new engine.

Selecting the PW306A engine would reduce the current dry engine mass by 130 kg (287 lb). This would also reduce the propulsion group mass, thereby reducing the aircraft empty mass. Assuming a cruise specific fuel consumption of 0.64 (as quoted for the equivalent CFE engine) reduces the fuel required to fly the 1000 nm ferry range from the previously estimated 1733 kg for the Adour engine to 1099 kg. This is close to the 900 kg (1985 lb) initially assumed for the fuel mass. The 2000 nm ferry range (assuming external tankage) would require 2328 kg of fuel. This is close to the combined fuel and weapon load (900 + 1360 = 2260 kg/4984 lb) originally specified. Therefore, it appears that by installing this type of engine it would not be necessary to request a reduction in the specified ferry range from originators of the design brief.

The design penalty for installing the higher-bypass type engine lies in the requirement for a larger rear fuselage diameter. The PW306 engine is 0.17 m (7 in) larger in diameter than the Adour. The extra fuselage mass required to house the fatter engine would be more than offset by the reduction in fuel tank weight. The higher bypass ratio engine will also suffer greater loss of thrust with altitude and speed than a pure jet engine.

For designers, the selection of an engine is always a difficult decision as many non-technical factors may intrude into the process (e.g. political influences, offset cost and manufacturing agreements, national manufacturing preference). Without a knowledge of these influences on this project it is recommended that the PW306A engine is installed. This decision will still allow the other competitor high-bypass engines listed above to be used if commercially advantageous. Alternatively, the Adour engine could be used but this would involve a substantial reduction in aircraft range capability unless external tanks are fitted.

5.11.3 Opportunities

Most of the successful training aircraft were originally designed over 20 years ago. Although many have subsequently been 'modernised' they still present old technologies for structure, engines and some systems. The capability of modern fast-jets in the same period has substantially changed and the nature of air warfare which has developed with these improved capabilities. This situation opens a wide gap in the effectiveness of old trainers to meet current demands. Here lies the major opportunity for a new trainer design.

Nearly all of the existing successful trainers have been developed into light combat variants for local area defence and ground attack. However, many of these aircraft are of limited capability due to the age of their systems and their inadequate performance. Our new trainer could be developed into an effective combat aircraft to compete with these existing older trainer aircraft variants.

There is therefore substantial worldwide potential for marketing a new trainer and its derivatives.

5.11.4 Threats

We are not alone in identifying the need for a new trainer. Two other countries have started to manufacture and develop new trainer aircraft over the past few years. These could present a serious commercial challenge to our project unless we can exploit our advanced technologies to produce a more effective and technically capable design solution.

5.11.5 Revised aircraft layout

The result of the study review has proposed significant changes to the existing baseline layout. These include:

- a two-seat cockpit,
- a change of engine,
- a requirement for less internal fuel volume.

Each of these changes will effect the aircraft mass and geometry. A revised general arrangement drawing of the new baseline layout is shown in Figure 5.24. Initial calculations showed that the increase in aircraft structural mass resulting from the addition of the second seat and larger diameter engine has been offset by the reduction in mass from the lighter engine and the reduced fuel requirement.

The single-seat derivative of the new aircraft would benefit from either a 230 kg/507 lb increase in weapon load, or by an increase in range from the equivalent 230 kg increase in fuel load. The single-seat variant is shown in Figure 5.25.

The detailed analysis of the new aircraft follows the same methods as outlined earlier in this chapter. To avoid repetition these calculations have not been included in this chapter.

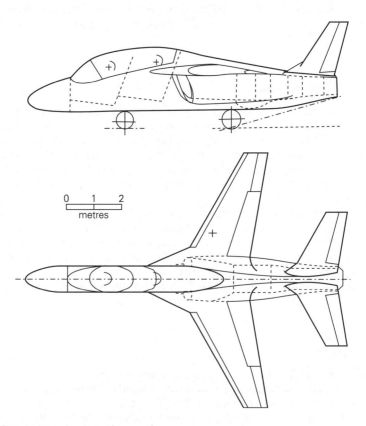

Fig. 5.24 Post-design review layout (two seat)

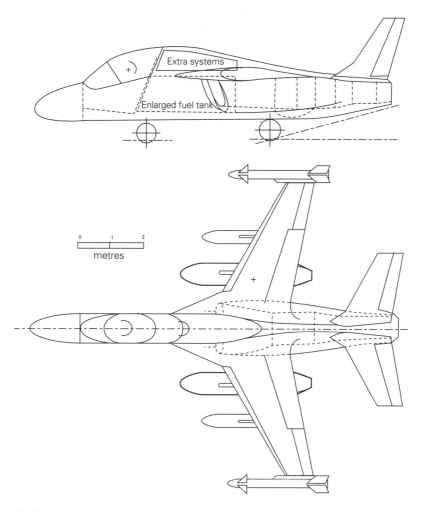

Fig. 5.25 Single-seat aircraft variant

5.12   Postscript

This study has demonstrated how project design decisions may change as the aircraft is more thoroughly understood. This demonstrates the iterative nature of conceptual design. It is possible for students to continue this project into the next iterative stage using the final aircraft drawings (Figure 5.25) as the starting point.

References

Textbooks for military aircraft design and performance:

1 Raymer, D. P., *Aircraft Design: A Conceptual Approach*, AIAA Education Series, 1999, ISBN 1-56347-281-0.

2 Brandt, S. A. *et al.*, *Introduction to Aeronautics: A Design Perspective*, AIAA Education Series, 1997, ISBN 1-56347-250-3.

3 McCormick, B. W., *Aerodynamics, Aeronautics and Flight Mechanics*, Wiley and Sons, 1979, ISBN 0-471-03032-5.

4 Nicolai, L. M., *Fundamentals of Aircraft Design*, METS Inc., San Jose, California 95120, USA, 1984.

5 Mattingly, J. D., *Aircraft Engine Design*, AIAA Education Series, 1987, ISBN 0-930403-23-1.

The following publication is also useful in collecting data on existing aircraft:

Aviation Week Source Book, published annually in January.

This handbook is a useful source of general aeronautical data:
AIAA Aerospace Design Engineers Guide, 1998, ISBN 1-56347-283X 1.

Project study:
electric-powered racing
aircraft

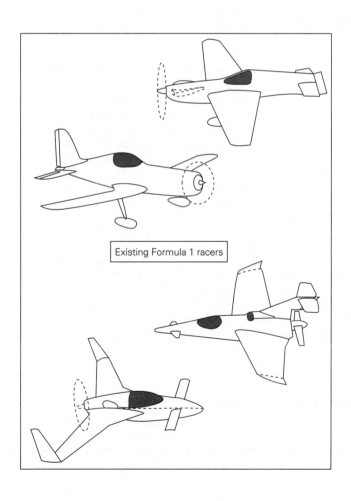

Existing Formula 1 racers

6.1   Introduction

This project is the direct result of collaboration between aeronautical and automotive research teams. Government requirements aimed at reducing the detrimental effects of emissions from automobiles on the environment have stimulated the automotive industries into investigating and developing alternative power sources for mass produced cars and light vans. Various types of electric propulsion systems have been studied in detail. These produce near-zero, harmless emissions. Future automotive legislation may require a substantial and increasing proportion of motor vehicles to be environmentally 'friendly'. It is expected that this will result in the development of lightweight and cheap electric propulsion systems. Such systems could be adapted for aircraft use. Although the reduction of emissions is not too significant for the short duration of a race, the development flights for this aircraft and the use of such propulsion systems in other applications must be considered. Investigating this possibility in a competitive environment that will stimulate rapid technical development is the main objective of this project. And, of course, the design of a fast racing aircraft should also be fun!

6.2   Project brief

From the earliest beginnings of powered flight, general/light aviation has modified automotive engines for powering aircraft. Even the famous Wright Brothers followed the principle in their epic first flights about a hundred years ago. As in the development of any new technology and innovation, it is necessary to introduce new concepts slowly and in a controlled environment. Sport aviation has traditionally been a suitable way of developing such technologies into commercial opportunities. Air racing is currently reported to be the fastest growing motor sport in the USA. Commercial sponsorship and television sports coverage of weekend race meetings have generated renewed interest in the sport. This environment offers the means by which we could gain flying experience with a new propulsion system in a highly controlled environment.

As we will be designing a new racing aircraft, it is important to investigate the current air-racing scene. At present, there are several classes of air racing. The two most closely controlled pylon-racing organisations are Formula 1 and Formula V (vee). The main difference between these lies in the specification of the engine type. Formula 1 relates to the 200 cu. in. Continental (0–200) engine and for Formula V to a converted Volkswagen engine (hence the significance of the vee). Using this pattern, we should project a new Formula (E) to relate to the electric propulsion.

Apart from the engine details, all other requirements should match the Formula 1 rules. In this way, the new formula will benefit from the many hours of successful racing experience. It will also ensure that the race organisers accept the new formula. The rules and procedures are available from the Formula organisers and are published on the Web.[1] The main features, and a brief history of air racing, are described below.

6.2.1   The racecourse and procedures

The race starts with a field of six to eight aircraft on the ground (runway) for a simultaneous take-off. The normal formation consists of three aircraft in front, two in the middle position, and three at the rear. As in motor racing, the positions on the starting grid are related to previous race performance. The fastest aircraft/pilots are at the front

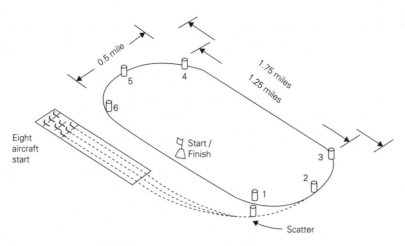

Fig. 6.1 Racecourse geometry

of the grid and have a 150-yard advantage over those at the back. As in earlier motor racing, the racing team ground crew assist in starting the engine, securing the pilot (etc.) and preparing the aircraft for the race but must leave the take-off area no later than one minute prior to the start. A green flag is raised about ten seconds before the 'off' at which point the pilots apply full throttle. When the green flag drops the race begins.

The racecourse consists of a two turn, three-mile oval as shown in Figure 6.1. The seven marker pylons that define the course are typically 60-gallon oil drums fastened on the top of short poles. The first pylon (outside the oval track) is called the scatter marker. Although the aircraft are racing from the take-off, the lap that includes the scatter pylon is not included in the race. The racing time starts when the first aircraft passes the start/finish line. Races usually last for eight laps (sometimes six depending on the number of heats that are required to sort out the field). Overtaking is the 'name of the game' but pilots should pass high and outside the flight path of the slower competitor. Stewards are positioned at each pylon to ensure that pilots do not 'cut' the track. Such indiscretions earn the pilot penalty time. This is two seconds per lap, which is more than can be won back in the race. It is therefore important to have clear visibility to ensure that such penalties are avoided. The 24 fastest aircraft/pilots from the heats are split into three groups. The slowest group competes for the bronze, the next for the silver and the fastest for the gold. The winner of the gold race is crowned the champion of the race. These victories build up points for the national championship. Prize money is earned in proportion to the success in the heats and, more profitably, in the finals.

6.2.2 History of Formula 1 racing (further reference can be found on the Formula 1 web site[1])

Prior to 1945, racing aircraft were mostly original designs specifically aimed at racing. They were unique creations that often advanced the field of aeronautics. Innovative designers of air racers consistently produced aircraft that outperformed the best military aircraft of the day. In the early days, these aircraft led to the development of monoplane wing layouts and introduced materials and construction methods that were lighter and more reliable. After World War II, there was a surplus of high-powered, mass-produced, ex-military aircraft. These were introduced into air racing but the purists argued that this was not in the spirit of the original sport. Such aircraft were raced in an open classification category that still exists. To return to the original racing concept and to make the sport more pilot-centred, the Formula rules mentioned earlier were created. In 1946, these special racing criteria were introduced to continue aerodynamic refinement within the financial means of many people.

The first racing class was referred to as the 'Midget' class due to the relatively small size of the aircraft. Initial rules limited the aircraft to a minimum wing area of 66 sq. ft, fixed-pitch propeller, and engines limited to a maximum displacement of 190 cu. in. The aircraft also had to have good pilot visibility, nose-over pilot protection and weigh at least 500 lb (227 kg) empty. These constraints forced designers to concentrate on minimising drag and structural weight. Sponsorship was initially with the Goodyear Company, hence the name 'Goodyear racers' was formed.

The first competition of the Midget class was held at Cleveland in 1947. Thirteen aircraft competed in this event. By 1949, the field had grown to 25 and the prize money to $25 000. This attracted the famous names of aviation enthusiasts and model builders. Unfortunately, it was also during this meeting that Floyd Odom, a renowned racer, crashed his ex-military fighter aircraft into a home killing a young mother and her baby. This accident and the sudden realisation that racing over populated areas was potentially very dangerous caused many in the population to call for a ban on all air racing and air shows. The multi-classes Cleveland National Air Races were abandoned. Fortunately, Continental Motors sponsored the 190 cu. in. class and kept the sport alive. During the 1950s and early 1960s, the Midget class air races were held throughout the country. Multi-class racing was finally resumed in 1964 with the first Reno National Championship races. Reno offered a remotely located site that minimised the risk to non-participants. The 1964 races were the first major racing event in 15 years. Six 190 cu. in. racers took part in this event with famous pilots such as Steve Wittman and Air Scholl participating.

Formula 1 racers have continued to evolve, reflecting the spirit for which the sport was formed. The most significant change in the rules occurred in 1968 when the maximum engine size was increased to 200 cu. in. and the name changed from 'Midget' to '1'. Radical designs such as the Miller Special pusher configuration and the all-composite Nemesis racer have recently 'pushed the envelope' further. Formula 1 and the more recent Formula V races continue to attract a healthy mixture of new enthusiasts and seasoned veterans each year.

6.2.3 Comments from a racing pilot

This section is added to offer guidance to prospective designers who have not had experience in air racing. Collecting such information from people who have operated the type of aircraft you are designing is a good strategy if it is easy to find.

> Constant vigilance is the number one rule when racing. You always have to be looking and thinking. It is essential to always be looking around for other aircraft and to be keeping abreast of the current situational awareness. In fact, you should start this process on the ground at the start of the race and understand which aircraft are likely to be accelerating to gain the early advantage. Know which way the wind is blowing as this will affect the racing line to be taken by your competitors. Watch out for shadows when you are turning around the pylon. Who is above you and likely to overtake you in the turn, and who is below and perhaps will be over-run by you.

In the race, you want to fly a smooth line at constant altitude but this will not be entirely feasible if you want to overtake and avoid contact with a competitor. I like to fly with my lowest wing tip about 20 feet above the top of the pylon. In this way, if I get blown off line or misjudge a turn I still avoid the pylon. It also leaves me with a way out if I think that I might get hit. Normally I like to see about half a circle looking through the pylon. If you try to see more than this you run the risk of a cut if the wind or turbulence suddenly alters your course. In a turn, we do not throttle back, as in motor racing, we just bank and add a little load factor on!

When it comes to overtaking, it is important to take care. This means seeing them and the course ahead at all times. The best way is to pass them above and on the outside. This requires good airmanship. You have to fly carefully but if you go too wide or high you will erode your speed advantage and you could interfere with those aircraft that are following and hoping to overtake. Passing a plane that is well piloted is fun. One that is flown erratically, or by someone who tries to climb in the turn, can be hazardous. You must know your opponents' flying abilities. Once you get past a slower aircraft some pilots wave, salute or make other gestures. You are close enough to see facial expressions and hand signals easily!

Turning is the fun part of the race. It is here that the skill in piloting and the technical advantage of the aircraft design are most clearly apparent. I try to look at the next two pylons at all times and keep updating my flight path to get the best out of the situation. Once a pylon is passed, I forget it and start lining up for the next one. Since the course is only about three miles long (mile straights and two half-mile turns) you can pick up the corner pylons as you enter the straight. Using big landmarks like hangars, road and mountain features help in quickly identifying the pylons from the ground clutter.

(with grateful acknowledgements to the Formula 1 web site,[1] June 2000)

The description above provides an excellent insight into the enthusiasm and skill of racing pilots. Once the race is over and the times have been adjusted for any 'cuts', the winner is paraded in front of the cheering crowd (often in a specialist automobile).

6.2.4 Official Formula 1 rules

Current Formula racing rules can be found on the Internet.[1] Most of the rules relate to the design and modification of the engine. The list below summarises the aircraft related aspects of the rules as they appeared in June 2000:

- wing area, minimum size = 66 sq. ft (6.132 sq. m),
- cockpit height, minimum = 30 ins (0.75 m),
- pilot visibility, minimum 25° aft, 5° over nose, 45° upwards,
- aircraft empty weight (mass) = 500 lb (227 kg) minimum,
- aircraft centre of gravity = between 8 per cent and 25 per cent of the wing mean aero. chord,
- main landing unit fixed (nose may retract),
- main wheels = 5.00 × 5 tyre size,
- fixed pitch propeller (metal props not allowed),
- structural design limits = +/−6g minimum.

There are several rules that relate to pilot experience and medical condition. The race committee has the right to revoke the licence if unsafe flying habits are demonstrated.

The flight test requirements include:

- perform aileron rolls in each direction with no more than 50 ft loss of height,
- perform a half roll to the left then recover with a half roll to the right with less than 50 ft height loss,
- fly in formation,
- fly the race course safely without climbing in turns,
- properly overtake on the racecourse,
- complete normal and aborted starts without deviating more than 10 and 20 ft (respectively) from the straight ground track,
- complete normal and engine-out landings.

Reference to the full set of rules should be made if unconventional concepts are to be considered.

Apart from the constraints described in the official rules there are a number of other design considerations that must be taken into account when configuring a new racer layout:

- It is common practice to tow the race aircraft to the air show in a specially built and prepared trailer. This trailer must comply with national construction and use regulations for highway vehicles. The aircraft will need to be partially dismantled to fit into the trailer. Reassembly of the aircraft at the race site must be easy and quick. More importantly, the aircraft reassembly must be reliable and safe. The description of the aircraft must detail how this is to be done and what special equipment and skills are required.
- Flying at speeds in excess of 200 kt at only 50 to 100 ft above the ground, and in competition with equally enthusiastic racing pilots, is not regarded as the safest of activities! If an accident happens it is important that the pilot is adequately protected. Crashworthiness is a significant consideration in the overall design philosophy of the aircraft.
- Even with our best endeavours, it may be necessary to make changes to the aircraft design after the first few races. This 'tweaking' of the aircraft is regarded as an essential part of the development process that matches the aircraft, pilot and course characteristics. A design that offers some flexibility will be more easily 'tuned' to achieve higher performance and therefore be more competitive.
- Although vision and cockpit dimensions are specified in the rules, these may be considered as the minimum safety standard. Pilots win races by overtaking competitor aircraft, often in a turn. To do this effectively they need to be able to see suitable opportunities quickly and as they arise. For example, a prone position for the pilot may be better aerodynamically as it reduces frontal area, and this position resists higher 'g' loading, but the pilot would find it too restrictive in racing conditions.
- Although formula racing is becoming 'commercial', enthusiastic amateurs still largely dominate the sport. Many aircraft are built from kits by such people. They often personalise them to suit their preferences and competitive spirit. Aircraft must be designed to be 'easy to make' and with the freedom to allow for small changes if required.
- Race duration is only about 10 minutes but, to account for possible delays and emergencies, a flight time of at least 30 minutes must be possible.
- Consideration must be given in the design for flight trials prior to the commercial launch of the project. This also extends to the need for pilots to become familiar with the aircraft prior to the racing season. Flight test programmes lasting up to 2 hours should be possible, perhaps with a modified engine if the electric system is

not feasible for this duration. Such a change must be carefully considered to avoid penalising the fundamental configuration of the aircraft.

6.3   Problem definition

For this project, the problem definition phase is relatively straightforward. Put simply, the aircraft is designed to win Formula E races. However, there are several associated criteria that must also be taken into account. As with most projects it is important to identify who is likely to be the 'customer' for the aircraft. Although the race pilot will be the ultimate 'user' of the aircraft, he may not be the owner or purchaser. As with automobile racing, it is possible to envisage that a corporate/sponsor involvement is desirable. Development of advanced technologies, like electric propulsion, requires substantial investment. It would be preferable to design the aircraft for a 'professional' team rather than an enthusiastic amateur. Such considerations would allow a more ambitious approach to the design and manufacture of the aircraft. This would permit more sophistication in the systems to be used. However, such possibilities are accompanied by the need to be ultra-confident in the performance and operational capabilities of the aircraft. Sponsors will soon evaporate if the aircraft is not providing the success that they will demand. Equally, safety (or more precisely the lack of safety) would be a paramount consideration. Successful companies will not want to be associated with failure of any kind. Such thoughts lead to a number of essential requirements for the aircraft:

- The aircraft must have performance that is at least as good as existing Formula 1 racers, or it will not be taken seriously by the existing competitor and viewing public.
- Crashworthiness must be a priority in the design layout.
- Excellent pilot visibility is essential.
- The aircraft must have significant development possibilities.
- Aircraft flying characteristics will be of particular significance. It is often difficult to achieve a suitable blend between racing trim and development flying characteristics.
- Aircraft development will require significant effort in flight testing.

Although the main intention of this project is to work alongside a major industrial sponsor, it must be appreciated that air racing is largely dominated by amateurs. In the event that a sponsor cannot be found, the design should be capable of development for amateur construction and operation.

Assessment of the 'viability' of the aircraft at the project stage for a racing aircraft is difficult as it obviously involves, first, the attraction of sponsors and then success in racing. Obviously both of these are interlinked but the main technical criterion is directed at superiority in competitive racing.

The initial brief, the mandatory airworthiness requirements, and the compulsory Formula racing rules describe the main definition of the problem. However, as described above, there are several other design considerations that must be included in the development of the aircraft specification:

- Consideration must be given to the methods to be used in the pre-race flight trials.
- The specification must allow for 2-hour flight duration in development aircraft.
- In racing trim, 30-minute flight duration is required.
- Better than average pilot visibility is required to allow tight turns around the course pylon to be executed with precision.
- Pilot crash-survivability must be built into the fuselage/cockpit structure.

- The aircraft must be capable of been 'trailered' to the race site.
- The aircraft must be capable of reassembly at the race site within 30 minutes and must not require special facilities, tools or skills.
- Safety of the aircraft structure, controls, trim and systems must be assured following reassembly.

For this aircraft, the principal area of innovation lies in the propulsion system. It may be possible to distribute the components of the electric system more advantageously than with traditional engines. For example, avoiding the blunt nose profile and cooling air intakes of a conventional internal combustion tractor engine layout. An electric propulsion system could be arranged to have the fuel cell components behind the cockpit yet still adopt a tractor propeller layout. Because of the technical uncertainty of the propulsion system, it may be a better overall design strategy to avoid developments in other technical areas (e.g. novel aerodynamic devices and complex structures). Such considerations may be delayed until after the integration of the electric propulsion system has been successfully achieved. This will ensure that the development of the new power system is separated from other, potentially troublesome, innovations. A consequence of this approach may be that the aircraft layout and structure follow similar lines to existing successful racing aircraft, except for the detail positioning of the propulsion system components.

6.4   Information retrieval

The sections above have set out the foundations for the initial aircraft specification. Before we move on to selecting a configuration for the aircraft, it is preferable to spend some time conducting a literature search to review competitor aircraft, understanding racing operations and researching the current electric propulsion systems.

Many of the aspects relating to air racing have been described above. These have included a description of the racing environment and a racing pilot's views on his flying requirements. More can be found from the various web sites related to air racing. No more information will be discussed here. The other two aspects of the information retrieval phase are concerned with existing aircraft and electric propulsion.

6.4.1   Existing aircraft

Existing aircraft can be used to guide us in the choice of configuration for our design. Many of these are home-built designs. Some of them were designed by their enthusiastic owners. A feature of this collection of aircraft is the uninhibited selection of unconventional layouts and novel details. The list below is not intended to be a recommendation of the 'best' designs or of preferred configurations. It is intended to give a flavour of the variety of aircraft to which our design will need to compete. (Apologies to those who feel that we have not included their favourite designs in this survey!) A brief introduction to each aircraft is given below and a list of the main characteristics is compiled for use in the initial sizing stage.

Nemesis

We start with an aircraft that is regarded by many as the 'state of the art' in Formula racing aircraft. It has been designed and developed by professional aeronautical engineers. It encompasses many features that have contributed to its outstanding success in national championships. The mid-fuselage mounted wing uses a natural laminar flow aerofoil. The wing tips are a noticeable and unique detail of the aircraft. They are of a novel 'cusp' design. They have been geometrically sheared rearward to reduce wing tip vortex power, thereby reducing aircraft induced drag. Recognising the need to reduce wing/fuselage interference effects, the wing is permanently fixed to the fuselage structure. To transport the aircraft, a fuselage joint behind the wing detaches the rear fuselage and tail controls. The aircraft structure is mainly constructed in moulded graphite composite (carbon/foam sandwich). To reduce weight and drag, the Nemesis designers have paid great attention to detail design (e.g. one piece wheel fairings). This is a good overall strategy that is worth considering by all new racing aircraft design teams.

AR-5

This aircraft is another example of a modern low drag design. Although powered by only a 65 hp motor it set a world record in 1992 of 213 mph (which gave a remarkable 3.28 mph per hp). It has a completely, all-composite, home-built structure. The conventional configuration achieves its outstanding performance by the designers' and manufacturers' attention to detail. This is particularly evident in the treatment of the wing (and tail) to fuselage junctions to reduce interference drag. The lower wing section profile is flat and arranged to be coincident with the bottom fuselage section. The upper surface of the wing has been carefully 'filleted' to the fuselage side profile notably at the nose and tail intersections. The wheels are faired and located in the airstream outside of the propeller wash. It is regarded as a tribute to the designers that the relatively old NACA 65 series aerofoil performs so well. The wing achieves almost 50 per cent laminar flow. The design strategy for the aircraft is of a simple (uncomplicated) layout, very clean aerodynamic shape, high-quality (smooth) surface manufacture and an almost obsessive attention to the avoidance of interference drag. As in the Nemesis aircraft, any successful racing aircraft is worth careful study by new aircraft teams.

Monnett Sonerai

This aircraft, a Formula V racer, shows how the configuration is affected by the requirement to fold the wings to lie along the side of the fuselage for transport (in this case tail-first towing). The resulting low aspect ratio wing planform will be 'draggy' in high-g turn manoeuvres but set against this deficiency is the low wing weight from the short span composite structure. This presents the classical aeronautical dilemma – savings in drag from low weight but increased drag from lower wing aspect ratio.

Option Air Reno Acapella

This aircraft illustrates a different configuration for the engine/airframe integration. The small sized fuselage houses a rear engine 'pusher' propeller. Twin booms mounted from the wing structure provide support for the rear fins, rudders, tailplane and elevator. The advantage of the rear propeller lies in the avoidance of prop-wash over the fuselage profile. This should reduce drag, but as the propeller is positioned in a more disturbed airflow, it will be less efficient. (Another aeronautical dilemma!)

Alpha Macro J-5

The Alpha is another example of a 'pusher' layout. The reason for the configuration is the desire to avoid the increased airspeed from the propeller flowing over the fuselage and increasing aircraft parasite drag. To avoid the reduction in propeller efficiency from the blockage of airflow into the propeller, the engine in this aircraft is mounted high on the cockpit 'capsule'. This layout allows a low mounted single tail boom to be positioned below the propeller disc. A 'butterfly' tail is positioned at the rear of the tail boom. Although overcoming some of the problems normally associated with a pusher layout, care must be taken with this configuration to avoid aircraft trim variation with power changes.

Holcomb Perigree

Although this is not a racing aircraft, it has been included in this review to illustrate another pusher propeller layout. In this case, the propeller is mounted on an extension to the rear fuselage. The extension is long enough and suitably shaped to reduce the propeller 'blanking' and to ensure that the tail surfaces are not too much affected by the inflow to the propeller. The fin is arranged in a ventral (downward) position to provide the mounting structure for the tail wheel. This gives the aircraft an approximately parallel attitude to the ground at take-off and avoids the high-drag, tail-down fuselage and propeller attitude of more conventional designs. It is expected that this will reduce drag during take-off but has the disadvantage of reducing wing incidence, therefore lift generation. However, it is claimed that the aircraft take-off performance is improved over the tail-down attitude of more normal configurations. Without the use of air brakes, the landing distance will be increased due to the streamlined layout of the aircraft on the ground run.

FFT SC01B: Speed Canard

Although this aircraft is also not a racing aircraft or even a single seater, it is an interesting example of the canard configuration that is used on some aircraft. This aircraft is based on the successful Rutan VariEze tourer aircraft. Its structure is a composite of GRP and foam materials. The wing is mid-mounted and swept back to provide wing tip mounted vertical control surfaces. The rear engine, pusher layout complicates the shape of the rear fuselage but has been carefully contoured to provide a good aerodynamic profile. This layout requires a tricycle undercarriage arrangement. The nose unit retracts into the front fuselage profile. The main units are fixed. To provide the required balance to the aircraft, forward swept wing root extensions are used as fuel tanks.

In summary, several aircraft were researched in the literature search. Some of these are potentially competitor aircraft while others are included to illustrate alternative configurational options. Some selected technical details of the aircraft are shown in Table 6.1.

The details in Table 6.1 will be useful when making the initial estimates for our aircraft.

6.4.2   Configurational analysis

In reviewing all the different types of aircraft that are similar to our expected design, it is clear that the main configurational decision to be made rests between the choice of tractor or pusher propeller position. Both have advantages and disadvantages associated with airflow conditions over the aircraft profile. As neither configuration has emerged in the preferred layout for modern racing aircraft, there seems to be no over-riding technical (racing efficiency) reason for the choice.

From the review, the conventional tractor layout is seen to have less variation in the overall aircraft layout. The traditional two-surface layout prevails with the mainplane

Table 6.1 Survey of existing aircraft

Name	Description	Long (m)	Span (m)	Area (sq. m)	AR	M_E (kg)	M_{TO} (kg)	M_E/M_{TO}	W/S (N/m²)	P (kW)	T/W
Nemesis	Formula 1 racer	6.71	6.41	6.22	6.6	236	340	0.69	536	75	0.16
AR-5	Formula 1 racer	4.42	6.4	5.12	8.0	165	290	0.57	556	49	–
Monnett Sonerai	Formula V racer	5.08	5.08	6.97	3.7	199	340	0.59	479	45	0.15
Optiori AirReno	Formula V – twin boom pusher	5.03	8.08	6.06	10.8	295	473	0.62	766	88	0.18
Alpha Macra J-5	Home-built – twin boom pusher	–	8.16	6.28	10.6	165	290	0.57	453	19	0.12
Perigree	Kit – pusher – high wing	4.78	8.53	7.57	9.6	172	326	0.53	422	26	0.13
FFT Speed Canard	Two seat – canard – sport	7.79	7.79	7.88	7.7	440	715	0.62	890	120	0.17
Cassult Special	Formula 1 – Home-built	4.88	4.57	6.27	3.3	227	363	0.63	568	64	0.17
Pottier P70s	Home-built – sport a/c	5.15	5.85	7.21	4.7	215	325	0.66	442	45	0.2
Monnett Money	Blended canopy sport a/c	4.67	5.08	4.27	6.0	191	295	0.65	678	60	0.19
Aerocar Micro Pup	High wing – pusher	4.57	8.23	7.49	9.0	118	238	0.50	312	22	0.15

ahead of the control surfaces. On the other hand, the pusher layout offers several options. These include either tail or canard control surfaces. If the tail arrangement is selected, this presents difficulties at the rear fuselage. Using a twin boom layout avoids the tail surfaces/propeller interference but complicates the wing and fuselage structure. Lifting the propeller line above the fuselage may cause trim changes with power and also complicates the rear fuselage profile.

The choice of landing gear geometry lies between the nose (tricycle) and the tail (tail dragger) arrangements. The tail wheel layout is lighter but introduces the possibility of ground looping. Current formula rules prohibit retraction of the wheels but our proposed Formula E rules will allow the auxiliary wheel to be retracted as this does not seem to overcomplicate the design yet improves aerodynamic efficiency.

In selecting the aircraft configuration, the most significant criterion is the requirement for high aerodynamic efficiency (i.e. low drag). This implies:

- smooth profiling of the external shape of the aircraft,
- avoidance of the canopy/windscreen discontinuity,
- fairing of the landing gear and other structural details,
- reduction of airflow interference areas (e.g. mid-mounting of the wing to fuselage), and
- avoidance of engine/propulsion system cooling drag.

Many of the low drag features would be considered during the manufacturing (surface smoothness and preparation) and operational (gap taping and surface cleaning) phases.

For this project, the most significant difference in configuration compared with conventional designs is the location of the various components of the propulsion system. Whereas conventional designs have the propeller and engine closely positioned, in an electric system only the electric motor is linked to the propeller. This motor is much smaller than a conventional internal combustion engine and can therefore be streamlined into the fuselage profile. All other components in the electrical system can be located in convenient positions in the aircraft. These options will create an installation that has potentially less drag and higher propeller efficiency. It is also envisaged that the electrical system will require less cooling than the equivalent internal combustion engine. This will also reduce aircraft drag.

6.4.3 Electrical propulsion system

Powering an aircraft with an electrical power system is not new. Several attempts were made in the early days of aviation to incorporate electrical power into aircraft. However, the main difficulty rested with the storage of electrical energy in bulky and heavy batteries. Even now, batteries are still too heavy to be used in general aviation aircraft as the main power source, except for aircraft with very short operating time. One aeronautical field in which electrical propulsion has become established over the past 20 years is for model aircraft flying. Progressive miniaturisation of electrical and electronic components has shown the advantages of the technology. The aircraft are environmentally cleaner, more reliable and easier to integrate the operating systems. More recently, a number of large aircraft have been designed and flown using solar panels as the power source. Reference to the Pathfinder, Solar Challenger and Penguin aircraft should be made to understand these developments (namely, NASA and Lawrence Livermore National Laboratory).

As described earlier in this chapter, the automotive industry has again become interested in electrical propulsion for environmental reasons. The search for clean power systems has led to the development of alternatives to the traditional lead–acid battery. Many of the leading companies have displayed prototype vehicles using fuel cells.

A fuel cell is a chemical and mechanical device to convert chemical energy stored in a source fuel into electrical energy without the need to burn the fuel. Theoretically, if the device can be made to work, it is potentially highly efficient, has almost harmless emissions, and is quiet. Such devices are not new. They have been successfully used in spacecraft and submarines for many years. There have also been some small-scale applications used in some industrial, power generation units.

The fundamental operation of a fuel cell matches that of a traditional battery. Electrons are freed from one element in order to create an electrical potential. The essential difference between a battery and a fuel cell lies in the ability of the fuel cell to perform the process of dissociation of the chemical components continuously, providing fuel is supplied to the cell. The fuel cell is fed with hydrogen. After the electrons have been removed, the spent hydrogen protons pass through an electrolyte to combine with oxygen to form pure water, an environmentally acceptable emission. Several types of electrolyte could be suitable for our application. As the 'solid polymer' type has been successfully developed for automotive applications this is the one that is recommended. Knowledge of the precise details of the construction of a fuel cell is not necessary for this project. For general interest, such details can be found in textbooks, technical papers and various automotive companies' web sites.

The fuel cell requires hydrogen and oxygen. The latter is easily obtained from the ambient atmosphere. Hydrogen is more difficult to feed to the system. In its pure form (as a gas or a cryogenic liquid), it would be a very efficient fuel. The problem with hydrogen in this form is the need to store and transport it in pressurised gas tanks or refrigerated liquid containers to reduce tank volume. Both of these options are heavy and bulky, seriously eroding the chemical efficiency of the system. Although not pure, a more convenient source of hydrogen supply for our application would be a hydrocarbon fuel. Methane is the preferred fuel as it is rich in hydrogen and can be easily reformed. Reformation is the process by which methane is mixed with water and vaporised to split it, with the assistance of heat and a catalyst, into hydrogen and carbon dioxide. Any carbon monoxide present in the emissions can be converted to carbon dioxide using another catalyst. While supplied with the two input gases, the fuel cell process is continuous. It is therefore not suitable for changes in energy demand. To provide higher power, for example on take-off and climb or some emergency condition, it would be necessary to supplement the fuel cell energy with a battery. The battery could be recharged by the fuel cell during low-energy flight periods. This feature may be less appropriate for a racing aircraft that continually uses full power.

Several components are required for a fuel cell system. These are shown diagrammatically in Figure 6.2, and described below.

The basic principles of the reformer (2) and fuel cell (5) have been described previously. The oxidiser unit (3) is more accurately referred to as the 'flue gas clean-up unit'. This reduces harmful (NOX) gases. The air compressor (4) provides the supply of oxygen to the oxidiser unit and fuel cell. The compressor is powered by the battery (8). The battery also provides the start-up energy for the system. This may take up to 15 minutes. It could be provided by an external power source. The internal battery will be recharged during flight. The DC/DC converter (6) transforms the low-voltage DC supply from the fuel cells to the inverter (7) and then to the electric motor (9). The controller (10) provides the overall system control. This includes:

• the input to the motor (via the converter and inverter),
• the start-up sequence,

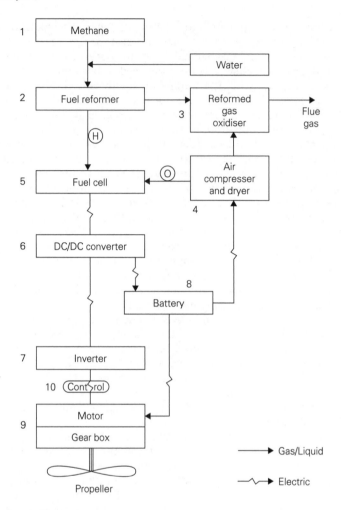

Fig. 6.2 Electric propulsion system

- the power-down control, system condition monitoring, and
- management of the various subsystems.

In a fast moving technology there is a risk that developments not envisaged at the time may arise to change the details described above. For example, the reformation process to create hydrogen from the methane is heavy and complex. It would be a substantial improvement if the methane could be fed directly to the fuel cell. Considerable research work is currently being conducted to make this possible. If this works, the system will be changed from that described but the fundamental process will be unaffected. With such thoughts in mind, it is wise to treat the details of the system presented below with a degree of scepticism. The technical details (e.g. weight, volume requirements and efficiencies) have been extrapolated from experimental automotive applications and suitably modified to include improvements expected in the timescale of the project. This process is not untypical when designing aircraft that incorporate advanced technology features.

Table 6.2

Component	Mass (kg)	Dimensions (m)
Fuel cell	28.8	$0.44 \times 0.20 \times 0.13$
Reformer	53.6	$0.61 \times 0.31 \times 0.31$
Compressor	4.5	0.08 dia. × 0.04
Inverter	5.0	$0.24 \times 0.24 \times 0.12$
Motor	26.0	0.23 dia. × 0.20
Battery	3.0	$0.15 \times 0.15 \times 0.15$
Total	120.9	112 litres

Table 6.3

Component	Mass (kg)	Dimensions (m)
Fuel cell	37.5	$0.64 \times 0.30 \times 0.20$
Compressor	4.5	0.08 dia. × 0.04
Inverter	5.0	$0.24 \times 0.24 \times 0.12$
Motor	26.0	0.23 dia. × 0.20
Battery	3.0	$0.15 \times 0.15 \times 0.15$
Total	76.0	65 litres

Technical specification of the fuel cell system

When assessing the performance of a rapidly advancing technology (e.g. fuel cells), it is necessary to define the year of adoption. In our project, we will assume 2010. We also need to define the size (power) of the propulsion system. We will assume 75 kW. For the methanol reforming system described above, the projections shown in Table 6.2 were made.[2]

This compares to a mass of about 65 kg for a conventional internal combustion engine. The estimated system mass of 121 kg will be used in the design of the aircraft.

In contrast, the potential mass and size of a direct methanol system are also given in the same reference which quotes the data shown in Table 6.3.

Notice the considerable savings in mass and volume to be gained if such a system can be made to work efficiently. However, the system is very speculative and therefore cannot be assumed for our design.

The methanol used in the half-hour duration of the race and warm-up period has to be added to the system mass. Estimates suggest that 60 kg will be required. This will need a tank size of $0.44 \times 0.44 \times 0.44$ m. Allowing about 13 per cent for propeller, system installation mass and other contingencies brings the propulsion group mass to 205 kg.

6.5 Design concepts

At this stage in the development of the project, it is difficult to decide between the tractor or pusher layouts. Both will benefit from the cleaner and more streamlined shape made possible by the fuel cell system and the electric motor. To avoid this decision it is

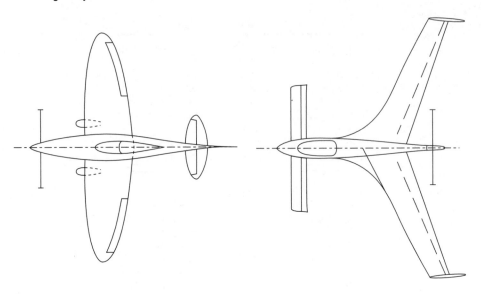

Fig. 6.3 Initial concept sketches

proposed to consider both configurations in the initial stages of the project and to compare them to establish the 'best' design.

The tractor layout is associated with the conventional monoplane and tail config-uration. The Nemesis aircraft is the best example of this configuration for a racer. For the pusher layout, either a canard or a tail control position is possible. For the latter choice either a twin boom or a high propeller position must be selected. To avoid the complications involved in the design of the rear fuselage to accommodate a tail surface we will select a canard layout for our aircraft. Initial concept sketches of both designs are shown in Figure 6.3. The appearance of the conventional design results from the elliptical wing and tail planforms, the tail-dragger landing gear, the mid-fuselage cockpit, and the mid-fuselage wing mounting. It resembles the conventional aircraft of the 1940s. The canard layout follows the FFT/VariEze configuration but is scaled down to suit the racer requirements. The aircraft has a swept back wing, wing tip fin/rudder control surfaces and a semi-retractable undercarriage (assuming that this would be permitted in the Formula E rules).

6.6 Initial sizing

Unlike most aircraft projects, the selection of wing area and engine power is not a problem as they are part of the Formula rules. The wing area is set at the minimum allowed by the rules (66 sq. ft/6.132 sq. m) and the engine power at 75 kW to the pro-peller is set to match the fuel cell performance. A constraint analysis could be conducted later in the design process to show the influence of these restrictions on the design. Cockpit size will be kept as small as practical within the formula rules. The shape of the canopy will be significantly different on the two aircraft due to the mid-fuselage position on the conventional and the forward position on the canard layouts.

Although structurally heavier, the 'canard' wing will have a higher aspect ratio (7.5) to provide for a greater fin arm. The 'conventional' will have a low aspect ratio (5.5) to reduce aircraft span and roll inertia. This will make the wing structure lighter giving a lower aircraft empty mass. Aerodynamically the high aspect ratio of the canard will provide a lower induced drag but the sweepback will reduce lift capability. Detailed analysis will be needed to determine the absolute effects of the layout differences.

In order to undertake the necessary performance calculations it is essential to estimate the aircraft mass (and balance) and the aerodynamic characteristics of the two aircraft. These calculations require an accurate (to scale) drawing of the aircraft (see Figure 6.4 for the conventional and Figure 6.5 for the canard layouts). The dimensions from these drawings are input into the mass and aerodynamic spreadsheets.

6.6.1 Initial mass estimations

As the aircraft propulsion system does not use conventional fuels, and since the flight duration of the race is short compared with non-racing aircraft, conventional methods of predicting take-off mass are not appropriate. In this case, it is necessary to make the initial calculations using known data to validate an acceptable estimation.

Data from the two closest existing aircraft to our designs, namely the Nemesis racer and the FFT Speed Canard, will be used to verify the estimation formula. Adjustments will need to be made in the case of the FFT, as this is a two-place touring aircraft.

For our aircraft the mass analysis will be considered in the following breakdown:

$$M_{gross} = M_{structure} + M_{propulsion} + M_{fixed\ equipmt} + M_{crew} + M_{disposables}$$

In this breakdown, any fuel (methane) will be considered in the 'disposable' mass section.

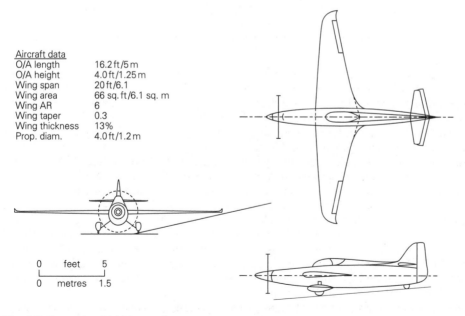

Aircraft data

O/A length	16.2 ft/5 m
O/A height	4.0 ft/1.25 m
Wing span	20 ft/6.1
Wing area	66 sq. ft/6.1 sq. m
Wing AR	6
Wing taper	0.3
Wing thickness	13%
Prop. diam.	4.0 ft/1.2 m

Fig. 6.4 'Conventional' initial general arrangement

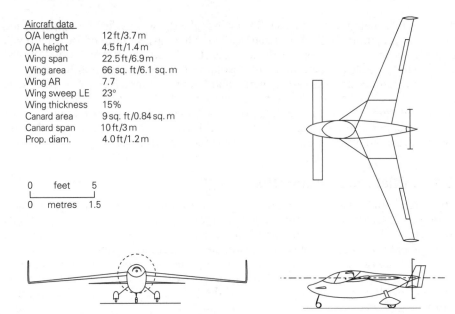

Aircraft data
O/A length	12 ft/3.7 m
O/A height	4.5 ft/1.4 m
Wing span	22.5 ft/6.9 m
Wing area	66 sq. ft/6.1 sq. m
Wing AR	7.7
Wing sweep LE	23°
Wing thickness	15%
Canard area	9 sq. ft/0.84 sq. m
Canard span	10 ft/3 m
Prop. diam.	4.0 ft/1.2 m

```
0     feet    5
L_____J
0    metres   1.5
```

Fig. 6.5 'Canard' initial general arrangement

Table 6.4

Mass	Actual	Method 1	Method 2	Method 3
Structure	104.7	93.7	90.9	117.7
Empty	235.8	237.0	206.1	251.7
Gross	340.1	341.3	310.4	356.0
% change		<1%	−8%	+5%

Details of the Nemesis aircraft were found in professional journals (e.g. SAWE paper 2343, June 1996) and from the Nemesis web site.[3] These include a detailed mass statement that can be used to compare the predictions from mass formulae, etc. in various textbooks. These comparisons are shown in Table 6.4.

Method 1 is a well-publicised general aviation method,[4] method 2 is a method used for civil aircraft design[5] and method 3 is mainly associated with military aircraft.[6,7] All the methods give reasonable results that generally bracket the actual values. The structural results seem to be more variable. A word of caution is appropriate here as the Nemesis is mainly built from composites and the formulae used are based on conventional metal structures. Method 3 allows the incorporation of 'technology factors' that accounted for composites. Although research shows that composite structures can reduce mass by 20 to 30 per cent, it has been found that such improvements are only achievable in large aircraft structures due to the connections required to feed loads into and out of the shell. Applying factors of 0.85 for the wing, 0.83 for the control surfaces, 0.90 for the fuselage and 0.95 to the landing gear result in the estimations (per cent MTOM) for three specimen aircraft shown in Table 6.5.

Method 3 appears to underestimate the wing structure and overestimates the fuselage of the Nemesis aircraft. Because of the difficulty of accounting for the wing to fuselage

Table 6.5

Component	Nemesis Actual	Nemesis method 3	FFT Canard method 3	VariEze method 3
Wing	12.3	10.0	12.3	9.4
Fuselage	10.0	12.5	5.9	6.9
Horiz. tail	1.3	1.1	1.0	1.0
Vert. tail	0.4	0.4	2.0	1.0
Main u/c	7.1	8.0	7.9	6.8
Aux. u/c	–	1.0	1.2	1.1
Structure	31.1	33.0	30.3	26.2
Engine	29.0	28.9	21.0	17.9
Propeller	1.2	1.2	0.8	0.9
Fuel system	1.6	0.9	1.8	2.1
Propulsion	31.8	31.0	23.6	20.9
Fixed equip.	6.4	6.7	3.7	4.3
A/C empty	69.3	70.7	57.6	51.4
Crew	26.7	25.5	28.2	32.4
Fuel	4.0	3.8	14.2	16.2
A/C gross	100	100	100	100
Gross (kg)	340	356	643	560

joint structure into either the fuselage or wing mass component, uncertainty always exists to the precise division between wing and fuselage components. The above analysis does give a reasonable estimate for the combined (wing + fuselage) structure ratio. This may be due to the design of the landing gear (small and with perhaps no brakes) for racing aircraft. The method seems to overestimate the mass of the landing gear. Note that the Nemesis empty mass fraction is much higher than the other two general aviation aircraft. At about 70 per cent, this is typical for the short range/duration, single-seat racer aircraft.

The analysis above was also done to indicate any variations in mass fractions due to the canard layout. Although both of these aircraft are much heavier than our proposed design, some general conclusions can be drawn. The wing structure for the FFT is seen to be about 2 per cent heavier than the Nemesis. This is largely due to the need to sweep the wing planform back to provide an acceptable fin control arm. For both of the canard aircraft, the fuselage mass is substantially less than the conventional layout. This is because the pusher layout shortens the fuselage length. In addition, the engine is mounted close to the wing/fuselage joint making all the heavy loading on the fuselage concentrated in the same area. The control surface mass is slightly higher for the canard designs.

The propulsion system for the two aircraft layouts will be assumed the same. The predicted electric propulsion mass (205 kg) is twice as heavy as a conventional petrol engine.

For all small aircraft, the landing gear represents a substantial weight penalty. The tricycle arrangement will be slightly heavier than the tail dragger type. The retraction system, which is to be used on the canard aircraft, will also add a little extra mass.

The estimated[8] mass statement for the two layouts is shown in Table 6.6.

The estimating method seems to correctly identify the higher wing mass and lower body mass for the canard layout compared to the conventional design. As both estimates for the maximum take-off mass are so close, the aerodynamic and performance analyses that follow will assume that both designs are at 470 kg.

Table 6.6

Component	'Conventional'		'Canard'	
	Mass (kg/lb)	%MTOM	Mass (kg/lb)	%MTOM
Wing structure	68.3	14.7	71.9	15.3
Body structure	41.5	8.9	37.3	7.9
Control surfaces	9.0	1.9	14.3	3.0
Landing gear	32.0	6.9	33.5	7.1
Propulsion group	205.0	44.2	205.0	43.6
Fixed equipment	18.0	3.9	18.0	3.8
Pilot	90.0	19.4	90.0	19.1
TO mass (MTOM)	463.8	100	470.0	100

6.6.2 Initial aerodynamic considerations

Aircraft speed is one of the most significant factors in racing. Therefore, the main aerodynamic analysis for racing aircraft focuses on the reduction of drag.

The layout details of the two aircraft will affect the aerodynamic calculations. The pusher propeller configuration will reduce the size, and therefore the wetted area, of the fuselage. The clean flow conditions over the nose will help to maintain laminar conditions over the forward fuselage profile. The smooth contours on the front fuselage and the forward position of the cockpit will allow the windscreen and canopy to be blended into the fuselage profile. This will substantially reduce drag. All these features are an advantage for the canard layout. The conventional layout will conversely suffer drag penalties from the disturbed airflow, from the propeller, over the front fuselage. The mid-mounted cockpit will force the adoption of a bubble canopy. This will be 'draggy'.

The conventional aircraft wing will produce a clean and efficient aerodynamic result; a low drag coefficient and good lift generation. The canard layout will suffer aerodynamic penalties due to the swept wing planform and the wing tip 'fins'. The canard surface will be 'flying' therefore producing lift that will help to off-load the main wing. The relatively close coupling of the fins will mean that larger surface areas are necessary and this will add to the aircraft drag.

Assuming a flight (racing) speed of 200 kt at sea level, the drag of each component of both aircraft has been calculated using classical aerodynamics. Based on the descriptions above the following assumptions have been made:

'Conventional' wing

- Reference area 6.14 sq. m (66 sq. ft).
- A modern, high-performance, general aviation wing section.
- Average thickness 14 per cent.
- No twist.
- Aspect ratio 6.0.
- Taper ratio 0.5.
- Smooth surface.
- 50 per cent chord laminar flow.
- Oswald efficiency factor 0.9 due to the elliptical planform.
- Wetted area twice the wing ref. area as the small penetration into the fuselage will compensate for the wing section curvature.

'Canard' wing

- Reference area 6.14 sq. m (66 sq. ft).
- A modern, high-performance, general aviation wing section.
- Average thickness 16 per cent.
- 23° sweep at quarter chord.
- Aspect ratio 7.7.
- Smooth surface.
- 25 per cent laminar flow due to the effect of spanwise drift caused by the sweep and the 30 per cent max. thickness of the wing section.
- Oswald efficiency 0.8 due to the wing tip/fin interference.

'Conventional' fuselage

- The complex fuselage profiles increase wetted area and interference factors.
- Tractor propeller position makes fuselage flow totally turbulent.
- Mid-fuselage wing position reduces interference factor.
- Fuselage wetted area 6.9 sq. m (74 sq. ft) with planform area 2.54 sq. m (27.3 sq. ft).
- Equivalent fuselage diameter 0.7 m (28 in).
- No fuselage base drag as rudder extends below the fin.
- Canopy drag effects calculated separately.

'Canard' fuselage

- Smooth profile.
- No base drag.
- 10 per cent laminar flow assumed (this is considered as conservative).
- Wetted area 4.5 sq. m (48 sq. ft).
- Width of fuselage 0.64 m (26 in), depth 0.88 m (35 in).
- Wing/fuselage interference factor 1.07.
- Turbulent flat plate coefficient 0.00245.
- Zero-lift drag reduced by 7 per cent due to the pusher propeller position.

'Conventional' empennage

- Thickness 10 per cent throughout.
- Wetted areas 1.52 sq. m (16.3 sq. ft) horizontal, 0.69 sq. m (7.4 sq. ft) vertical.
- Fin sweep at quarter chord 27°.

'Canard' control surfaces

- Wetted areas 1.68 sq. m (18 sq. ft) horizontal, 0.88 sq. m (9.5 sq. ft) vertical (total).
- Thickness 18 per cent horizontal, 15 per cent vertical.
- Flat plate skin friction coefficient 0.00375.

Canopy (both aircraft)

- 'Conventional' wetted area 0.057 sq. m (0.62 sq. ft).
- 'Canard' blended profile therefore no extra drag.

Trim (both aircraft)

- A value of about 6 per cent of total drag is common but as the flight duration is short and the aircraft can be pre-race adjusted for trim reduction, no extra drag is assumed in the race condition.

Table 6.7 summarises the detailed drag calculations:[8]

Table 6.7

Component	Parasite	Induced	Total	Per cent
'Conventional'				
Wing	0.00463	0.00052	0.00515	26.8
Body	0.00564	0.00007	0.00571	29.8
Controls	0.00177	0.00006	0.00183	9.5
Canopy	0.00284	zero	0.00284	14.8
L/gear	0.00366	zero	0.00366	19.1
Total	0.01854		0.01919	100.0
'Canard'				
Wing	0.00694	0.00038	0.00732	38.4
Body	0.00359	zero	0.00357	18.8
Controls	0.00425	0.00002	0.00427	22.3
Canopy	zero	zero	zero	zero
L/gear	0.00391	zero	0.00391	20.5
Total	0.01869		0.01907	100.0

For this class of aircraft, interference drag will be kept low in the racing trim. A contribution has been added to each of the component drag calculations shown in Table 6.7.

The effect of the tractor propeller on the fuselage skin-friction drag can clearly be seen by the fact that this is the largest drag component on the conventional aircraft. (It has been reported elsewhere that a 7 per cent increase can be expected.) Also, the influence of the blister canopy on drag is seen to add about a further 10 per cent to the total drag. For the canard design, drag is seen to be predominantly affected by the wing and the large contribution from the control surfaces (wing tip fins and canard). As expected, on both configurations the landing gear represents a substantial drag penalty (about 20 per cent in both cases). Fairing the main gear would seem to be a sensible option for these aircraft.

It is interesting to note that the predicted drag of both aircraft is approximately the same. This confirms the view that a choice of the preferred configuration cannot be made on the basis of aerodynamic and mass efficiency (a view borne out by the fact that both types of aircraft are currently used in formula racing).

For both wing layouts, a non-flapped lift coefficient of 1.0 can be assumed. The swept wing of the canard design will suffer a reduction in lift generation but the canard surface will contribute to the overall lift and so reduce this disadvantage. A simple stall speed calculation can now be done:

$$\text{Stall speed} = [\text{aircraft weight}/(0.5\rho S C_{Lmax})]^{0.5}$$
$$\text{Stall speed} = [470 \times 9.81/(0.5 \times 1.225 \times 6.14 \times 1.0)]^{0.5}$$
$$\text{Stall speed} = 35\,\text{m/s}\ (64\,\text{kt})$$

This is regarded as a little too high for a light aircraft. Either the wing area needs to be increased (this will increase aircraft drag) or a flap will be required. To reduce the stall speed to 60 kt (making the approach speed $1.3 \times 60 = 78\,\text{kt}$) will demand a lift coefficient of 1.29. This could be easily achieved with a simple plain or split flap. Careful detail design of the wing trailing edge and flap hinges, to minimise drag increases, should be possible.

As the aircraft will be pulling g in the tight turns, it is necessary to determine the stall speeds in relation to the load factor (n). Using the equation above the following results are obtained:

Load factor (n)	Aircraft stall speed (m/s)
2	49.5
3	60.6
4	70.0

6.6.3 Propeller analysis

For light aircraft, propeller performance is the most difficult parameter to accurately assess. The diameter of the propeller must be limited to avoid sonic flow over the tips. This would generate noise and be aerodynamically inefficient. Most prototype light aircraft have to be refitted with a different propeller after the initial flights because it is virtually impossible to predict accurately the aircraft drag and thrust values. A fixed-pitch propeller produces its best performance at a specific combination of aircraft forward speed and engine rotational speed. The lower the number of blades, the better as the preceding blade disturbs the airflow for the following blade. One blade would be aerodynamically best but . . . ! The formula rules dictate the use of a propeller with fixed pitch. This creates a problem for racing aircraft, as a fine pitch propeller will be most efficient at low aircraft forward speeds and a coarse pitch at high speed. In a race, it is important to have good take-off performance in order to achieve a good position at the first turn on the circuit. Being ahead of the field allows the pilot to choose his racing line (height and position) and avoids flying in the turbulent airstream from other aircraft. A clear view with a preferred racing line is a significant advantage. However, the take-off and early race represents only a small proportion of the total competition. As airspeed builds up during the race, a fine pitch propeller will be a serious handicap. Aircraft with a coarse pitch propeller with the same engine will fly faster and may eventually overtake the early leaders.

The choice of propeller size (diameter) may be dictated by the geometric constraints of the layout. If the diameter is too large the landing gear will need to be longer and the aircraft ground clearance high. This will make it more difficult to climb into the cockpit and the increased height of the aircraft centre of gravity above the ground may make ground manoeuvring over rough ground unstable. If the diameter is small, the inefficient hub area will form a larger proportion of the total disc area reducing the propeller overall efficiency. To make towing easier, a two-blade layout is best. The blades can be stopped in a horizontal position, parallel to the road.

For the electric propulsion system, the electric motor speed can be varied to better match the propeller requirements. This is not as easy to achieve with a conventional internal combustion engine. This feature is potentially very useful and should be investigated in more detail in later stages of the project (after the preliminary design phase has been completed). For example, it may be possible to adopt a higher motor speed for the take-off phase than used in the race condition. This would effectively produce a thrust boost for take-off if the propeller geometry/performance can account for such a change.

For initial design considerations, typical propeller details would be:

- Tip diameter 1.2 m.
- Spinner diameter 0.24 m.
- Rotational speed 2000 rpm (racing), 3000 rpm (take-off).

- Advance ratio 2.0 (racing), 0.5 (take-off).
- Efficiency 82 per cent max.

6.7   Initial performance estimation

For racing aircraft, performance is the key issue in the design. As there is little difference in mass, drag and thrust between the two proposed configurations, their performance will be similar. At this initial stage, it will not be possible to distinguish between the two aircraft and identify the best design. It may be necessary to build, test and then race both types to decide which is the best! Very small differences in performance are always to be expected between competitive racing aircraft. Pilot ability will be exaggerated and the best flyers will be successful.

Notwithstanding the above comments, it is necessary to determine the overall performance to establish the viability of the aircraft and the new racing formula. The following estimates are required:

- maximum level speed,
- climb performance,
- turn performance,
- field performance.

6.7.1   Maximum level speed

As the drag and propeller parameter estimates are made with several crude assumptions (e.g. extent of laminar flow over the surfaces), and as the aircraft profile and induced drag coefficients are similar for both aircraft, an average between the two aircraft types will be used. To reflect the variability in the estimation of the coefficients, a $+/-5$ per cent range will be applied to show the sensitivity of optimistic and pessimistic estimates. We will also apply the same variability to the propeller efficiency.

The values used in the analysis are shown in Table 6.8.

Two curves (fine and coarse pitch) for propeller efficiency against aircraft forward speed are shown in Figure 6.6. Aircraft drag and thrust curves are shown in Figure 6.7. The effect of propeller pitch selection on aircraft performance is clearly seen in this graph. The extra thrust provided at low speed by the fine pitch propeller is eroded as speed increases. The aircraft maximum level speed is seen to be 96 and 102 m/s for the fine and coarse propellers respectively. The $+/-5$ per cent variation shown above results in a $+/-2$ to 3 m/s change in maximum speed. Although seemingly not very much this change would result in either a 'dog' or a 'pearl' of a racing aircraft. This confirms the essential requirement to get the aircraft parameter estimation as accurate as possible in the early stages.

Table 6.8

	Pessimistic	Mean	Optimistic
Profile drag coeff.	0.0195	0.0186	0.0177
Induced drag factor	0.0488	0.0465	0.0442
Aircraft gross mass	470 kg	470 kg	470 kg
Prop. efficiency	0.78	0.82	0.86

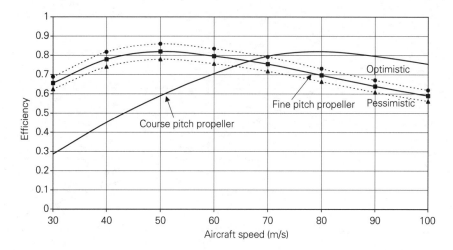

Fig. 6.6 Propeller efficiency versus aircraft forward speed

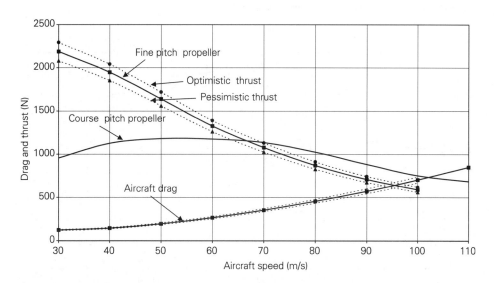

Fig. 6.7 Drag and thrust versus aircraft forward speed

The difference between the thrust and drag curves shows the energy available for aircraft manoeuvre. For the sea level, straight and level, flight performance the (thrust–drag) versus aircraft forward speed is shown in Figure 6.8.

Dividing the aircraft drag at a given speed into lift ($=Mg$ for straight and level flight) gives the aircraft lift to drag ratio (L/D) variation. Figure 6.9 shows the L/D ratio with speed. For economical flight it is necessary to fly at the speed close to maximum L/D. For our aircraft, this speed is very slow due to the very low drag characteristics but fuel economy in racing aircraft does not have a high priority.

All the above calculations have assumed that the aircraft is not manoeuvring (i.e. structural load factor (n) = 1.0). Pulling extra 'g' will increase the lift on the

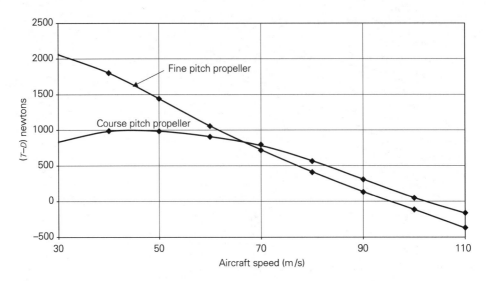

Fig. 6.8 (Thrust–drag) versus aircraft forward speed

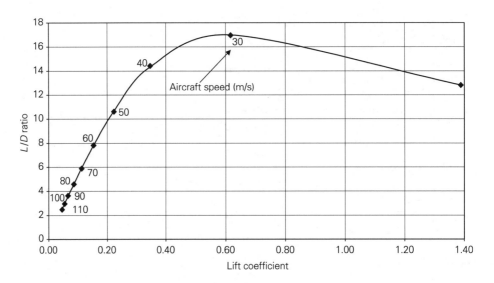

Fig. 6.9 Lift/Drag ratio versus lift coefficient

wings and thereby increase induced drag. Figure 6.10 illustrates the change of aircraft drag with manoeuvring load factor. Notice how the minimum drag speed progressively increases with load factor. The pilot will not want to fly the aircraft 'on the back side of the drag curve' as this results in unstable and difficult handling and will prefer to only pull 'g' at higher speeds. The extra drag will slow the aircraft. The relationship between aircraft speed and manoeuvre is considered further under the climb and turn performance below.

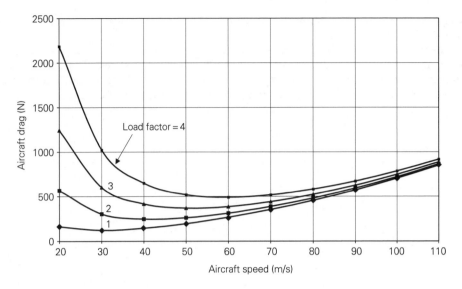

Fig. 6.10 Drag and load factor versus aircraft forward speed

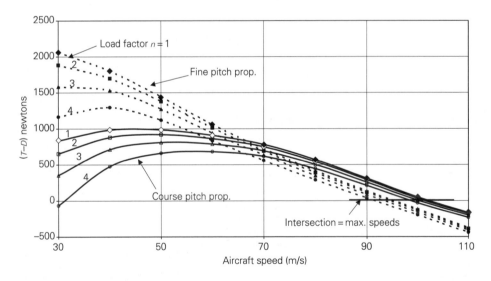

Fig. 6.11 $(T-D)$ and load factor versus aircraft forward speed

6.7.2 Climb performance

As mentioned above, the difference between the thrust and drag curves, at a specific speed, represents energy that is available for the pilot to either accelerate (kinetic energy increase) or climb (potential energy increase) the aircraft. The excess force available (thrust–drag) at various aircraft speed, and with the aircraft pulling 'g', is shown on Figure 6.11. This figure also shows the advantage of fine pitch at low speed and coarse pitch at high speed. Using all the available extra energy to gain height provides the maximum rate of climb. Multiplying $(T - D)$ by aircraft speed and dividing by aircraft weight gives the max. climb performance of the aircraft at constant aircraft forward speed (i.e. with zero acceleration).

The term $[V(T-D)/W]$ is referred to as the specific excess power (SEP). At sea level the maximum rate of climb versus aircraft speed is shown in Figure 6.12. Drag increase in manoeuvring flight, as mentioned above, has a significant effect on the aircraft SEP. Figures 6.13 and 6.14 illustrate the effect of choice of propeller pitch.

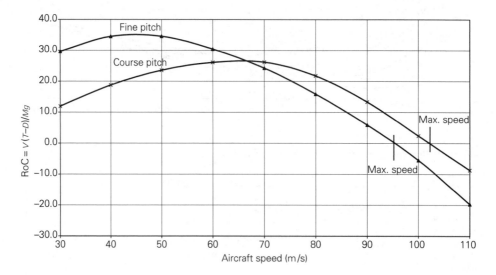

Fig. 6.12 Rate of climb versus aircraft forward speed

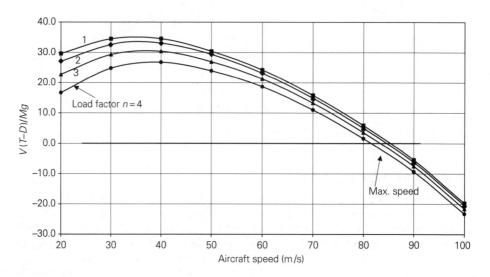

Fig. 6.13 Specific excess power (SEP) versus aircraft forward speed (fine pitch)

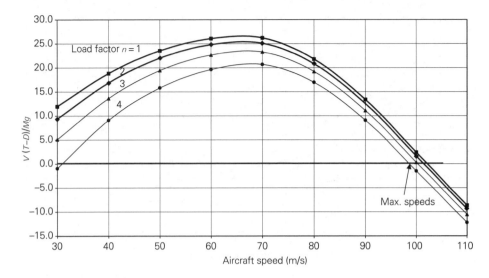

Fig. 6.14 Specific excess power (SEP) versus aircraft forward speed (coarse pitch)

6.7.3 Turn performance

Racing aircraft fly an oval circuit; it is therefore necessary to investigate the aircraft turn performance in some detail to establish the optimum racing line. Good turning performance will allow the aircraft to fly a tighter turn and therefore cover less distance in the race. The pilot faces a dilemma. Pulling a tight turn will increase drag and therefore reduce aircraft forward speed. This loss of speed will have to be made up along the straights. Alternatively, flying gentle (larger radius) turns will maintain speed but extend the race distance. Figure 6.15 shows the basic relationship between aircraft forward speed, manoeuvring load factor (*n*) and aircraft turn rate. Tight turns (high '*g*') are achieved at low speeds. Race pilots do not like high '*g*' and slow speed. They like to fly fast and gentle.

To achieve a balance of forces on the aircraft in a turn, it is necessary to bank the aircraft. The angle of bank is related to the aircraft load factor as shown in Figure 6.16. Although the loads on the aircraft in a correctly banked turn are balanced, it is necessary to instigate the turn from a straight and level condition and then to return to it. The application of the control forces required to change these flight conditions creates extra drag. To avoid these complications, a race could be flown in a fully balanced and constant attitude if a circular, or near circular, path outside of the pylon was selected. This would result in a much longer flight distance that would penalise the pilot unless a higher average race speed could be achieved to offset this disadvantage. The best strategy to adopt for the race is not obvious. Here lies the essence of good racing technique.

Not all of the aircraft parameters can be considered in the performance analysis. For example, sighting and aligning the pylons is an important element in successful racing. The mid-fuselage cockpit position of the conventional layout may be regarded as less effective than the forward position on the canard. Also, the canard control surface may offer the pilot a reference line to judge his position more accurately. 'Cutting a pylon' carries a substantial time penalty but flying a line that is too wide may present an opponent with a passing opportunity. These are features that are difficult to assess in the

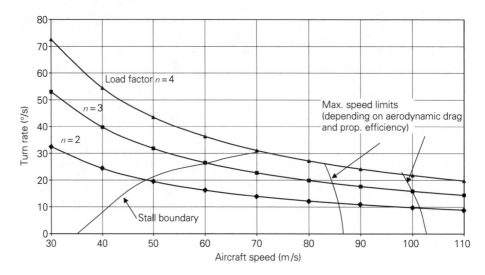

Fig. 6.15 Turn performance

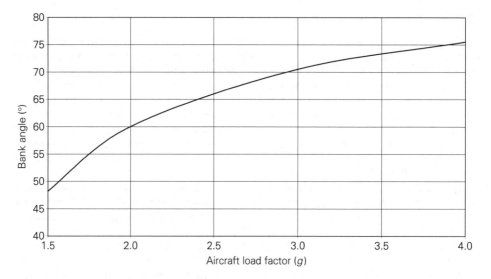

Fig. 6.16 Aircraft bank angle (balance turn) versus load factor

initial design stage. The combination of turn performance and flight path strategy offers a good example of the application of computer flight simulation in the early design stages. In this way, it is possible to test the external (visual) and internal (handling) features of the aircraft in a synthetic racing environment. Unfortunately, the initial aerodynamic, mass, propulsion and performance predictions do not hold sufficient fidelity to make accurate judgements from such simulations. However, some crude assessments are possible.

6.7.4   Field performance

As described previously, Formula racing starts with a grid of eight aircraft that have won the previous heats. The pole positions are awarded to the fastest aircraft in previous races. Take-off performance is therefore a significant aspect of the race. Obviously, there is an advantage to the first aircraft to reach the scatter pylon and avoid the congestion of other competitors. As mentioned in the propeller section, the designers must make a difficult choice between compromising race speed for take-off advantage, or vice versa. Short take-off performance and initial climb ability demands good lift generation at low speed. This implies a thick wing section profile, a cambered chord line, a low wing loading, efficient flaps and a fine pitch propeller. Conversely, maximum race speed will be achieved with high wing load, thin unflapped wing section and a coarse pitch prop. This is a difficult choice for the designers that will involve compromises to be made. Of all the parameters mentioned, the propeller selection is the easiest to change after the aircraft is built. In the early stages of the design all that can be done is to analyse the aircraft in a generalised method.

Estimation of field performance comprises both take-off and landing manoeuvres. In race conditions, the aircraft will not follow generalised procedures. For example, a racing pilot may hold the aircraft down in ground effect to build up energy before starting the climb. Disregarding such aspects, we will analyse the field performance using established design methods. Using average values for the aerodynamic coefficients, a sectional max. lift coefficient of 1.0, simple landing flaps, and aircraft gross (race) mass gives:

> *Take-off to 50 ft at* $1.2 \, V_{stall}$ *(with max. lift coeff.* $= 1.0$*)*
> Ground run $= 340$ m $(1114$ ft$)$
> Climb to 50 ft $= 136$ m $(446$ ft$)$
> Total take-off distance $= 476$ m $(1560$ ft$)$
>
> *Landing from 50 ft at* $1.3 \, V_{stall}$ *(with flapped max. lift coeff.* $= 1.3$*)*
> Approach distance $= 406$ m $(1330$ ft$)$
> Ground distance $= 117$ m $(384$ ft$)$
> Total landing distance $= 523$ m $(1714$ ft$)$

These values appear to be acceptable for this type of aircraft.

6.8   Study review

Design of racing aircraft is different to most design projects in that the main objective is simply to win competitive races. As these are set in a highly controlled design and operational environment, the design process is made easier. For the designer, the Formula rules and the racing conditions provide a very narrow focus to the selection of the design criteria and a simplification of technical decisions. Some of the normal design procedures (e.g. constraint analysis and overall operational trade-off studies) are not appropriate. The 'rules' set the wing area, engine type and power so the main design drivers become:

- reduction of aircraft mass (down to the specified minimum allowed by the rules),
- making the configuration aerodynamically efficient (reducing drag and generating lift),

- selecting a propeller geometry that is 'matched' to the race requirements,
- ensuring that the aircraft is easy to fly in the competitive racing environment,
- ensuring that the aircraft is reliable and serviceable at the race location,
- enabling the aircraft to be transported to the racecourse and easily reassembled.

Many of the detailed developments involved in the above will only be possible during the racing season. The 'fine tuning' of the aircraft is an established feature of a successful race team. Such late changes to the aircraft arise because it is not possible to model the aircraft using the analytical methods that are available in the design stages. Races are won by very small margins in aircraft performance between aircraft. These differences are much smaller than the accuracy of our design calculations. All that can be done in the design stages is to provide the best starting point for the race development process.

This illustrates a tenet of aircraft design:

> Analytical methods will only provide a starting point for the aircraft design which will subsequently only be improved by detailed design, empirical trimming and flight test work.

However, this should not be used as an excuse to avoid quality in the preliminary design phase, as subsequent improvements will not overcome inherent weaknesses in the basic design.

This project has provided a good example of the strengths and limitations of the conceptual design process. It should serve as a reminder that good design relies on excellence in each phase of the total design and development process. Ineptitude in any of the parts of the design work will only produce a poor quality aircraft.

References

1 Formula 1 web site (www.if1airracing.com/Rules).
2 Warner, F., 'An investigation into the application of fuel cell propulsion for light aircraft', Final-year project study, Loughborough University, May 2001.
3 Nemesis web site (www.nemesisnxt.com).
4 Stinton, D., *The Design of the Aeroplane*, Blackwell Science Ltd, 2001, ISBN 0-632-05401-8.
5 Jenkinson, L. R. *et al.*, *Civil Jet Aircraft Design*, AIAA Education Series and Butterworth-Heinemann Academic Press, 1999, ISBN1-56347-350-X and 0-340-74152-X.
6 Raymer, D., *Aircraft Design – A Conceptual Approach*, AIAA Education Series, ISBN 1-56347-281-0, third edn, 1999.
7 Brant, S. A. *et al.*, *Introduction to Aeronautics: A Design Perspective*, AIAA Education Series, 1997, ISBN 1-56347-250-3.
8 Tully, C., 'Aircraft conceptual design workbooks', Final-year project study, Loughborough University, May 2001.

7

Project study: a dual-mode (road/air) vehicle

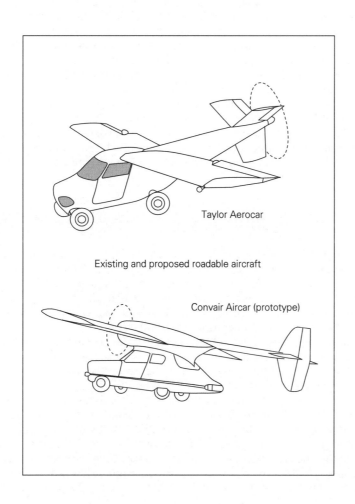

Taylor Aerocar

Existing and proposed roadable aircraft

Convair Aircar (prototype)

7.1   Introduction

'Flying car', 'roadable aircraft', 'dual-mode vehicle' and other terms are used to describe the all-purpose vehicle that can fly like an airplane and drive on the highway like an automobile. Make it amphibious and we have the perfect all-purpose vehicle! Nevertheless, this might be taking our ideas a bit too far.

It has long been the dream of aviation and automobile enthusiasts to have a vehicle that will bring them the best of both worlds. Many drivers stuck in rush hour traffic have fantasies about being able to push a button and watch their car's wings unfurl as they lift above the stalled cars in front of them. Just as many pilots who have been grounded at an airport far from home by inclement weather have wished for some way to wheel their airplane out onto the highway and drive home. This yearning has resulted in many designs for roadable aircraft since as early as 1906.[1]

A designer of a flying car will encounter many obstacles, including conflicting regulations for aircraft and automobiles. As an automobile, such a vehicle must be able to fit within the width of a lane of traffic and pass under highway overpasses. It must be able to keep up with normal highway traffic and meet all safety regulations. It must also satisfy vehicle exhaust emission standards for automobiles. (Note: these regulations are easier to meet if the vehicle could be officially classed as a motorcycle.) Therefore, the wings must be able to fold (or retract) and the tail or canard surfaces may have to be stowable. The emission standards and crashworthiness requirements will add weight to the design. The need for an engine/transmission system that can operate in the stop and go, accelerate and decelerate environment of the automobile will also add system complications and weight.

For flight, the roadable aircraft must be lightweight and easy to fly. It must have a speed range at least comparable to existing general aviation airplanes. Conversion from aircraft to car or vice versa must be doable by a single person and the engine must be able to operate using either aviation fuel or auto fuel. Ground propulsion must be through the wheels and not via propeller or jet which would present a danger to nearby people, animals or other vehicles.

7.2   Project brief (flying car or roadable aircraft?)

While some people use the above terms interchangeably, or use the latter term to bypass the science fiction connotations of the former, they are explicitly two quite different concepts. One wishing to design such vehicles must first decide which approach is appropriate. The 'flying car' is primarily a car in which the driver has the option of taking to the air when desired or necessary. The 'roadable aircraft' is an airplane that also happens to be capable of operation on the highway.

In the past, most designs[1] have actually been for roadable aircraft. They started out looking like conventional airplanes but with wings and possibly with tails that could be retracted or folded. Alternatively, they may be removed and towed in a trailer when the vehicle is operated on the road. Several such vehicles have been designed and built. A few, such as the Taylor Aerocar[1] or the Fulton Airphibian,[2] have been certified for use in flight and on the highway. Both types of vehicle have been sold to the public. The roadable aircraft is meant to be primarily an airplane but with the capability of being driven on roads to and from the airport. It must also be capable of getting the pilot and passengers to their desired destination on the highway when the weather prevents flight. As such, it is a vehicle primarily sold to licensed pilots. They would use its on-road capabilities in a limited manner, and not as a substitute for the family automobile for everyday trips to the supermarket. Typical problems with such designs have been their poor performance both in the air and on the road. Also, there has been in the past a reluctance of insurance companies to write policies which will cover their operation in both environments.

The 'flying car', unlike the roadable aircraft, has proved to be more of a fantasy than an achievable reality. A key element in the development of a successful flying car is designing a control system that will enable a 'driver' who may not be a trained pilot to operate the vehicle in either mode of travel. This virtually necessitates a 'category III capable' automated control system for the vehicle. This must provide a 'departure-to-destination' flight control, navigation and communication environment. Many experts feel that such a design is possible today, but only at high cost. Ideally, if the 'flying car' is to become the family car, it must have a price that is at least comparable to a luxury automobile (preferably less than 25 percent of the cost of the cheapest current four passenger general aviation aircraft).

Both the flying car and the roadable aircraft concepts usually assume a self-contained system capable of simple manual or even automated conversion between the car and airplane modes. A third choice is the dual-mode design which is capable of operation on the road or in the air but does not necessarily carry all the hardware needed for both modes with it at all times. One such vehicle was the Convair/Stinson CV-118 Aircar.[2] Designed in the 1940s, it combined a very modern looking fiberglass body car with a wing/tail/engine structure that could be attached to the roof of the car for flight. This design successfully flew, and operated well on the highway, but was a victim of high cost and changing corporate goals for its manufacturer.

Another decision facing the designer of any airplane/automobile hybrid vehicle is whether to attempt to meet government standards for both types of vehicles. Unless one wishes to go to the extreme of developing a very light weight flying motorbike which will operate under ultra-light regulations, one must meet FAR or JAR requirements for general aviation category aircraft. On the other hand, there is a choice when one considers the automotive aspects of the design.

Automobile safety and emission control requirements necessitate structural and engine designs that are heavier than one would ordinarily need for an aircraft. There is, at least under United States law, a 'loophole' in the regulations under which any roadable vehicle with fewer than four wheels can be classified as a motorcycle and not an automobile. This allows those who wish to avoid the extra weight and expense of meeting automobile design standards to develop a three-wheeled vehicle and classify the resulting design as a flying motorcycle, a vehicle that officially is an airplane in the air and a motorcycle on the road. Motorcycles have very few safety or emission design requirements beyond the specification of lighting, horn and engine muffler. Three-wheeled road vehicles do have operational speed restrictions in the United States.

Another decision that must be made is the extent to which the vehicle will meet the 'luxury' standards of automobile buyers that are not normally seen in general aviation aircraft. A typical modern American automobile lists in its 'standard' equipment package air-conditioning, electric window controls and door locks, automatic transmission, CD/tape players and similar items. None of these are usually found in most general aviation aircraft and all add (sometimes considerable) weight to the aircraft.

7.3   Initial design considerations

This design was developed by a single team of students from two universities in the United States and in Britain to satisfy the requirements for an aircraft design class.

The final design was to be entered in an American design competition sponsored by NASA and the FAA. As such, there were no initial customer requirements other than the above-mentioned regulations for the design of aircraft and automobiles in both the US and the EU. The student team had to decide which of the above design approaches to take and had to determine their own specifications for things like range, endurance, rates of climb, and cruise (on land and in the air) speed. For this study, the designers selected a 'roadable aircraft'. This is defined as a vehicle which is primarily meant for air travel but which, when pressed into duty in its automobile mode, will be able to fully meet the requirements for travel on high-speed motorways as well as city streets. It was designed to meet all EU and US requirements for both automobiles and aircraft. The initial assumptions were that the vehicle, as an aircraft, had to match the performance of current four-place, piston-powered, general aviation. As a car, it must have performance similar to a family sedan type of vehicle.

The general goals agreed upon at the start of the design process were for an aircraft with a cruise speed of 150 knots and a range of between 750 and 1000 nautical miles (1388 to 1850 km) at a cruise altitude of about 10 000 ft (3048 m). It must be able to take off and land in less than 2000 ft (610 m) and carry four people. As an automobile, it must be able to cruise at 70 mph (113 km/hour), have a reasonable acceleration capability, a range at highway speed of at least 300 miles (482 km), and handling qualities comparable to a family sedan. In addition, the design had to meet all FAR (JAR) regulations for airworthiness and meet both American and EU requirements for automobiles. There was considerable discussion about opting for a three-wheel design in order to eliminate many of the automotive design constraints but this was rejected. The team accepted the challenge of meeting US and EU automobile safety and emission requirements in order to have a vehicle that would handle like a car on the highway.

Additional challenges noted by the team at the beginning of the project included:

- the need to have acceptable in-flight wing aerodynamics while being able to retract, fold, or detach and stow the wing for road travel,
- the need to 'rotate' on take-off,
- the need to find an engine/transmission combination which could meet the conflicting demands of ground and air travel,
- the need for dual-mode control systems, and the need to meet rigorous stability and performance requirements in both modes of travel.

The design of a satisfactory wing is a dominant part of any roadable aircraft layout. As a 'car' the vehicle must fit into standard roadway widths. The resulting vehicle footprint (aspect ratio) is less than unity. This is regarded as inefficient for an aircraft wing planform. A wing of reasonable aspect ratio must then be capable of being extended from the body (fuselage) for flight and somehow stowed for highway use. There are many ways to do this including folding wings, rotating wings, telescoping wings, and detachable wings. These could be stored in, under, or over the car configuration. Alternatively, they could be towed behind the car.[1] All such designs impose structural compromise and weight penalties. The use of the wing for a fuel tank location would also be ruled out.

The take-off problem reflects the differing stability requirements of automobiles and airplanes. Most modern aircraft are designed with a tricycle landing gear arrangement with the rear or main wheels placed only slightly behind the center of mass (center of gravity). This allows easy rotation in pitch to a reasonable take-off angle of attack after ground acceleration. Placement of the rear wheels in the optimum location for the main gear of an aircraft would result in a very unstable car. It would have a tendency for its front wheels to lift off the road at highway cruise speeds near the desired take-off speeds for the aircraft. Cars are designed to minimize the likelihood of the wheels lifting off the road at highway speeds! Some roadable aircraft designs have attempted to solve this problem by having a conventional aircraft tail section that is removed for road travel. This effectively moves the center of mass further forward between the front and rear wheels. Others have employed a car type suspension with wheels or axles that can be extended or retracted to give the needed angle of attack for take-off.

Further complications arise due to the need for the wing on the airplane to develop some lift during the take-off run while the automobile must produce as little lift as possible at highway cruise speed. Removing or retracting the wings for the car layout will obviously solve most of the highway lift problem.

Aircraft piston engines are designed to be run at constant rpm for long periods of time. Automobile engines are designed to operate over a wide range of rpm and are coupled to a transmission to make possible combinations of torque and power suitable for a variety of operational needs. Aircraft engines must also be capable of efficient operation over a wider range of altitude than car engines. Air-cooling is normally used with aircraft engines while water-cooling is usually used for automobile engines. Both a water-cooling system and a transmission system will add extra weight not common in most aircraft designs. Some flying car designs have proposed using separate engines tailored to each mode of travel. This is on the assumption that two optimized engines may not weigh much more than a single dual-mode engine and drive train, and that the improved efficiencies may allow lower fuel consumption. Other designers have suggested the use of an engine and transaxle from a small 4WD automobile with the drive for one set of car wheels attached to the wheels and the other to the propeller.

The extent to which the controls for flight and ground operation can be merged is also a design concern. Do the in-flight rudder pedals become the accelerator and brake pedals on the road? Does the car steering wheel, with a release to allow it to move toward and away from the driver/pilot, become the in-flight control yoke, or can a 'stick' replace the wheel and be used in both modes of travel? Moreover, how are these controls coupled to the rest of the vehicle? Can a fly/drive-by-wire system work in both modes or must the controls be mated to two separate mechanical or hydraulic systems?

Finally, there is the question of 'roadability'. Beyond the question of tip-over angles (or ground loops during taxi, take-off, and landing), this is an issue that does not normally face the aircraft designer. The vehicle's wheel placement and suspension system and even the choice of tires must take into account the need for comfortable, stable handling on the highway as well as be able to absorb the sudden shock of landing.

7.4 Design concepts and options

Given the above challenges, constraints, and goals, the design process employed in this study was different from the usual systematic approach to aircraft design. It was not possible to generate a specific set of aircraft performance targets from studies of 'comparable' aircraft designs, or to initially 'size' the vehicle and refine it by the use of constraint diagrams and plots.

The design process began with each member of the team proposing an initial concept. The perceived merits of each concept were evaluated and compared and three general configurations from each of the two collaborating universities were brought to the table

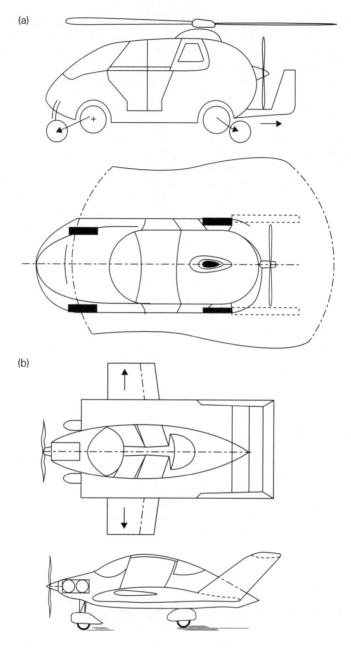

Fig. 7.1 Three initial concept sketches

at the first formal meeting of the complete team. Figure 7.1 shows sketches of three of these 'intermediate' concepts:

- a gyrocopter,
- a lifting body design with telescoping wings, and
- a car with ducted fans and folding wings.

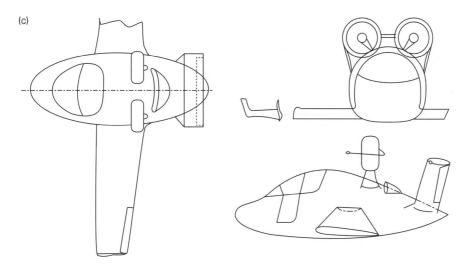

(c)

Fig. 7.1 Continued

The team then developed a 'decision matrix' with which to evaluate these six proposals. The decision matrix included assessments of the following features:

- the structural design,
- performance, and control aspects,
- propulsion system(s),
- 'roadability',
- cost,
- complexity of manufacturing, and
- ergonomics and human factors considerations.

They then divided themselves into six smaller groups, each rating all six concepts based on one of the above criteria. These ratings were subjective in nature since none of the designs had been developed beyond the stage of an initial sketch and concept. The resulting matrix is shown in Table 7.1.

Based on this matrix analysis, a decision was made to merge some elements from the second and fifth of the preliminary concepts. This resulted in a design with a lifting fuselage, a dual ducted fan propulsion system, and retractable (telescoping) wings as illustrated in the sketch in Figure 7.2.

The design employed a conventional four-place seating arrangement. The ducted fans were felt to solve the problems presented by using either a conventional tractor or pusher propeller, either of which would probably have to be removed for road travel to preclude accidental damage. This design had the engine located in front of the passenger cabin to provide protection to passengers in a crash. However, this led to the need for a complex drive train to couple the engine to the drive wheels and propulsive fans.

7.5 Initial layout

The concept initially selected employed an inner wing section that blended, to some degree, with the fuselage. It also had outboard wings which could be retracted into the

Table 7.1 Aircraft concept selection

	1. GA craft with transmission in wings	2. Lifting body with telescoping wings	3. Lifting body with folding wings	4. Gyrocopter	5. Car with ducted fans and folding wings	6. Cessna with rotating wings
Structures						
1. Wing position	−2	−1	−2	0	2	2
2. Aspect ratio	1	2	1	2	1	1
3. Sweep/Taper	−2	2	1	0	1	2
4. Number of moving parts	0	1	0	1	0	1
5. Size of moving parts	−2	1	−2	1	−2	−1
6. Wing loading	1	0	−1	−1	2	1
7. Weight distribution (moments)	0	1	−1	−1	1	1
8. Crashworthiness	−1	0	0	2	1	1
Subtotal 1/6	−5	6	−4	4	6	8
Stability and control/aerodynamics/performance						
1. Control in all aspects (air)	0	1	1	1	1	1
2. Aft CG for rotation (air)	2	0	1	0	−1	1
3. Crosswind effects (road)	−2	0	−1	1	−1	0
4. Low CG (road)	1	0	1	1	2	−2
5. Central longitudinal CG (road)	2	1	1	−1	2	2
6. Reduced lift (road)	1	−1	0	2	1	2
7. Clean flow over surfaces and props (air)	1	2	1	0	0	2
8. Streamlined frontal cross-section (air)	1	1	0	−1	1	1
9. High aspect ratio (air)	1	−2	1	0	2	−1
10. Wing placement-mid wing (air)	2	0	2	0	0	1
11. Low profile drag – after conversion (road)	1	1	1	1	1	2
Subtotal 1/6	10	3	8	4	8	9

Propulsion						
1. Power	−1	0	0	−1	−1	−2
2. Size, weight of engine and transmission	−1	−2	0	0	−1	1
3. Engine type (fuel)	0	0	0	0	0	−1
4. Fuel efficiency and range	−1	0	1	0	0	0
5. Cost	0	0	0	−1	0	1
6. Location of engine and transmission. Easy access	0	0	0	0	−2	−1
Subtotal (1/6)	−3	−2	1	−2	−2	−2
Car						
1. Stability	−2	1	−2	2	1	0
2. Crashworthiness	−2	0	−2	1	1	−1
3. Driver visibility	0	2	2	2	2	1
4. Road friendly	−2	1	2	2	2	−1
5. Ease of conversion	−2	0	−2	0	0	1
6. Aesthetics	−2	1	0	2	2	−2
7. Access	−1	1	2	2	1	2
Subtotal (1/6)	−11	6	0	11	9	0
Cost/manufacturing						
1. Market	0	1	0	1	0	0
2. Development (outsourcing)	0	−1	−1	−1	−1	2
3. Simplicity of design	−1	1	−1	−2	−1	1
4. Service and running costs	−1	−1	1	−1	0	0
Subtotal (1/6)	−2	0	−1	−3	0	3
Human factors						
1. Safety	−2	0	1	1	0	−1
2. Ingress/Egress	−2	−1	2	2	2	2
3. Visibility	−2	2	0	2	1	1
4. Conversion ease	−2	2	0	2	1	2
5. Aesthetics/Noise	−2	1	−1	0	2	−1
Subtotal (1/6)	−10	4	2	7	6	3
Total	−3.50	2.83	1.00	3.50	4.50	−3.50

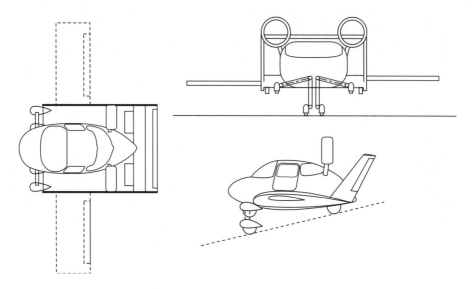

Fig. 7.2 Sketch of agreed configuration

inner wing when in the road configuration. This would give a vehicle width of less than 8 ft (2.44 m) for highway use. This is within normal highway lane-width limits for main roadways and is no wider than some automotive vehicles already in use. The vehicle's height and length were 8 ft (2.44 m) and 17 ft (5.18 m) respectively; both selected to enable roadway travel without restriction due to length or height. The weight (mass) was estimated to be 3500 lb (1591 kg).

The design of the wing was a crucial part of the concept. The inner wing was to have a chord of 11.32 ft (3.45 m) with its 8 ft (2.44 m) span, giving an aspect ratio of 0.707. An end plate that expanded into a vertical stabilizer/winglet was to be used at the tip of these inner wings in expectation of improving performance and providing a clean separation between inner and outer wing sections in flight. A horizontal stabilizer connected the vertical stabilizers. The 6 ft (1.83 m) chord outer wings were each to be made of either three or four sections that would telescope out of the inner wings to a final span of 23 ft (7.01 m) for flight. The telescoping mechanism was to be similar to that patented in the United States by Branko Sarh.[1]

For highway stability, the center of gravity (mass) is located midway between the front and rear wheels. This makes rotation for take-off difficult, if not impossible. To overcome this problem, the inner wing was positioned on the vehicle such that it is at a negative angle of attack when operating on the highway. For take-off and landing, the front wheels were to be extended to raise the nose of the vehicle. This gives a positive angle of attack for both the inner and outer wings and allows take-off in a reasonable distance without rotation.

The twin vertical tails and the horizontal stabilizer are used to provide pitch and yaw stability and control. Due to the relatively short moment arm, all these surfaces are larger than normal. Flaperons on the outer, telescoping wing sections are used for roll control and to provide extra lift in landing.

As noted above, the vertical tails were extended around the inner wing tip to give an end-plate effect. This was done to improve the performance of its very low aspect ratio planform. It was later decided to twist these slightly to provide a winglet effect, providing slightly more thrust. The aerodynamic analysis of the wing included an optimization of the winglet angle for these vertical tail sections.

During the aerodynamic analysis of the vehicle, it was found that the thrust from the twin-ducted fans was insufficient for the desired cruise speeds. The design was therefore changed to employ a single, large, unducted, pusher propeller placed behind the fuselage over the trailing edge of the inner wing. This change was accompanied by a relocation of the engine to a position aft of the cockpit. This placed it closer to the propeller and rear drive wheels. This change necessitated the addition of a rear firewall designed to force the engine downward, under the cabin, in case of an accident. This shift in engine location moved the center of gravity (mass) aft, requiring an increased sweep of the vertical stabilizer to provide an additional moment arm for the horizontal tail. The resulting inner wing, tail, propeller configuration represents a variation on a 'channel' wing where the large propeller enhances the flow over the top of the inner wing and thus increases lift. The winglet-like capabilities of the vertical stabilizers are also enhanced by the repositioning of the propeller. This configuration is shown in Figure 7.3.

The baseline configuration had a slightly smaller width on the highway with the inboard wing-span now at 7.48 ft (2.28 m) and had a chord of 8.2 ft (2.5 m). The outer wings were redesigned to telescope in four sections with a chord averaging 5.74 ft (1.75 m). They extend to a span of 27.16 ft (8.28 m), giving a gross wing area of 174.4 ft² (16.2 m²) and an aspect ratio of 4.23. The resulting unusual wing planform with its partial span located 'winglets' (vertical stabilizer/end plates) would require careful analysis and testing to ensure that the vehicle performs as required in flight.

The housings for the wheels were also modified from the initial concept to make the front wheel enclosures integral with the body/fuselage of the vehicle. The drive wheels were located at the rear of the inboard wing section and enclosed in housings that projected from the inside corner formed by the inboard wing and the vertical stabilizers. The front wheels were designed to retract tightly into their housings in flight and to extend to both a 'highway' position and a take-off position. Observers noted that the resulting vehicle, in its highway configuration, looked like a propeller-powered, turn of

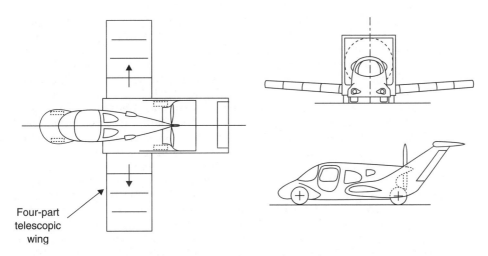

Four-part
telescopic
wing

Fig. 7.3 Baseline aircraft layout

the twenty-first century Volkswagen 'Beetle' sitting on a stubby wing and employing a large racing spoiler.

7.6   Initial estimates

Let us now further examine the technical details and performance (flight and roadway) of this unusual vehicle.

7.6.1   Aerodynamic estimates

This design incorporated a unique combination of aerodynamic concepts including:

- a lifting fuselage,
- an inboard 'channel' wing,
- 'inboard' winglets, and
- a telescoping outboard wing.

These made the analysis of vehicle performance a challenging prospect.

The aerodynamic examination needed to consider both in-flight and highway modes. These operating conditions presented contradicting aerodynamic requirements. The analysis was described in detail in the final project report.[3] A summary of the main findings is given below.

The analysis of the in-flight aerodynamics required detailed examination of the new concepts incorporated into the design. The vehicle body/fuselage, the inboard wing, the vertical tails and the propeller ducting accounted for a substantial part of the lift and drag. The fuselage was shaped with a flat bottom and curved top in the hope of producing some lift in flight. The flow over the fuselage was enhanced by the pusher propeller. These combined with the inboard wing and vertical tails to form a modified 'channel wing'. The channel wing was proposed by Custer[4] in the 1940s. In the take-off phase, the channel wing design enables the inboard wing to develop extra lift at low speeds due to flow augmentation from the propeller. The performance of the low aspect ratio inboard wing can be improved by as much as 15 percent by using the vertical stabilizers as winglets. This effect should also be enhanced by the propeller if the 'winglets' are properly designed. In the aerodynamic analysis, the outboard wing sections can be treated as separate, low aspect ratio surfaces. However, attention needs to be given to assessing the effect of the inboard/outboard wing junction on the spanwise loading.

Seven airfoil profiles were considered for the wing: the NACA 2412, 4412, 63_1-412, 63_2-415, 65_2-215, the NASA LS 0417 (GA (W) 1) and LS 0413 (GA (W)-2). The initial selection was to use the 4412 airfoil with its almost flat lower surface for the inner wing and the LS 0417 for the telescoping outer panels. The outboard wing was to be mounted with its chord set at an angle 2° higher than the inboard wing to give enhanced lift on take-off, but this produced stall control problems in flight. Consequently, the design was altered to use the LS 0417 for both parts of the wing with both at the same angle. To gain the needed lift on take-off the front undercarriage legs are extendable to give an angle of attack of 8° prior to the take-off run. At the desired 150 kt (77.2 m/s) cruise speed the ideal angle of attack for the airfoil was essentially 0°, hence, that angle was selected as the mounting angle of the wing to the fuselage. The telescoping outer wings were given a dihedral angle of 5° for roll stability in yaw. In order to counteract the nose-down pitching moment inherent in the LS 0417 airfoil, the horizontal stabilizer was moved rearwards.

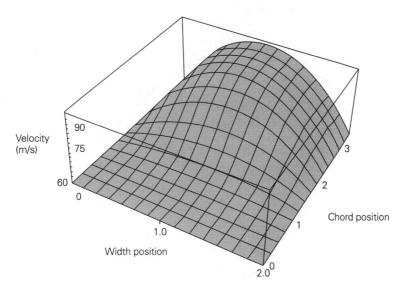

Fig. 7.4 Velocity distribution over 'scoop wing'

The NACA 0012 airfoil profile was selected for the horizontal tail. Based on winglet thickness recommendations in Raymer[5] an NACA 0008 was used for the vertical stabilizers/inboard winglets.

The aerodynamic characteristics of the channel wing were analyzed using the conventional actuator disk model for propellers. This simple momentum theory assumes a continuous acceleration of the flow forward and aft of the propeller disk. Since the wing was flat and not wrapped around the bottom of the propeller disk (as in the true channel wing configuration), this design was termed a 'scoop' wing. This term arose due to the resemblance to a rectangular scoop in frontal profile. The analysis of the 'scoop' wing performance produced the plots of velocity and pressure coefficient over the wing upper surfaces shown in Figures 7.4 and 7.5. Using this data, it was estimated that the propeller could induce a lift of 522 pounds (2248 N) at zero forward speed. As the speed of the vehicle increases this lift enhancement decreases.

The extent to which two vertical stabilizers serve as winglets and a fence between the outboard and inboard wing sections was analyzed using a model of the vortex generated at the intersection of two semi-infinite wings of different chords. A winglet works by using the cross flow from the tip (or in this case the junction) vortex combined with the free stream flow to create a 'lift' with 'spanwise-inward' and forward (thrust) components. This vortex-induced cross flow decreases with increasing distance from the vortex core. The angle of the local flow to the vertical stabilizer/winglet can be found as a function of that distance, allowing the calculation of the optimum local 'angle of attack' or twist angle of the winglet for 'thrust' production. This result is shown in Figure 7.6.

Based on the above it was found that at cruise conditions each winglet could generate 9.32 pounds (41 N) of thrust. Although the winglet only makes use of the inner/outer wing-junction vortex and not the full wing tip vortex, this corresponds to a 5.5 percent increase in L/D for the entire wing (including the outer wing lift and drag). This boost in performance may seem small but on a vehicle with such a low aspect ratio, any such help is welcome.

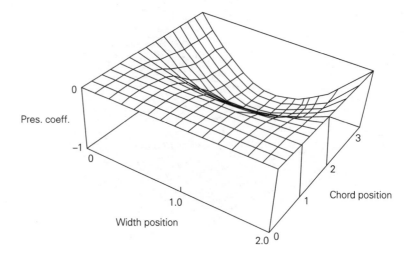

Fig. 7.5 Pressure coefficient distribution over 'scoop wing'

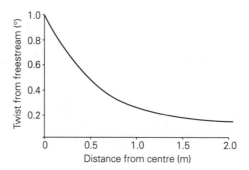

Fig. 7.6 Optimum wing twist distribution

The unique shape of the wing planform with the outer wing extended required a detailed three-dimensional aerodynamic analysis. Using methods in Raymer[5] and accounting for the effect of the winglets, lift curve slopes of 4.97 and 4.92 were calculated for the cruise and take-off/landing phases of flight respectively. The associated maximum lift coefficients were estimated at 1.818 and 1.755 respectively.

Using methods outlined in Raymer[5] and Torenbeek[6] to calculate flap effectiveness, it was found that full span, 20 percent chord flaps on the outer extended wing would increase C_{Lmax} at landing to 2.092 with 10° deflection and to 2.534 with 60° of flap.

There was some concern about the possible need to twist the wing to assure that the outboard sections of the wings stalled later than the inboard wing to provide adequate roll control in post-stall flight. Calculations using methods in Torenbeek[6] indicated that twist would not be needed. It was also found that the vortex at the inboard/outboard wing junction would help assure that stall progressed outboard along the wing-span. Later wind tunnel tests confirmed that no twist was needed for acceptable spanwise stall progression.

Similar thorough analyses were performed on the vertical and horizontal stabilizers to assure their effectiveness in providing the required aircraft trim and control.

Drag was estimated using methods of Torenbeek.[6] This involved totaling the drag of all aircraft components (wings, flaps, winglet/stabilizer, horizontal stabilizer, fuselage, engine cooling, windscreen, and the undercarriage). Corrections for fuselage/wing interference, the low mounting of the wing, surface roughness, and control gap effects were applied. The drag coefficient based on the extended wing planform (gross) area was found to be 0.0241 in cruise, 0.0718 at take-off, and 0.0830 in landing. The increased drag in take-off and landing configurations comes from flap deflection and the extension of the landing gear.

Finally, a similar aerodynamic analysis was made for the vehicle in the highway mode to determine its highway aerodynamic drag characteristics. The outer wing being retracted to assure that the lift would not endanger controllability at a cruise speed of 70 mph (31.3 m/s). These results were used in the vehicle road handling analysis.

7.6.2   Powerplant selection

Propulsion represented a unique problem for this design since the selected engine must provide power for both airborne and highway use. Alternatively, two separate engines could be used. Based on the drag calculations above with a cruise at 150 kt (77.2 m/s) at 3000 m (9843 ft) altitude, and an estimated maximum gross weight (maximum take-off mass) of 3308 lb (1500 kg), it was found that a cruise engine power of 207 hp (155 kW) was required.

The use of two engines was considered, one to power the propeller in flight and the other to drive the wheels on land. Weight and the capability of the propulsion system to operate efficiently in both modes of travel are the critical issues. For road travel, an engine plus a transmission is required to provide the needed power over a wide range of wheel speeds and the needed acceleration in stop and go traffic. For flight, no transmission may be needed and the engine would operate at a constant speed. While it is conceivable that separate, optimized systems could be found which would weigh less than a single, perhaps more complex engine and transmission system, it is not likely. The design team made the decision to employ a single, aircraft type engine, optimized for constant rpm operation, and to combine it with a transmission/drive train that would allow operation in highway traffic. The decision to design around the aircraft engine was made because the primary mode of travel for the vehicle was to be flight. The automobile role was considered as secondary; i.e. it would not be a vehicle used in day-to-day commuter highway use.

Several engines were considered as shown in Table 7.2. Two of these are standard general aviation piston engines, one is an automobile engine which has been popular as a 'conversion' engine for home-built general aviation aircraft and two are engines which were in development at the time of the design project.

Table 7.2 Engine properties

Engine	Current use	Power, hp (kW)	Weight (mass), lb (kg)
Continental IO-520	Beech C-55	310 (231)	436 (198)
Lycoming IO-540	Cessna 182	290 (216)	437 (198)
Subaru EJ-22	Conversion	160 (119)	295 (134)
Dyna-Cam	In development	200 (149)	300 (136)
Wilksch	In development	250 (186)	287 (130)

The selected engine was the Wilksch five-cylinder turbo diesel with intercooler. It was the closest to the required power output, gave an excellent thrust to vehicle weight ratio (0.3), and could run on both diesel and jet fuel. Turbocharging gives the engine sea-level performance at the desired cruise altitude of 3000 meters (9843 ft). This engine had a small size, $3.45 \times 1.48 \times 2.1$ ft ($1.05 \times 0.45 \times 0.64$ m), and at maximum power it produced 486 ft-lb (660 N-m) torque at 2700 rpm. Its specific fuel consumption was 0.441 lb/hp-hr (0.270 kg/kW-hr). The engine had an internal fuel pump that would pull the fuel from an 80 US gallon (303 liter) tank in the fuselage below the engine.

Aircraft engines such as the selected Wilksch diesel are designed for hour after hour of constant rpm operation while automobiles, especially in urban traffic, may encounter continuously varying rpm requirements. Automobile transmissions use a variety of gear ratios to provide a set range of drive shaft speeds and torque within the rpm range of the engine. This utilizes the torque capability of the auto engine over a range of rotational speeds. While some car engines, like the Subaru engine in Table 7.2, have been used in aircraft, they tend to be heavier than comparable power aircraft engines. The solution to this dilemma was found in the use of a 'continuously variable transmission' or CVT that permits constant rpm operation of the engine while varying the amount of power transmitted to the wheels in road use. The selected system was the Audi Multitronic CVT that has a weight (mass) of 220 pounds (99.8 kg) and measures 1.3×0.82 ft (0.40×0.25 m). The quoted weight (mass) includes the needed transmission fluid and is slightly lower than standard automobile transmissions with comparable capabilities.

A gearing system was also incorporated into the drive train design to allow coupling of the propeller and wheel drives. A 'dog' clutch is used to switch power between the wheels and the propeller and is designed to prevent simultaneous operation of both systems.

The 6.56 foot (2 meter) diameter, three-bladed propeller was designed using a propeller design program from the Internet web site of Leonard Newnham.[7] The program required the input of engine power, engine rpm, aircraft speed, number of blades, and the propeller diameter and blade angle of attack in order to output an 'optimum' blade design. The result promised to give a high propulsive efficiency.

The location of the liquid-cooled engine requires a cross-flow radiator and an electric fan similar to that used in modern automobile engines. Size constraints of the design limited radiator area to 1.5 ft^2 (0.14 m^2) and calculations showed that a 0.5 hp (373 W) electric motor would be more than adequate for cooling under the most adverse conditions.

7.6.3   Weight and balance predictions

Initial sizing (weighing) of the vehicle was attempted using published statistical curve fit methods (Roskam[8] and Raymer[5]) but these were of questionable value, given the unconventional nature of the 'aircraft'. Hence, a combination of traditional sizing methods and actual system weight (mass) data was used to produce Table 7.3.

Based on the estimated weights (masses) the balance of the aircraft could be analyzed and the center of gravity (mass) excursion determined. This is shown in Figure 7.7.

7.6.4   Flight performance estimates

Based on the above aerodynamic analysis and powerplant performance estimates, the flight performance of the vehicle in cruise could be calculated using the following

Table 7.3 Component weight (mass) estimates

Component	Weight (lb)	Mass (kg)
Structure		
Wing	215	97.7
Horizontal tail	30	13.6
Vertical tail	50	22.3
Fuselage	350	159.1
Main gear/rear wheels	85	38.6
Nose gear/front wheels	85	38.6
Subtotal	815	370.5
Propulsion		
Engine	400	181.8
Transmission	305	138.6
Propeller	50	22.7
Fuel system	45	20.5
Subtotal	800	363.6
Systems		
Controls	20	9.1
Electrical	190	86.4
Avionics	100	45.5
Anti-icing system	80	36.4
Subtotal	390	177.3
Cabin furnishings	115	52.3
Variable weights (masses)		
Fuel	480	218.2
Front passengers	320	145.5
Rear passengers	320	145.5
Luggage	60	27.3
Subtotal	1180	536.4
Total	3300	1500

characteristics:

- a wing aspect ratio of 4.46,
- a calculated Oswald efficiency factor of 0.92,
- an aircraft parasitic drag coefficient $C_{D0} = 0.025$,
- a propeller efficiency, η_p, of 88 percent giving a constant thrust of 1012 lb (4500 N),
- a specific fuel consumption of 0.441 lb/hp-hr (0.270 kg/kW-hr), and
- a maximum gross weight (mass) of 3308 lb (1500 kg).

The power plot, Figure 7.8, shows a cruise speed (at 80 percent power, at 9843 ft (3000 m) altitude) of 157.5 kt (81 m/s) and a maximum speed at this altitude of 179 kt (92 m/s).

Using take-off at 1.2 stall speed from a hard surface gave a take-off ground roll of 689 ft (210 m) and a 50 ft (15.24 m) obstacle clearance take-off distance of 920 ft (280 m). With touchdown at 1.3 stall speed, which can be achieved with less than 10° flap deflection, and braking at 80 percent of touchdown speed, the landing ground roll was calculated at 755 ft (230 m). This gave a total distance of 1148 ft (350 m) after clearing a 50 ft (15.24 m) obstacle at an approach sink rate of 787 ft/min (4 m/s). With 30° flap deflection, this distance is reduced to 1066 ft (325 m).

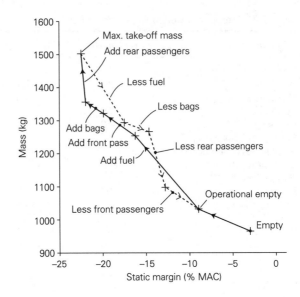

Fig. 7.7 Aircraft center of gravity excursion diagram

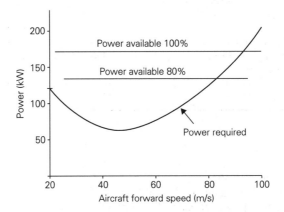

Fig. 7.8 Performance envelope at 3000 meters

The aircraft maximum rate of climb at sea level was found to be 1460 ft/min (445 m/min), and 755 ft/min (230 m/min) at the cruise altitude of 9842 ft (3000 m). The absolute ceiling was determined as 21 650 ft (6600 m). In normal 80 percent power cruise conditions at 9842 ft (3000 m) the range was calculated to be 825 nm (1528 km) with a 5.7 hour endurance. Flying at minimum drag conditions gave a maximum range of 960 nm (1778 km). At the speed for minimum power required the maximum endurance was found to be 9.5 hours. The design had proved to exceed all performance goals in the aircraft operation. It would be possible to re-optimize the aircraft configuration to better match the operational specification at this point but time was not available to do this in this project.

7.6.5 Structural details

There is an essential difference in structural design considerations for aircraft and cars. For aircraft, low weight with strength is paramount, while automobile designers need to add a focus on structural stiffness to improve handling and suspension performance. For this project the structure was designed to meet both general aviation aircraft and automobile requirements (FAR 23 and US National Highway Transportation Safety Advisory respectively).

The aircraft loads and their distributions over the lifting surfaces were developed based on the information shown in the flight envelope (*V-n* diagram), Figure 7.9.

The general structural layout of the vehicle is shown in Figure 7.10 with the major structural members numbered on the figure and identified in Table 7.4.

The structural design was evaluated in three parts:

1. at the fuselage/inner wing combination,
2. at the telescoping outer wings, and
3. at the tail.

The fuselage/inner wing structure consists of four regions:

1. the crumple zone forward of the cockpit,
2. the passenger compartment,
3. the wing box, and
4. the engine compartment.

The crumple zone was designed with an aluminum substructure covered by a composite skin. The skin is only lightly stressed and the aluminum frame is designed for controlled deformation in a crash using v-shaped indentations, termed 'fold initiators'. The forward wheels (landing gear) and their structure are mounted to the first bulkhead at the rear of the crumple zone. The aluminum substructure continues through the passenger and engine compartments. The passenger compartment skin is fabricated with carbon composite for stiffness and deformation resistance. The aluminum bulkhead at the rear of the passenger compartment transfers the loads between the forward spar of the inboard wing and the fuselage. Attached to this bulkhead is a fiberglass firewall coated with sperotex and phenolic resin. The firewall is mounted to the bulkhead at a

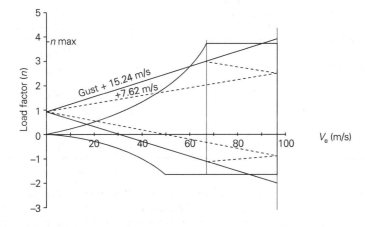

Fig. 7.9 Aircraft structural flight envelope

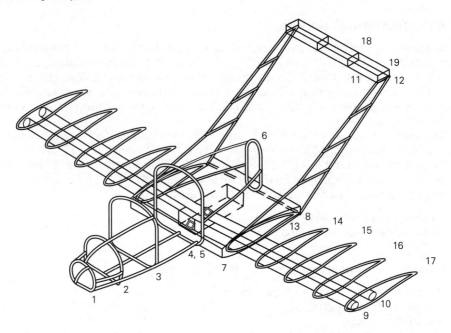

Fig. 7.10 Structural framework

Table 7.4 Location and identification of major structural members

Component	Member	Number in fig.	Fuselage sta.	Wing sta.
Crumple zone	Rib 1	1	107.87	n/a
	Bulkhead 1	2	128.35	n/a
Passenger compartment	Rib 2 (doorframe)	3	164.96	n/a
	Bulkhead 2	4	202.36	n/a
Engine compartment	Firewall	5	202.36	n/a
	Bulkhead 3	6	275.20	n/a
Inboard wing spars	Forward spar	7	202.36	n/a
	Rear spar	8	275.20	n/a
Telescoping wing spars	Forward spar	9	213.19	n/a
	Rear spar	10	229.96	n/a
Horizontal tail	Forward spar	11	350.79	n/a
	Rear spar	12	362.20	n/a
Telescoping wing	Rib 1	13	n/a	44.09
	Rib 2	14	n/a	73.62
	Rib 3	15	n/a	103.15
	Rib 4	16	n/a	132.48
	Rib 5	17	n/a	161.81
Horizontal tail	Rib 1	18	n/a	14.72
	Rib 2	19	n/a	44.33

slight angle. This configuration, combined with fold initiators, is designed to drive the engine below the passenger compartment in a collision.

The most complex part of the structural framework is the telescoping wing. The loads on the telescoping outer wing section were first approximated by examining the aerodynamic behavior of the combined inner and outer wings. The discontinuity

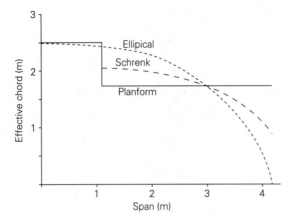

Fig. 7.11 Schrenk spanwise lift distribution

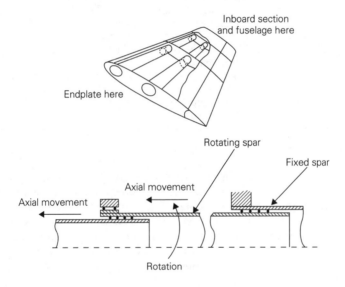

Fig. 7.12 Diagram of telescopic wing mechanism

in wing chord at the inner/outer wing junction makes load analysis a challenge. An approximation based on Schrenk's method[5] was used to estimate the loads over the entire span. The result is shown in Figure 7.11.

Each of the outboard wings consists of four sections. These telescope outward from their stowed position inside the inboard wing. The mechanism used to deploy and retract the outer wings is based on a patented design[9] as illustrated in Figure 7.12. Each of the telescoping outer sections from tip to root is slightly larger than the inner ones, allowing it to slide in over its neighbor. The telescoping sections are driven by threaded, rotating spars supported by bearings and powered by a 12-volt motor in the central wing box. To prevent accidental deployment/retraction of the outboard wings, the motor can only be operated when the wheels/landing gear are in their extended position supporting the weight of the vehicle, and when the wheels are not turning.

Table 7.5 Structural material selection

Structure	Component	Material
Crumple zone	Rib 1	Al 7075
	Bulkhead 1	Al 7075
Passenger compartment	Rib 2 (doorframe)	Al 7075
	Bulkhead 2	Al 7075
Engine compartment	Firewall	Fiberglass coated with sperotex and phenolic resin
	Bulkhead 3	Al 7075
	Engine mounts	Steel
	Fuselage skin	Carbon fiber
	Windows	Plexiglas
Inboard wing	Forward spar	Al 7075
	Rear spar	Al 7075
	Top skin	Al 7075
	Bottom skin	Al 2024
Telescoping wing	Rotating spars	Stainless steel
	Non-rotating spars	Carbon fiber
	Spar attachments	Al 7075
	Ribs	Carbon fiber sandwich
	Skin	Carbon fiber sandwich
Horizontal tail	Forward spar	Al 7075
	Rear spar	Al 7075
	Skin	Glass/Carbon fiber hybrid
Vertical tail	Skin	Glass/Carbon fiber hybrid
Landing gear	Struts, supports, etc.	Steel
	Wheels	Al 7075

The rotating spars are made of stainless steel for strength and stiffness. The rest of the outboard wing is mostly manufactured in carbon fiber composite construction.

The twin, vertical tail sections are designed to be manufactured entirely of carbon-glass-epoxy resin, composite materials. Material thickness is greater toward the root of the vertical stabilizer/winglets where the greatest bending moments would exist. The number of composite fiber layers will be reduced toward the horizontal stabilizer. The spars in these elements will also be composite in construction. The horizontal tail has aluminum spars.

The structural analysis included an extensive investigation of materials, strengths, and certification requirements for the composite structures. Table 7.5 lists the materials used in the various parts of the vehicle.

7.6.6   Stability, control and 'roadability' assessment

A wide range of factors must be considered when examining the stability and control needs of a vehicle that operates as either a car or an airplane. These include:

- the sizing and design of aircraft control surfaces and the resulting static and dynamic flight stability,
- the ease and predictability of handling in the automobile operating mode, and
- the internal systems needed to operate both the automotive and flight control systems.

Despite the somewhat unusual configuration of this vehicle, its flight control system and the requirements placed on that system are fairly conventional. The design is different from most general aviation aircraft in its use of a twin vertical tail and in its telescoping wing. The adoption of the large twin vertical tails resulted in the need for relatively small rudder size, as a percent of tail chord. The telescoping wing design led to the need for simplicity in flap/aileron systems and, ultimately, to the use of a plain 'flaperon' system, combining the role of conventional flaps and ailerons.

The static and dynamic control and stability requirements were calculated using methods of Raymer,[5] Thurston,[10] Etkin and Reid,[11] and Render.[12] The resulting tail volume coefficient was 0.35 and both rudder and elevators were sized at 35 percent of their respective stabilizer areas. Full span, 25 percent chord flaperons were used on the outer, telescoping wings. Calculations showed that with these controls the aircraft was able to meet Military Specification 8785C, level-one dynamic stability requirements for all cases except Dutch roll mode, which met level-two requirements. A complete analysis of the flight stability is presented in the final design report[3] but is not included here.

In highway use, this vehicle was not designed to be a high performance automobile. The emphasis was on handling and control, safety and predictability, and passenger comfort. All US and EU transport regulations related to safety and environmental impact had to be met. An added consideration was the requirement that a vehicle designed to fly does not do so on the highway!

7.6.7 Systems

One of the major decisions in the design process was to integrate the car and airplane control systems as much as possible. This has been achieved by using electronic rather than cable or hydraulic actuation of both automotive and aeronautical control systems. In this fly/drive-by-wire system, a joystick would replace both the automobile steering wheel and the aircraft yoke or stick. On the road the vehicle would have an automatic rather than a manual transmission and thus would have two foot controls, the brake and the accelerator pedals. In the air, these would serve as conventional rudder pedals. Both of these controls (floor pedals and joystick) would be attached to a fully electronic, fly/drive-by-wire control system. This would include a feedback to the pedals and joystick designed to give normal feel in both flight and the highway operation.

The instrument panel would have a large liquid crystal display (LCD) which would show a conventional automotive instrument array on the road and a modern aircraft flight control system display in the air. Required mechanical back-up instruments would be placed on the perimeter of the LCD panel. Switching from aircraft to automotive (or reverse) control and instrument display systems would be accomplished manually with system locks that would prevent any changeover when the vehicle was in motion.

The joystick controls are side-mounted, simulating the practice in many modern transport and military aircraft. The throttle control when in the aircraft mode is mounted on a center panel. Numerous studies of joystick type control systems for automobiles have shown that such systems are easy to use for most drivers and other studies of drive-by-wire automobile control and steering systems have proven their feasibility. Table 7.6 illustrates the way in which the driver/pilot would use the joystick and pedals for control of the vehicle in both operational modes.

There will also be a four-way toggle switch on top of the joystick. This will operate either the elevator trim or the headlight beam position when moved forward and aft, and either the rudder trim or the turn signals when moved left or right.

Table 7.6 Control system actions

Action	Aircraft mode	Automobile mode
Left rudder/brake pedal depressed	Yaw to left	Four wheel braking
Right rudder/accelerator pedal depressed	Yaw to right	Vehicle accelerates
Move joystick to left	Roll to left	Steer to the left
Move joystick to right	Roll to right	Steer to the right
Stick pushed forward	Lower aircraft nose	No action
Stick pulled back	Raise aircraft nose	No action

The wheels/tires and suspension system represented a unique challenge. The suspension system had to meet requirements for all three modes of operation:

- highway use (normal extension),
- flight (full retraction into wheel wells),
- take-off and landing (normal extension of rear wheels, full extension of front wheels for increased take-off roll angle of attack).

The system had also to be designed to absorb the vertical and horizontal impact forces encountered in landing and to handle the side force loads associated with cornering in the automotive mode. This required a careful specification of tire type and size as well as a good design of the suspension system itself.

The tires will need to possess characteristics that represent a hybrid of normal aircraft and car tires in terms of cornering stiffness and impact deflection. These properties are primarily a function of the tire aspect ratio (height to width). Low aspect ratio gives increased cornering stiffness and high aspect ratio gives better impact deflection. Different tire widths were specified for front and rear units to provide greater cornering stiffness at the rear (main) gear location. The front suspension uses an upper wishbone configuration with the lower arm attached to a longitudinal torsion bar. A screw jack is used with a damper (shock absorber) to attach the suspension wishbone to the vehicle frame, allowing extension or retraction of the wheel into the wheel well. The rear suspension is a trailing arm configuration with a spring/damper unit between the wheel and the vehicle frame. An extensive analysis of this suspension system and its behavior under all conditions was undertaken using methods of Gillespie.[13] This was presented in the design final report.[3]

7.6.8   Vehicle cost assessment

An analysis of the projected cost of an airplane is always difficult and such an evaluation for a combination automobile/airplane is necessarily based more on guesses than technical methods. Cost estimation began with standard methods outlined by Roskam.[14] Such methods are heavily based on past experience of general aviation aircraft. There are few, if any, vehicles comparable to this design. However, based on an admittedly optimistic production estimate of 1000 vehicles per year over a ten-year period and on assumptions of modern manufacturing techniques, an estimated cost per vehicle is $276 627. This figure is based on the cost components outlined in Table 7.7.

Table 7.7 Summary of estimated costs per vehicle

Research, development, testing, and evaluation cost	$15 000
Program manufacturing costs	
Airframe engineering and design	$1688
Aircraft production	$215 935
Flight test operations	$400
Overhead and indirect costs	$21 802
Profit	$21 802
Total	$261 627
Aircraft estimated price	$276 627

Fig. 7.13 Wind tunnel test model

This projected cost is at the high end of a range of four-place aircraft with comparable performance. However, our aircraft provides a 'roadable' option. It would be interesting to see if there is a viable market for such a design.

7.7 Wind tunnel testing

An eighth scale model of the vehicle was constructed of wood, plastic foam with aluminum wing spars. It was tested in a wind tunnel with a 6×6 (1.83 m $\times$ 1.83 m) test area cross-section. The model was mounted in the wind tunnel on a six-component strain gauge balance and tested through a range of angle of attack (from -6 to $+16°$). Test results consisted of force and moment data as well as photographic flow visualizations. Figure 7.13 shows the model being tested with wool tufts for flow visualization.

Although, due to time constraints, testing was limited in scope, the results did confirm the viability of the design. Stall was quite manageable and the outboard wings were shown to have attached flow after the inboard wing stalled, allowing control in stall. The horizontal tail also exhibited attached flow after stall of the inboard wing. Despite the somewhat unusual design of the vehicle, there was no evidence of separated flow areas at the rear of the fuselage, even with the propeller not operating. The tests also confirmed a rather broad range of angle of attack for near maximum lift to drag ratio showing that cruise efficiency is not very sensitive to angle of attack.

Tests were also run with the outboard wings removed from the model, simulating the on-road configuration. These confirmed that this gave a lift coefficient low enough to avoid unintended 'lift-off' while in use on the road.

7.8   Study review

The design of the roadable aircraft proved a challenging but successful student project. The design report was entered in the 2000 NASA/FAA General Aviation Design Competition and won first prize. Details of the final design are given in Table 7.8. While it may remain unlikely that a truly roadable aircraft will ever be successfully marketed, this exercise, like several designs for 'flying cars' that have been built and introduced in the past, shows that such a vehicle is feasible. There continues to be strong interest in such vehicles among inventors and dreamers. In the future, a design with many of the features described here may finally fulfill these dreams. As illustrated in Figure 7.14, a car/plane that will give its owners and operators a freedom of transport that does not exist with present-day aircraft or automobiles must one day be a reality.

Table 7.8 Aircraft description

Aircraft type:	General aviation four-place radable aircraft		
Propulsion:	Wilksch 250 hp (186 kW) diesel engine		
Aircraft mass:	Empty = 1568 kg	3457 lb	
	Max. fuel = 480 kg	1058 lb	
	Payload = 800 kg	1764 lb	
	Max. TO = 2848 kg	6280 lb	
Dimensions:	Overall length	= 4.25 m	14.0 ft
	Overall height	= 1.30 m	4.2 ft
	Span (wing extended)	= 4.14 m	13.6 ft
	Span (wing retracted)	= 2.16 m	7.1 ft
	Wing area (total)	= 15.88 sq. m	170 sq. ft
	Aspect ratio (total)	= 4.46	
	Wing taper ratio	= 1.0	
	Wing profile	NASA GAW-1	
	Wing thickness	= 17%	
	Wing sweep	= 0°	
	Wing dihedral (outbd)	= 5°	
	Horizontal tail area	= 2.85 sq. m	30.6 sq. ft
	Vertical tail area	= 3.18 sq. m	30.6 sq. ft
	Tail profile	NACA 0012	
Performance:	Stall speed	= 28 m/s	54 kts
(at max. TO mass)	Cruise speed	= 77 m/s	150 kts
	TO speed	= 33.6 m/s	65 kts

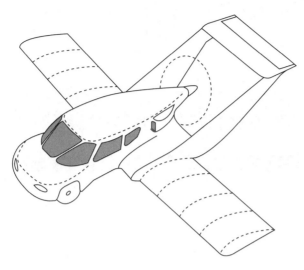

Fig. 7.14 Computer simulation of vehicle in flight

References

1 Stiles, Palmer, *Roadable Aircraft, From Wheels to Wings*, Custom Creativity, Melbourne, FL, 1994.
2 Mertins, Randy, *Closet Cases*, Pilot News Press, Kansas City, MO, 1982.
3 Gassler, R. *et al.*, *Pegasus, the First Successful Roadable Aircraft*, Virginia Tech Aerospace & Ocean Engineering Dept., Blacksburg, VA, 2000.
4 Anon. 'The wing that fooled the experts', *Popular Mechanics*, Vol. 87, No. 5, May 1947.
5 Raymer, D. P., *Aircraft Design: A Conceptual Approach*, 2nd edition, AIAA, Washington DC.
6 Torenbeek, Egbert, *Synthesis of Subsonic Aircraft Design*, Delft University Press, Delft, 1981.
7 Newnham, L., http://helios.bre.co.uk/ccit/people/newnhaml/prop.
8 Roskam, Jan, *Airplane Design, Part IV*, DARcorporation, Lawrence, KS, 1989.
9 Czajkowski, M., Clausen, G. and Sahr, B., 'Telescopic wing of an advanced flying automobile', SAE Paper 975602, SAE, Warrendale, PA, 1997.
10 Thurston, David B., *Design for Flying*, 2nd edition, McGraw-Hill, New York, 1994.
11 Etkin, Bernard and Reid, Lloyd, *Dynamics of Flight, Stability and Control*, Wiley & Sons, New York, 1995.
12 Render, Peter M., *Aircraft Stability and Control*, Aeronautical & Automotive Engineering Dept., Loughborough University, UK, 1999.
13 Gillespie, Thomas D., 'Fundamentals of Vehicle Dynamics', SAE, Warrendale, PA, 1992.
14 Roskam, Jan, *Airplane Design, Part VIII*, DARcorporation, Lawrence, KS, 1989.

Project study: advanced deep interdiction aircraft

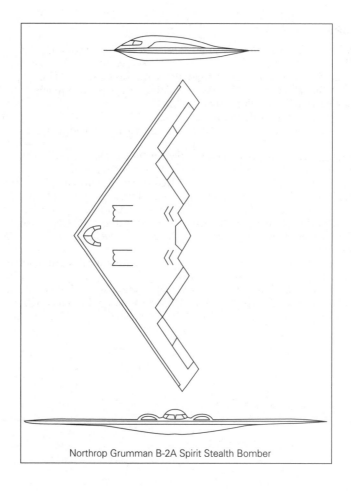

Northrop Grumman B-2A Spirit Stealth Bomber

8.1   Introduction

This project formed the basis of the American Institute of Aeronautics and Astronautics (AIAA)[1] annual undergraduate team aircraft design competition in 2001/02. Teams of three to ten students from the best aeronautical courses compete for prestige and cash prizes. The Request for Proposal (RFP) published by AIAA is based on recent industrial project work. Judges look for a thorough and professional submission from the team, which demonstrates a specific and complete understanding of the problem. This competition provides a useful source of current projects and operational data that can form the basis of undergraduate design projects even if the designs are not to be submitted for the competition.

The background to the project, as described in the original RFP,[2] is given below:

> When the F111 was retired from service in 1996 it was partially replaced by the F-15E. The balance of USAF deep-interdiction capabilities are provided by the F-117, B-1 and B-2 aircraft. All of these aircraft are expected to reach the end of their service lives in or before the year 2020. The need exists for a new aircraft which can effectively deliver precision guided tactical weapons at long range and which can rapidly deploy with minimum support to regional conflicts world-wide. Improved threat capabilities dictate that this new aircraft have signatures in all spectra comparable to or less than those of the F-117. The capability to super-cruise (fly supersonically without the use of afterburner) will allow these aircraft to respond to crises around the world in half the time required for current strike assets. Approximately 200 aircraft are needed to replace the F-15E, F-117, B-1 & B-2 aircraft.

The complete AIAA description of the problem[2] includes some detailed operational requirements, mission profiles and some engine and weapon design data. These are incorporated and discussed in the problem analysis and aircraft specification sections below.

8.2   Project brief

Recent conflicts in the Middle East, Eastern Europe and Central Asia have displayed the military strategy for modern warfare. The first objective of a new offence is to 'neutralise' the command and control centres of the enemy and to degrade their air defence facility. This is termed 'interdiction'. For the Airforce, this is a difficult and dangerous mission. In the initial attacks, the aircraft are expected to engage well-defended targets lying deep inside enemy territory. The range of the mission may be beyond the operational range of protective fighter aircraft and other support. The interdictive-role aircraft must therefore be self-supporting and able to evade, or protect themselves against, all the defensive systems of the enemy.

8.2.1   Threat analysis

Interdictive strike aircraft are expected to operate early in the conflict. This is at a time when the enemy's defensive systems have not yet been degraded. To avoid threats, the traditional tactic relied on fast, low-level approach under the protective screen of the enemy radar. Improvements in radar technology and the introduction of relatively cheap surface-to-air missiles (SAM) eventually made this tactic ineffective. Modern practice relies on aircraft stealth and high-altitude penetration. This avoids low and medium height threats from small-arms fire and low-technology SAM which now makes flight at altitudes below 20 000 ft very dangerous. A high-altitude mission profile ensures that the aircraft can only be attacked with much more sophisticated defensive weapons. The development of effective precision guided munitions and accurate target designation makes the high-altitude operation effective. Providing the aircraft with a high-speed capability, reduces the duration of the mission over the target area and thereby lowers the exposure to enemy defensive systems. The adoption of stealth means that the aircraft is more difficult to detect. However, this means that it must act without close air support that is any less stealthy. Defensive missile systems are becoming more effective at high altitude and such threats are also getting harder to detect and counteract. To rely on self-defence weapons and systems in future manned aircraft may be regarded as too optimistic. It is likely that even small countries will be able to afford such defence systems. Stealth, speed and height, which will make the defensive task more difficult, are likely to be the best forms of protection in future interdictive operations.

In order to strike deep inside enemy territory, from friendly airfields, requires a long operational range capability. The AIAA specification called for a combat radius of 1750 nm (3241 km) without refuelling. This long-range, high-altitude performance demanded an aerodynamically efficient aircraft configuration.

The two most significant design drivers for this project are identified as 'stealth in all spectra' and 'high aerodynamic efficiency at supersonic speed and high altitude'.

8.2.2 Stealth considerations

In recent years, the technical and popular press has focused so much attention on radar detection (radar cross-section, RCS) that it would be easy to forget that there are several other ways to identify and target an intruding aircraft. These include, infrared emissions (IR), electronic radiation, sound (aural signature) and sight (visual signature). Traditionally, the last of these led to the development of camouflage (the original stealth solution!). In modern warfare, it is important to make sure that each identifier is reduced to a minimum. None of the signatures should be more significant than the others. For example, we all are aware that in civil aviation the noise is much more intrusive than the visual characteristic. Similarly in military aircraft, the RCS or the IR characteristic must not dominate. Detailed technical information on stealth can be found in textbooks[3] and in technical papers. These textbooks and papers give advice on the analysis methods used to design for stealth. The methods used to predict RCS from the geometry of the aircraft are complex and beyond the scope of undergraduate preliminary design projects. However, generalised guidance on the selection of layout and profiling of the aircraft to minimise RCS is available.

Stealth issues influence the design of our aircraft in several different ways.

Radar

The AIAA specification required the RCS to be less than -13 dB. It is felt that with the expected technical improvements in radar performance in the period up to first flight (2020) this RCS may be too large. A value of -30 dB, if achievable, may be a better target for this aircraft. To achieve this figure will require as much help as possible from new technologies and the development of existing techniques. Existing methods include 'edge alignment', avoidance of shape discontinuities, elimination of flat surfaces, using radar absorbing structures (RAS), coating the external profile with radar absorbing material (RAM), and hiding rotating engine parts from direct reflection of radar waves. Attention must also be given to the avoidance of radar scattering caused by the aircraft profiles and from the edges of access panels. All of these methods have been demonstrated and proved on the B-2 aircraft. However, the main objective of such techniques is to reduce radar reflectivity. This is important when the radar transmitter and receiver are at the some location.

New defensive radar systems now displace the two parts of the system. This makes it more important to absorb the radar energy into the structural framework and the materials covering the aircraft profile.

Passive stealth techniques are currently being developed. These use plasma generation to 'assimilate' the radar energy. Another method attempts to displace or disguise the returning radar signature. This is intended to confuse defensive systems and make targeting more difficult. Obviously, for security reasons, published information on these developments is scarce. Therefore, little account can be taken of these new methods when currently deciding on our aircraft configuration. It is encouraging to note that research is identifying methods to reduce the radar threat. These are likely to be operationally mature for this next-generation aircraft.

Infrared

Infrared radiation is a natural consequence of heat. It is more pronounced at higher temperatures therefore the best way to reduce the exposure is to lower the temperature of the hot parts of the aircraft. The engine exhaust gases and surrounding structure give rise to the main source of IR radiation. A pure-turbojet engine exhaust is obviously easier to detect than that of a bypass engine. In the bypass engine, the hot core airflow is mixed with the cooler bypass air before leaving the engine. This substantially reduces the exhaust stream temperature and therefore the IR signature. Another way of reducing the IR signature is by shielding the hot areas from the potential detector. For example, if the IR detector is likely to be below the aircraft (a good assumption for our high flying aircraft) it would be possible to use the colder aircraft structure to hide the engine nozzle location. Positioning the engine exhaust forward and above the rear wing structure would provide this protection.

For aircraft travelling at supersonic speeds for long duration, the disturbed airflow will cause kinetic heating of the structure. The stagnation temperature resulting from aerodynamic heating is directly related to aircraft speed and ambient air parameters. For an aircraft in the stratosphere, travelling at M1.6 the stagnation temperature is estimated at over 100 °F (38 °C). It is difficult to estimate the actual skin temperatures that would result from this heat input as this will be dependent on the conductive properties of the structure and the heat radiation to the surrounding airflow. The temperature will be higher at positions of flow stagnation. Because of this, the leading edges of the flying surfaces and the nose of the aircraft will be affected more than the rest of the structure. This could present a potential problem as infrared radiation will naturally occur. If this is regarded as a serious problem, it would be necessary to cool these areas. In the case of the wing structure, it may be possible to use the fuel from the wing tanks to conduct the heat from the structure into the cold fuel mass. In other areas, it may be necessary to use ceramic coatings or other materials to improve the conductive path.

Other observables

For most of us, aircraft noise is the most noticeable characteristic. Exhibitionists at air shows try to make as much noise as possible to attract attention. For missions over enemy territory, the opposite strategy is advised! For our aircraft, there are mainly two sources of noise emission: the sonic boom and the engine.

For many years, researchers, mostly working on civil supersonic airliners, have been trying to reduce or eliminate the noise from the sonic boom. For example, it may be possible to mitigate the intensity of the noise energy by subtle shaping of the aircraft configuration, or by system innovations. However, the sonic boom is a natural consequence of the pressure changes as the ambient airflow is accelerated and then decelerated over the aircraft profile. The double-boom 'explosion' heard on the ground will alert the enemy defences to the presence of the aircraft. They will look upwards to confirm the sound. Hence, visual and aural observations are intrinsically linked.

Whereas noise heard on the ground from the sonic boom is transient, that from the engines is constant. This gives the observer more time to identify the aircraft visually. Engine noise is generated mainly from the impingement of the internal airflow on the rotating machinery and by the intermixing of the exhaust airflow into the atmosphere. The exhaust noise is affected by the jet velocity to the seventh power. Mixing exhaust velocity core airstream with a slower bypass stream before leaving the nozzle reduces engine generated noise significantly. The overall effect is to change the noise spectrum to increase higher frequency sound waves. As these are more rapidly dissipated with distance, there is a reduction of noise heard on the ground compared to a pure jet engine. The bypass engine will also provide better fuel consumption. From both of these aspects, this makes it a good choice for the interdiction mission.

Human sight is a very effective sensor. During day time, we can all see airliners high in the sky against a clear blue sky. If condensation trails or reflections (glints) from the aircraft are present the observation is even easier. Avoidance of visual detection was the first stealth technology. Camouflage has now become a natural strategy in warfare. Research has shown that it is possible to reduce the condensation and reflections from aircraft. At night, the main source of light comes from the exhaust glow. As the observer will usually be below the aircraft, shielding this glow from the observer, as previously recommended to reduce IR signature, will be an appropriate countermeasure.

The descriptions above provide some useful guidelines for the choice of configuration for our aircraft for stealth, but this is not the only requirement to be considered.

8.2.3   Aerodynamic efficiency

For the specified mission, the aircraft will spend nearly all of the flight time at supersonic speed. Therefore, it is important that the aerodynamic design concentrates on the reduction of wave drag. For a given size of aircraft, the longitudinal distribution of the cross-sectional area of the aircraft volume has a considerable influence on wave drag. Several aerodynamic and design textbooks (e.g. reference 4) describe the Sears–Haack analysis. They show that a smooth progression (i.e. following a statistically normal distribution) produces the minimum wave drag. The minimum increase in drag area due to wave drag is calculated using the formula below:

$$(S \times C_{Dwave}) = 14.14[A_{max}/L]^2$$

where S = aircraft reference (usually gross) wing area
A_{max} = maximum aircraft cross-sectional area
L = aircraft overall longitudinal length less any constant section segments

The equation shows that wave drag can be minimised by reducing ($A_{\max}$) and increasing (L). However, care must be taken to avoid adding significantly to wetted area and thereby increasing parasitic drag.

Shaping the aircraft body profile to try to achieve the Sears–Haack distribution is called 'area ruling'. Many existing aircraft exhibit this design strategy. It involves 'waisting' the fuselage of the aircraft to reduce the cross-sectional area at the intersection of the maximum wing profile and bodyside. This area ruling is shown clearly in the shape of the Northrop F-5 aircraft. The most cited case in aeronautical history books for the advantages of area ruling is that of the Convair F-102 which was originally designed with a straight fuselage but could not achieve supersonic speed until the shape was changed.

No practical aircraft can achieve an exact match to the Sears–Haack recommendation even by area ruling. The ratio of the actual wave drag to the minimum drag from the Sears–Haack prediction can vary from about 1.2 to 3.0. The lowest value would relate to a pure blended body with an area-ruled profile. The higher value would be for an aircraft not principally designed to minimise wave drag (e.g. a supersonic fighter that needs good pilot visibility and combat turn manoeuvrability). In this case, the main design driver would probably be combat effectiveness. For our project aerodynamic efficiency is paramount so every effort must be made to reduce wave drag. The adoption of a blended body configuration looks attractive. However, it must be realised that the main contributor to wave drag is the value of maximum aircraft cross-sectional area ($A_{\max}$). This term is squared in the Sears–Haack equation. The aircraft layout must focus on reducing this to the minimum value possible to hold the payload. Alternatively, or in combination, as the aircraft length is also squared, an increase will reduce wave drag.

At supersonic speeds, a Mach wave is formed which surrounds the aircraft. The angle of this wave cone relative to the longitudinal axis of the aircraft is known as the Mach angle (μ). This angle is a function of the aircraft forward speed (Mach number)[5] such that:

$$\mu = \sin^{-1}(1/M) \quad \text{With the specified cruise speed of M1.6: } \mu = 38.7°$$

To avoid discontinuity in airflow regions, it is desirable to keep the aircraft geometry, particularly the wing planform, within the Mach cone (i.e. keeping the wing leading edge sweepback angle greater than $(90 - \mu)°$). For our aircraft this dictates a **wing leading edge sweep angle greater than 51.3°**. As the air velocity in this region is substantially lower than free-stream, this also reduces wave drag.

The wing planform will be designed to fit within the Mach cone therefore the wing span will be restricted. This will increase lift induced drag but at the high cruising speed the lift coefficient will be relatively low which will make induced drag less significant from this effect.

The wing section profile will need to be of the 'supercritical' type to reduce the strength of shock in transonic flight. As the wing planform will be within the shock cone it would be possible to use a rounded wing leading edge profile. This will improve low-speed lift generation over the wing especially if a leading edge flap is used. Effort must be made to generate laminar flow over as much of the profile as possible to reduce parasitic drag. There is a potential conflict here between the preferred sharp wing leading edge profile for minimisation of radar signature and the rounded profile for aerodynamic efficiency. A choice will have to be made.

The body will need to be contoured to suit the area ruling mentioned above. In the region of the cockpit there are conflicting requirements. A smooth cross-section distribution in the forward part of the body may not provide the visibility requirements demanded of a strike aircraft. Good pilot visibility is also an advantage for the landing. Systems, including artificial vision and computer controlled imagery, will offer scope for innovation to overcome this problem in an aircraft designed for 2020. This aspect of layout and systems integration will require careful consideration.

8.3 Problem definition

The project description specifies a two-place advanced deep interdictor aircraft. The entire long-range mission will be flown at supersonic speed. The exact mission definition is shown in Figure 8.1. The long-duration, high-intensity flight conditions, much of which is over enemy territory, demands the security of twin-pilot operation. The long work periods and high manoeuvre load environment imposed on the pilots requires careful design of the cockpit. The workload related to flight safety and weapon delivery must be reduced by system design. Such systems must be made reliable and safe.

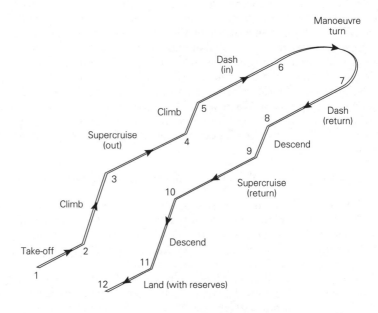

Segment	Description	Height	Speed	Distance/duration
1–2	Warm-up, taxi and take-off	Sea level		NATO 8000 ft, icy
2–3	Climb to best supercruise alt.			
3–4	Supercruise to conflict area	Opt. alt.	M1.6	1000 nm
4–5	Climb to 50 000 ft			
5–6	Dash to target	50 000 ft	M1.6	750 nm
6–7	Turn and weapon release	50 000 ft		180°
7–8	Dash out	50 000 ft	M1.6	750 nm
8–9	Descend to supercruise alt.			
9–10	Supercruise return	50 000 ft	M1.6	1000 nm
10–11	Descend to base			
11–12	Land (with reserve fuel*)			NATO 8000 ft, icy

*Diversion and hold at sea level with 30 min fuel at economical flight conditions.

Fig. 8.1 Mission profile

The aircraft must be capable of 'all-weather' operation from advanced NATO and other bases. Aircraft shelter dimensions may impose configurational constraints on the aircraft. Aircraft servicing and maintenance at austere operational bases demand minimum support equipment and skill. Easy access to primary system components must be provided.

Closed-loop, static and dynamic stability and handling flight characteristics must meet established military requirements. A digital flight control system will be necessary for a longitudinal unstable aircraft configuration. All systems must be protected against hostile damage and inherent unreliability.

In addition to strict stealth criteria, the AIAA problem description sets out several required design capabilities and characteristics. These include:

- The aircraft must accommodate two pilots but should be capable of single pilot operation. For such a long-range mission, pilot workload must be reduced by suitable design and specification of flight control and weapon delivery systems. Crew safety systems must be effective in all flight modes.
- The design layout should allow for easy maintenance. Minimum reliance on support equipment is essential for off-base operations.
- Structural design limit load factors of $+7$ to $-3g$ (aircraft clean and with 50 per cent internal fuel) are required. An ultimate design factor of 1.5 is to be applied. The structure must be capable of withstanding a dynamic pressure (q) of 2133 lb/sq. ft (i.e. equivalent to (q) at 800 kt) and be durable and damage tolerant.
- All fuel tanks must be self-sealing. Aviation fuel to JP8 specification (6.8 lb/US gal) is to be assumed.
- Stability and handling characteristics to meet MIL-F-8785B subsonic longitudinal static margins to be no greater than $+10$ per cent and no less than -30 per cent.
- The aircraft must be 'all-weather' capable. This includes operation from and on to icy 8000 ft runways.
- The aircraft must operate from austere bases with minimum support facilities. On these bases the aircraft will be required to fit into standard NATO shelters.
- The flyaway cost for 200 aircraft purchase must not exceed $150 M (year 2000 dollars).

In addition to the high-altitude, supercruising mission shown in Figure 8.1 and described in section 8.2 above, the design specification sets the following manoeuvring targets (specific excess power, SEP, is defined as P_S in Chapter 2 (section 2.7.1)):

- SEP ($1g$) military thrust (dry), 1.6 M at 50 000 ft = 0 ft/s.
- SEP ($1g$) maximum thrust (wet), 1.6 M at 50 000 ft = 200 ft/s.
- SEP ($2g$) maximum thrust (wet), 1.6 M at 50 000 ft = 0 ft/s.
- Maximum instantaneous turn rate, 0.9 M at 15 000 ft = 8.0°/s.

(all the above performance criteria are specified at aircraft manoeuvre weight (defined as 50 per cent internal fuel with two AIM-120 and four 2000 lb JDAM)).

The design specification calls for five separate weapon capabilities:

- Four Mk-84 LDGP + two AIM-120.
- Four GBU-27 + two AIM-120.
- Four 2000 lb JDAM + two AIM-120.
- Four AGM-154 JSOW + two AIM-120.
- Sixteen 250 lb small smart bombs.

(the AIAA specification gave details of the size, weight and cost of all government furnished equipment. This data is used in the layout, mass and cost estimations).

When details like those shown above are not provided with the initial specification, it is always necessary to spend time gathering the data before moving on to the next stage. In this case, we are now ready to consider initial aircraft design concepts.

The details below suggest several potential design requirements:

- The field take-off requirement, particularly with regard to the icy runway conditions will require a high thrust/weight ratio.
- Initial climb performance will require good specific excess power to reach the supercruise altitude and speed in reasonable time.
- Supercruise will require low overall drag to give a good lift/drag ratio and thereby a lower fuel requirement.
- The rear movement of the centre of lift in supersonic flight may require fuel transfer to balance the aircraft and reduce trim drag.
- The climb from supercruise altitude to 50 000 ft for the dash phase may require a burst of afterburning to offset the low SEP at high/fast operation. Stealth may be compromised by either the use of afterburning or from the long-duration climb from supercruise altitude to dash without the extra thrust.
- The aircraft must be able to drop the weapons without significant trim changes.
- The SEP requirements and the turn performance may require the use of manoeuvring flaps although this may compromise stealth.
- Landing will require low wing loading to avoid high approach speed and to reduce aircraft energy on the ground.
- Icy conditions may demand aerodynamic braking assistance (parachutes and lift dumping).
- Compatibility with NATO shelter size will limit the aircraft to a span of less than 20 m (65 ft) and length to less than 30 m (98 ft).

8.4   Design concepts and selection

Although initially many design layouts were envisaged, the three design concepts described below were selected for investigation.

- Conventional, straight wing
- Pure delta/diamond
- Blended delta

The conventional, tapered-wing layout (Figure 8.2) was selected as this offers less technical risk to the project. The design processes for this layout are well understood and the configuration can be easily developed for alternative roles.

The pure arrow-wing layout (Figure 8.3) results from considerations of stealth and aerodynamic efficiency. The main drawbacks of the diamond planform centre on the unorthodox control arrangement and the difficulty of developing the layout to accommodate alternative roles.

The blended arrow-wing configuration (Figure 8.4) can be regarded as either offering the best of the other options, or the worst of both types! The blended body can be configured to give lower wave drag than the straight wing and could be more easily developed than the pure delta.

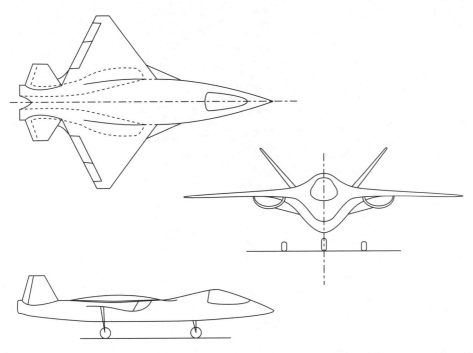

Fig. 8.2 Design concept – conventional straight wing

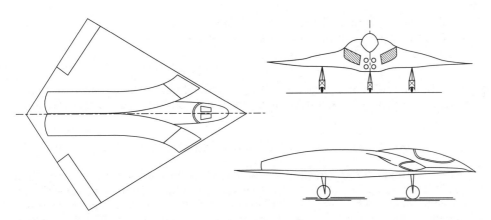

Fig. 8.3 Design concept – delta/diamond

A decision matrix method was used to analyse the different options on a consistent basis. The criteria used to assess the options in the selection process are listed below together with (in brackets) the significance (weighting) to the overall assessment.

Effectiveness of incorporating stealth technology into the layout (5)
Aerodynamic efficiency (mainly L/D ratio) of the layout (5)
Potential for low-weight design (4)
Technical difficulties (ease of analysis) and risk (3)

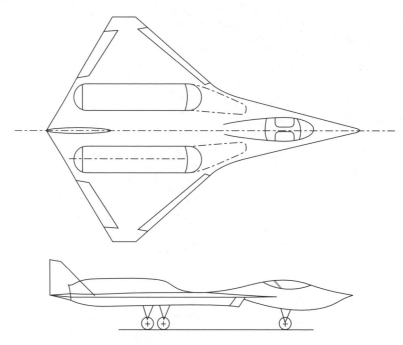

Fig. 8.4 Design concept – blended delta

> Field performance and rough ground handling (2)
> Maintainability and operational dependability (2)
> Survivability and ease of repair (2)
> Multi-role capability (1)

Naturally, the choice of criteria and the relative weightings is highly subjective but a group response tends to smooth the assessment process. The result of the 'voting' on the criteria above is shown below:

> Conventional option (56), Delta/diamond layout (72), Blended body (58)

The necessity for high L/D ratio and improved stealth were the key issues in the selection of the delta/diamond layout. It was also agreed that as much effort as possible should be given to the use of blending the profiling of the body (as on the B-2 aircraft). It was also decided that an advantage would be gained if the aircraft length was increased. These issues led to changes in the original configuration. To reduce aircraft maximum sectional area and effectively lengthen the aircraft, tandem seating and tandem weapon stowage was employed. This resulted in the concept sketch shown in Figure 8.5.

The basic structural framework consists of a continuous (tip-to-tip) wing box. The weapons and main landing gear are suspended below this and housed in profiled fairings with radar reflective door and hinge edgings. Forward of the weapon bay, the profile is extended to accept the engine intakes which sweep up in S-bends to the top wing surface. This duct-profiling protects the intake profile against radar reflections from the engine compressor face. It also ensures clean airflow to the engines with the aircraft at high-incidence attitude. The nose landing gear is retracted into the space between the separate intake ducts. The twin engines are supported in cradles above the wing

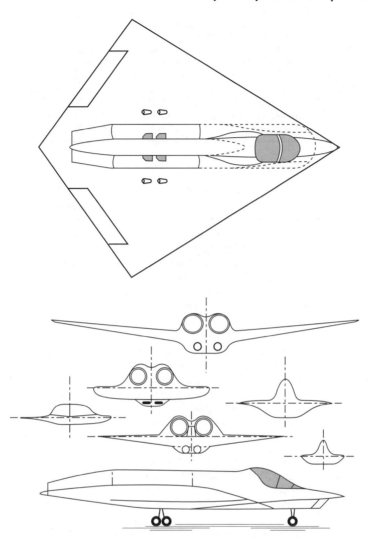

Fig. 8.5 Selected and revised concept sketch

structure. Nozzle exhaust ducts terminate forward of the wing trailing edge to shield the aircraft from downward infrared emissions. Fuel tankage is provided between the engine support cradles and intake ducting. The pilot and equipment bays are located in the aircraft centre line fuselage profile forward of the fuselage fuel tanks. The upper body is profiled to blend smoothly into the wing surface and to give an advantageous Sears–Haack volume distribution.

8.5   Initial sizing and layout

The initial sizing of the preferred configuration requires estimates of the main aircraft parameters. Instead of just guessing these values it is a good idea to investigate the values

Table 8.1

Parameter	Fighters	Strikers	Bombers
Empty mass ratio (M_E/M_{TO})	0.45–0.60	0.41–0.54	0.37–0.42
Fuel mass ratio (M_F/M_{TO})	0.21–0.33	0.17–0.33	0.40–0.62
Payload ratio (M_{PAY}/M_{TO})	0.21–0.28	0.18–0.37	0.14–0.19
Wing loading (M_{TO}/S) kg/sq. m	262–467	315–544	447–516
Thrust/Weight ratio (dry)	0.65–1.29	0.56–0.88	0.26–0.40

associated with existing aircraft of the same type. It is possible to compile a list of design data for existing military aircraft using published data.[6] The problem with using this approach for our project is the unique nature of the specified mission requirements of the design. It does not follow the 'fighter' class of aircraft because of our need to fly a longer range and carry a heavier weapon load than is normal for fighters. It does not fit into the 'bomber' class due to the higher speed and lower weapon load of our aircraft. 'Multi-role' and 'strike' aircraft may have some comparable features but these usually have much better manoeuvring ability and are not expected to supercruise for long periods. Using data on appropriate military aircraft from reference 6 (with extreme values ignored), it is possible to assess the variation of some design parameters (Table 8.1).

It is clear from this analysis that there is wide variation in the aircraft used in the study. Also, as with all published data, the definition of aircraft parameters (e.g. empty weight) may not be consistent from each manufacturer. The data therefore only provides a crude guide to the selection of parameters for use in the initial estimates. This implies that the initial estimates will be unreliable. It will be necessary to adopt a more refined analysis as quickly as possible.

Some thoughts about our design that might help us to select suitable starting values:

- Most of the aircraft in the survey are not supersonic in dry thrust so our design is likely to require a higher thrust to weight ratio than the bomber values.
- Travelling for long distance at supersonic speed will require more fuel than is seen in the fighter and strike classes above but not as much as the max. bomber (B-52) value.
- The fuel capacity required will be larger than on equivalent size aircraft so it may be advantageous to have a larger wing area to provide extra tankage.
- A large wing size (low wing loading) will help in meeting the icy runway requirements.
- The payload carried by our design, as defined in the specification, will give a relative low useful load ratio and the range flown at supersonic speed will give a high fuel mass ratio.
- The empty mass ratio would also be reduced due to the large fuel mass but to account for the stealth requirement extra structure mass (radar absorbent materials) will be required. With no better information these two effects will be assumed to cancel each other, giving a conventional empty mass ratio.

With these thoughts in mind, our initial estimates are shown below:

- **Empty mass ratio = 0.44**
  (this is low for fighters but high for bombers).
- **Fuel mass ratio = 0.46**
  (this is outside the range for fighter/striker aircraft but about average for bombers).

- Using the above assumption would make the **payload ratio = 0.1**
  (as predicted, this falls below all aircraft classes).
- **Wing loading = 390 kg/sq. m** (about 80 lb/sq. ft)
  (which is low for bombers, high for fighters and about average for strike aircraft).
- **Thrust loading = 0.60**
  (this is low for strike and fighter aircraft but it is not clear from the collected data how many of the sample have quoted afterburning (wet) thrust. It is outside the range for bomber aircraft).

It is now possible to use the assumed values to make our first 'rough' predictions of the size of the aircraft:

- From the problem specification we can predict that the payload (including two crew) is 6600 kg (14 550 lb). As we assume above that this represents $0.1 M_{TO}$, the aircraft **maximum take-off mass must be 66 000 kg** (145 500 lb).
- With an empty mass ratio of $0.44 M_{TO}$,
  the **empty mass = 29 000 kg** (64 000 lb).
- With a fuel mass ratio of $0.46 M_{TO}$,
  the **fuel mass = 30 360 kg** (67 000 lb).
- With a wing loading of 390 kg/sq. m = 3826 N/sq. m (about 80 lb/sq. ft),
  the **gross wing area = 170 sq. m** (1827 sq. ft).
- With a thrust loading of $0.6 W_{TO}$,
  the total **engine thrust (sea level, static, dry) = 388 kN** (87 300 lb).
  This equates to 194 kN (43 700 lb) per engine.

This makes our aircraft heavier and larger than any of the fighter and strike aircraft surveyed but much smaller than the existing bombers.

The diamond planform (area, $S = 170$ sq. m, 1830 sq. ft) which is limited in span (b) to 18.3 m (60 ft) (to keep within the hangar width) will have a centre line chord = $(2S/b) = 18.6$ m. For a symmetrical planform (90° at the tip) the wing sweep is only about 45° and we must have at least 51.3 (see section 8.2.3). It is also advantageous to maintain a long overall length to reduce wave drag. Both of these requirements can be met by reducing aircraft span to 17 m (55.7 ft). In this case the centre line chord will be increased to 20 m (65.6 ft). Providing a 90° angle between the leading and trailing edges at the tip gives a 60° wing leading edge sweep angle.

This geometry may need to be changed later in the design process if more fuel tankage is required.

Using the concept sketch (Figure 8.5) and the values above we can now produce our first scale drawing of the aircraft (Figure 8.6).

8.6   Initial estimates

With an accurate drawing of the aircraft (Figure 8.6) it is possible to estimate the component masses and drags (and lift). The predicted thrust will allow us to select a suitable engine or scale an existing design to provide engine performance data at all flight conditions. With mass, aerodynamic and propulsion data it will be possible to perform initial performance calculations and draw our first constraint diagram.

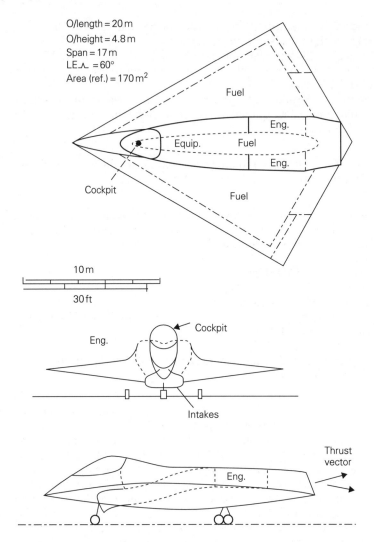

O/length = 20 m
O/height = 4.8 m
Span = 17 m
LE$\wedge$ = 60°
Area (ref.) = 170 m^2

Fig. 8.6 Initial baseline aircraft general arrangement

8.6.1   Initial mass estimations

The initial mass estimates can be calculated by using published empirical equations based on existing aircraft designs.[4] As our aircraft has a unique operating envelope, such methods may be regarded as crude. At this stage in the design process, the analysis is likely to be more accurate than the 'guesstimates' made from the survey used above. Using our knowledge of the aircraft specification, some corrections to the method can be applied. All the required input data for the method can be gleaned from the initial layout drawing, the project specification and common sense. Applying such data to the equations in reference 4 gives the mass statement shown in Table 8.2.

This initial estimate of MTO is substantially less than previously predicted. The main reason for this reduction is due to the lower prediction of aircraft empty mass. Although, as expected, the propulsion system mass is large due to the high thrust

Table 8.2

Component	Mass (kg)	(lb)	% MTO
Wing (inc. controls)	1 888	(4 163)	3.0
Body (inc. engine cowls)	7 070	(15 589)	11.3
Undercarriage (all units)	1 138	(2 509)	1.8
Total structure	10 096	(22 262)	16.1
Propulsion system	12 077	(26 630)	19.2
Fixed systems	3 673	(8 100)	5.8
Aircraft empty	25 846	(56 990)	41.1
Crew (two pilots)	500	(1 100)	
Weapons	6 100	(13 450)	
Zero fuel mass	32 446	(71 543)	51.7
Fuel*	30 360	(66 944)	48.3
Max. take-off mass	62 806	(138 487)	100.0

*The fuel load is retained at the value estimated from the higher MTO mass originally predicted. This will need to be checked when the mission analysis is completed

to weight ratio, the aircraft structure and fixed systems masses are low. This could have been expected as the compact and stiff structure framework will provide a light structure. However, for our high-tech, modern weapon system, the low systems mass must be treated as suspicious. As the project develops, and more detail is known about the aircraft systems, it will be necessary to reassess this estimate.

As the aircraft empty mass estimation was based mainly on the original value of MTO it is expected that the aircraft mass and size could be reduced. Before any changes are contemplated, it is advisable to continue with the aerodynamic and performance estimations using the original design. In this way, all the design modifications can be assessed at the end of the initial estimation process.

8.6.2   Initial aerodynamic estimations

The initial aerodynamic estimations concern the prediction of aircraft drag and lift. For this aircraft the main focus of drag will be on the supersonic wave drag (C_{Dw}) estimation. Using the wave drag equation in reference 4, with the following input values, gives the first estimation of C_{Dw}:

Aircraft cruise Mach number, $M = 1.6$
Aircraft max. cross-sectional area, $(A_{max}) = 10.06$ sq. m
Reference wing area, $S_{ref} = 170$ sq. m
Wing LE sweep $= 60°$
Aircraft overall length (less any constant sections), $L = 20$ m
An adjustment factor to relate the actual cross-section distribution to the Sears–Haack perfect shape, $E_{WD} = 1.4$ (assuming a smooth distribution from the blended body)
Gives, $C_{Dw} = 0.02104$

This is a very large drag increment that will substantially penalise the design. Somehow, we will need to either reduce the cross-sectional area or increase the aircraft length. The area cannot be changed significantly unless we alter the internal requirements. It is relatively easier to increase the length (see later aircraft drawings). Assuming that it is possible to stretch the aircraft to 28 m (92 ft) the calculation above would change to: $C_{Dw} = \mathbf{0.01408}$.

The parasitic drag will be estimated by using an equivalent skin friction coefficient of 0.0025 (representative of a smooth fast transport aircraft).

Hence, with an estimated aircraft wetted area of 400 sq. m (4300 sq. ft),

$$C_{DO} = 0.0025 \ (400/170) = \mathbf{0.00588}$$

This gives a total 'clean aircraft' zero-lift drag coefficient of (0.02692), in cruise condition.

At the start of the initial dash phase, the aircraft weight will be less than the take-off value due to the fuel used during take-off, climb and supercruise. As we do not know the fuel used yet we will assume that weight is at 80 per cent of the take-off value:

$$66\,000 \times 9.81 \times 0.8 = 518\,\text{kN} \ (116\,450\,\text{lb})$$

Therefore, the cruise $C_L = 518\,000/(0.5 \times 0.1864 \times [295 \times 1.6]^2 \times 170) = 0.147$
From reference 4, at M1.6, the induced drag factor $(K) = 0.3$
Hence, induced drag coefficient, $C_{Di} = 0.3 \times 0.147^2 = \mathbf{0.00648}$

Therefore the total drag coefficient at start of cruise $= \mathbf{0.02644}$
Hence, drag at 50 000 ft and dash speed of M1.6,

$$= 0.5 \times 0.1864 \times (295 \times 1.6)^2 \times 170 \times 0.02644 = 93.3\,\text{kN} \ (20\,982\,\text{lb})$$

Hence, the lift to drag ratio will be (518/99.5) = **5.56**

The reciprocal of (L/D) is equal to the (T/W) required in the initial dash. In this case $(1/5.56) = 0.18$. This has to be multiplied by the engine thrust lapse rate appropriate at the cruise condition (height and speed) and the weight reduction, to obtain an equivalent static, sea-level (SSL) value. Although a high bypass ratio engine would have a cooler exhaust temperature which would give a lower IR signature, it would be substantially larger. This would make the aircraft much bigger which would be less stealthy in other ways. Therefore, the engine we will select will be of the lower bypass type. For this type of engine the effect of speed on thrust lapse rate can be ignored for initial estimates. Reference 7 quotes the following expression to determine the lapse rate:

$$\text{Thrust at altitude/SLS thrust} = \sigma^x$$

where SLS denotes sea-level static condition, σ is the relative ambient air density and the exponential x has the value of 0.7 in the troposphere and 1.0 in the lower stratosphere.

At 50 000 ft the lapse rate is $(0.428 \times 0.51 = 0.22)$
As above, the weight ratio (W_{dash}/W_{TO}) is 0.8
Therefore, to achieve a cruise T/W of 0.18, requires an SLS value of 0.8 (0.18/0.22) = 0.654
This is higher value than the 0.6 value originally assumed

The above calculations have highlighted a potential problem area for the design. The high drag in cruise reduces the aircraft L/D ratio which will have a direct effect on the fuel required to complete the mission. The high T/W value will require a larger engine and corresponding propulsion system. Both of these effects will seriously compromise the effectiveness of the design. Hence, it is important to reduce the aircraft drag.

To illustrate the overall effect, if we could reduce wave drag by 20 per cent, a similar analysis to that above would yield:

$$\text{Drag in cruise} = 83.0\,\text{kN} \; (18\,668\,\text{lb})$$
$$\text{Lift/Drag at the start of cruise} = 6.24$$
$$\text{SLS thrust/weight} = 0.58$$

This looks more encouraging but a 20 per cent drag reduction may be difficult to achieve without a significant changes in aircraft layout. Obviously, the influence of wave drag is paramount in the drag estimation. For example, if the ($A_{\max}$) value could be reduced from 10.06 to 9.8 and the length increased from 28 to 30, the wave drag coefficient would lower to the 20 per cent required. This shows the significance of the parameters input to the equations and the need to carefully assess the values used. From the standpoint of layout, it is clear that in reviewing the existing design we must reconfigure the aircraft shape to reduce $A_{\max}$ and increase vehicle length, if this is possible.

In the evaluation of wave drag, the equation shows a direct proportionality to the factor (E_{WD}) which, as defined above, relates the actual aircraft longitudinal cross-sectional area distribution to that of the perfect Sears–Haack distribution. A value typical of fighter aircraft optimised for supersonic flow has been assumed (i.e. 1.4). It is impossible to achieve a value of 1.0 with realistic shapes but it may be possible[4] to achieve 1.2 for an optimum blended-fuselage, delta-wing configuration. Our aircraft layout fits into this category so we could consider reducing the originally assumed value. Lowering the 1.4 value to 1.2 generates a 14 per cent reduction in wave drag. The graph for cross-sectional volume for our current configuration is shown in Figure 8.7.

To avoid the profiling problems at the front of the aircraft associated with pilot windscreen and canopy shaping, artificial vision systems are proposed for the cockpit layout. This means that the cockpit could then be positioned away from the nose profile. With a reduction of the maximum cross-sectional area it should be possible to reduce the cruise drag coefficient to 0.0232, giving $L/D = 6.34$ and $(T/W)_{\mathrm{TO}} = 0.58$. Although this may be optimistic, it does represent a good initial estimate as it indicates the direction that future design decisions must take.

Drag of the aircraft in other flight conditions and in different configurations must also be estimated. The most significant of these are the take-off and landing phases. In addition to the aircraft clean condition we must add landing gear, flap and any aerodynamic

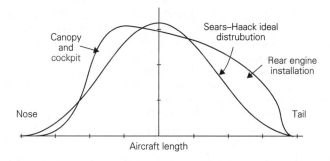

Fig. 8.7 Sears–Haack cross-sectional area distribution

retarding devices (e.g. lift dumpers and braking parachutes). In this initial stage it is sufficient to make sensible guesses, in terms of 'drag area' (drag/dynamic pressure $= D/q$) for each addition.

For the undercarriage (including interference effects) we will assume $D/q = 0.5$ sq. m. For our reference area of 170 sq. m this gives a ΔC_{DO} of $(0.5/170) = 0.003$.

At this stage in the evolution of the aircraft it is not known if conventional flaps will be required on the aircraft. Flaps may only be required for landing as at take-off the afterburner (if fitted) could be used. To assess the effect of flaps on aircraft drag, plain flaps with only 20° deflection will be assumed. (If it is possible to avoid flaps, the inboard trailing-edge surfaces could act as pitch control surfaces.) We will assume D/q for flaps if used to be 2.0 sq. m, giving $\Delta C_{DO} = 0.0118$. In later drag estimations it will be necessary to take into account changes in drag from the cruise condition. These arise from the reduced Reynolds number at the lower speeds in take-off and landing phases, and the effect on lift-induced drag due to the disturbed spanwise lift distribution caused by the flap lift. At this time, we can assume the clean drag coefficient is unchanged from the cruise value shown above, i.e. 0.00589. At take-off, we will assume that there is no flap deflection. The zero lift drag coefficients, based on the reference area of 170 sq. m, is shown below:

$$C_{DO} \text{ for take-off} = 0.00889 \quad \text{and} \quad C_{DO} \text{ for landing} = 0.02069$$

Using reference 8, the position of the aerodynamic centre can be calculated. For our aircraft ($\lambda = 0$, LE sweep $= 60°$, wing aspect ratio $= 1.7$) the supersonic position (X_{ac}/C_{root}) is 0.55 and subsonic 0.45. As this aircraft will spend most of the mission at supersonic speed, it would be sensible to trim the aircraft for this condition. The forward movement of the lift in the take-off and landing phases will need to be balanced by fuel mass transfer or control surface trim forces.

For the determination of the lift capabilities of the aircraft there are two principal characteristics; the lift curve slope and the maximum lift. For any aircraft configuration these are notoriously difficult parameters to predict accurately. In practice, many aircraft have required 'fixes' after flight tests, to correct lift characteristics that were not predicted.

The prediction of the lift curve slope is required to determine the best (low drag) angle between the fuselage and wing (this is not necessary on our blended body layout). It is also used in the prediction of drag due to lift, and in the stability and control analysis. When more accurate geometrical information on the aircraft layout is available, it will be possible to use computational fluid dynamic (CFD) methods to provide more accurate estimations. At this stage expending such effort would be inappropriate as the aircraft shape will be under continuous revision. For a bi-convex section of aspect ratio less than 2, the shape of the C_L versus angle of attack (α) graph is shown in Figure 8.8.

For moderately swept, high aspect ratio wings (typical of transport aircraft) the C_{Lmax} of the unflapped wing will be close to the infinite span (2D) aerofoil value but our aircraft is not of this planform. For highly swept, very low aspect ratio planforms, the airflow over the wing surfaces will be significantly affected by vortex generation over the leading edge. These leading edge vortices add both lift and drag and ensure that the flow over the upper surface remains attached well above the normal stall angle of higher aspect ratio trapezoidal planforms. This vortex formation (see Figure 8.9) will be most prominent for wings with a sharp leading edge. These conditions are expected to be found on our wing planform.

Vortex-generated lift will stay attached to the wing up to the point of vortex burst at high angles of attack. The traditional stall characteristic by which we predict maximum

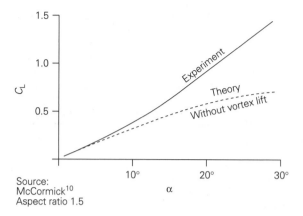

Fig. 8.8 Section lift coefficient versus angle of attack

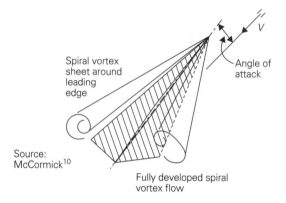

Fig. 8.9 Vortex induced flow

lift capability is not appropriate in this situation. As shown on Figure 8.8, the max. lift coefficient will not be reached until exceptionally high nose angles have been pulled. When the aircraft is on or close to the ground (e.g. on take-off and landing) such high angles will cause the aircraft rear fuselage structure to scrape the runway. The geometry of the aircraft will limit the max. attitude to about 15°. At this angle Figure 8.8 shows that the C_L is approximately 0.52. Therefore, the limit of lift generation on or close to the runway will be set by the aircraft tail scrape angle of 15°. For flight away from the ground, the max. lift coefficient will be set by the limit of controllable angle of attack.

8.7   Constraint analysis

As the initial mass (weight) and aerodynamic estimates have now been made, it is possible to conduct a constraint analysis to determine if the original choice of thrust and wing loading values are reasonable. As these were derived from data on other aircraft, it is likely that a better selection can improve the design. This process will also indicate which of the constraints on the problem are most critical.

The equation below, as developed from the specific excess power relationship in Chapter 2, is the general form of the constraint function:

$$(T_{SSL}/W_{TO}) = (\beta/\alpha)[\{(q/\beta)(C_{DO}/(W_{TO}/S)\} + \{[k_1 \cdot n^2 \cdot (W_{TO}/S)]/(q/\beta)\}$$
$$+ (1/V) \cdot dh/dt + ([1/g] \cdot [dV/dt])$$

In order to draw the constraint diagram (thrust versus wing loading) it is necessary to determine the values of the coefficients (etc.) to be used in the equation.

These are defined below:

List 1
T_{SSL} = engine static sea-level thrust
W_{TO} = aircraft take-off weight
S = aircraft reference wing area
W = aircraft weight at the condition under investigation
T = engine thrust at the condition under investigation

List 2
β = aircraft weight fraction for the case under investigation = (W/W_{TO})
α = thrust lapse rate at the altitude and speed under investigation = (T/T_{SSL})
q = dynamic pressure at the altitude and speed under investigation = $(0.5\rho V^2)$
V = aircraft speed at the condition under investigation
h = aircraft altitude at the case under investigation
ρ = air density at height h
C_{DO} = aircraft zero-lift drag coefficient
k_1 = aircraft-induced drag coefficient
n = aircraft normal load factor = L/W
dh/dt = aircraft rate of climb at the case under investigation
g = standard gravitational acceleration = $32.2 \, \text{ft/s}^2$ (or $9.81 \, \text{m/s}^2$)
dV/dt = aircraft acceleration at the case under investigation

For each constraint case, the analysis requires all the values for the parameters in the second list above to be substituted into the equation for (T_{SSL}/W_{TO}) above. Selected values of wing loading (W_{TO}/S) are then used to determine corresponding values for thrust loading (T_{SSL}/W_{TO}). These values are then plotted to indicate the constraint boundary for the case. This process is repeated for all constraints.

In the design proposal, there are several performance requirements:

• Take-off from 8000 ft (2440 m) runway, on standard day with icy runway.
• Climb to optimum supercruise altitude.
• Supercruise at optimum altitude at M1.6 for 1000 nm (less climb distance).
• Dash at M1.6 at 50 000 ft (min.).
• Manoeuvre with specific excess power (SEP), at specified weapon load and 50 per cent fuel:
  – at 1g, M1.6, alt. = 50 000 ft with SEP = 0 ft/s with no afterburning
  – at 1g, M1.6, alt. = 50 000 ft with SEP = 200 ft/s with afterburning
  – at 2g, M1.6, alt. = 50 000 ft with SEP = 0 ft/s with afterburning
• Land onto 8000 ft runway, on standard day with icy runway.

Before the analysis can be made there are several assumptions that must be made:

• Take-off from icy* conditions will be with afterburning (called maximum thrust).
• Take-off in normal conditions will be with no afterburning (called military thrust).

- As some of the constraints are related to military thrust, it is necessary to define the increase in thrust from afterburning. We will initially assume $(T_{max}/T_{mil}) = 1.5$.
- Initial climb to supercruise with final rate of climb of 1000 fpm (our requirement).
- Supercruise starts with 90 per cent MTOM.
- Dash starts with 80 per cent MTOM.
- Manoeuvres are at aircraft mass empty + crew + weapons + 50 per cent fuel $(25\,846 + 500 + 4000 + 15\,180 = 45\,526\,\text{kg}\ (100\,385\,\text{lb}))$.
  Basing all of the constraint analysis on our original mass estimate of $66\,000\,\text{kg}\ (145\,530\,\text{lb})$ gives $\beta_{manoeuvre} = (W/W_{TO}) = 0.69$.
- Landing approach speed less than 160 kts (82 m/s) at 95 per cent MTOM.
- Landing on an icy* runway with fuel dumping and possibly emergency braking parachute.
- Landing in normal conditions will be determined at 95 per cent MTOM with emergency braking ($\mu = 0.5$).

(*Operation from icy runways requires directional control that is not reliant on tyres.)

Aerodynamic surfaces and engine thrust mechanisms are the only alternatives. Lateral thrust vectoring will be available for take-off but not for landing. Operation from icy runways may be difficult unless other solutions can be found.

The last three constraints dictate maximum vales for (W/W_{TO}). The approach speed is only affected by the aircraft minimum speed. As we will not have a reverse-thrust capability on the aircraft, the landing distance calculations will be independent of engine thrust. The appropriate calculations are shown below and the results plotted in Figure 8.10.

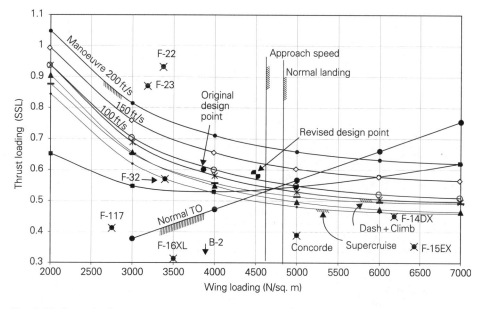

Fig. 8.10 Constraint diagram

(a) For the approach speed

$$(W_{TO}/S) = 1/\beta\{0.5\rho(V_{approach}/1.2)^2 C_{Lapproach}\}$$

where $\beta = 0.95$
$\rho = 1.225\,\text{kg/m}^3$
$V_{approach} = 82\,\text{m/s}$
$C_{Lapproach} = 0.52$

We are assuming that the approach speed is $1.2 V_{stall}$. This is slower than normal. This gives $(W_{TO}/S) = 1566\,\text{N/sq. m}$ (max.) (For reference $1000\,\text{N/sq. m} = 20.9\,\text{lb/sq. ft.}$)

This is much too low. It will create a large wing area which will be inefficient in the cruise phases. (For reference, the initial estimate for wing loading is 3880 N/sq. m.) It will be necessary to generate more C_L from the wing. The value used above was consistent with an unflapped delta wing limited to a maximum angle of attack of 15°. For an aircraft of our layout it may be possible to adopt a high angle of attack (HAA) approach (see Figure 8.11) as demonstrated by the X31 vector technology demonstrator.[9] This uses an HAA to provide a slow speed flight at low decent rate for most of the approach. Obviously, the aircraft must be stable in such a flight attitude and must be capable of maintaining its heading. As the aircraft gets near to the ground the angle of attack is raised to the maximum value to slow the aircraft. Just prior to the tail scrape, the incidence is rapidly reduced (nose-down). This will cause the aircraft to effectively have a controllable crash landing onto the runway threshold. This manoeuvre will demand an extra strong landing gear to withstand the high loads required to absorb the vertical energy. This flight profile requires automation as the pilots will not be capable of reacting to such a landing manoeuvre. (Most large civil aircraft landings are automatic these days, although not like this profile!)

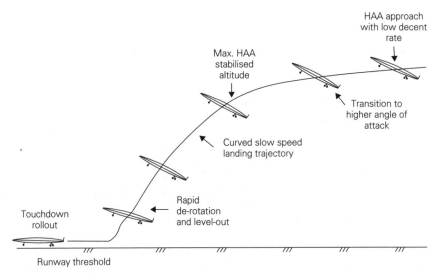

Source: S. W. Kandebo[9]

Fig. 8.11 High angle of attack approach profile

As we have already decided that artificial vision and automatic landing systems would be incorporated into the aircraft to avoid the forward cockpit profiling, the unusual aircraft attitude should not present a problem. Increasing the angle of attack to 35° would raise the C_L to 1.3. The wing could be fitted with a leading edge (vortex) flap to increase C_L to 1.5. Even when not deployed the additional mechanisms and systems needed to deploy the flaps would affect the stealth image of the aircraft so may not be desirable. Vortex flaps will not be included but will provide some insurance if the flight profile is seen in flight tests to require extra lift capability.

We could also assume some fuel dumping or burn-off before landing. This would reduce β to 0.8.

These changes would increase the maximum wing loading (N/sq. m) as shown below:

(a) baseline = 1566,
(b) with HAA profile = 3914,
(c) HAA plus fuel dumping = 4648.

The aircraft conditions to be adopted will be decided when all the constraints have been assessed.

(b) For normal landing

$$(W_{TO}/S) = (s_L \cdot \rho \cdot C_{Llanding} \cdot g \cdot \mu)/(1.69 \cdot \beta)$$

where s_L = available runway length = 2440 m (8000 ft)
$C_{Llanding}$ = 0.52 (see above)
$\mu = 0.5$
$\beta = 0.95$

This gives a maximum value of $(W_{TO}/S) = 4748$ N/sq. m.

Note that an approach speed of 1.3 times minimum speed has been assumed above (factor 1.69). This is typical of conventional aircraft to protect from stall due to sudden changes in atmospheric conditions. As the delta planform flying at 15° angle of attack is well away from the max. lift angle it may be argued that this factor could be ignored. If so, the maximum value of wing loading would be 8024 N/sq. m.

(c) For icy landing
The same formula as above is applicable if braking parachutes (etc.) are not used with input values of:

s_L = available runway length = 2440 m (8000 ft)
C_{Lland} = 0.52
$\mu = 0.1$
$\beta = 0.95$
Giving: $(W_{TO}/S) = 950$ N/sq. m
Or without the factor = 1605 N/sq. m.

These are obviously too low, therefore extra retardation is required. As we are likely to need thrust-vectoring and afterburning on the engine, it is unfeasible to expect thrust reversal to be available. Braking parachutes, air brakes, runway-retarding devices and ice removal offer some possibilities. As all of these devices complicate the analysis, it is not appropriate to get too involved in detail design at this early stage in the design of the aircraft.

(d) Normal take-off

For take-off conditions the constraint equation reduces to:

$$(T_{SSL}/W_{TO}) = [(1.44 \cdot \beta^2)/(\alpha \cdot \rho \cdot C_{Lto} \cdot g \cdot s_{TO})] \cdot (W_{TO}/S)$$

where s_{TO} = available runway length = 2440 m (8000 ft)
α and β = 1.0
C_{Lto} = 0.52

As the equation is a straight line through the origin, it is only necessary to evaluate it for one value of wing loading. For (W_{TO}/S) = 5000 N/sq. m, giving (T_{SSL}/W_{TO}) = 0.472.

The same argument as outlined above for landing can be made for the avoidance of the 1.44 factor in the take-off equation. In this case, the (T_{SSL}/W_{TO}) reduces to 0.328.

(e) For icy take-off

The calculation requires the estimation of the balanced field length using the maximum thrust for the flight condition but not for the braking condition to determine the decision speed. The braking part of the calculation involves the same difficulties as described in the icy landing description above. As with landing, it is too early in the design process to perform these calculations in sufficient detail. We will need to return to this subject later in the design process.

(f) Supercruise at optimum altitude

For a parabolic drag polar the condition for maximum range can be shown[4] to be:

$$C_{Do} = (3 \cdot k_1 \cdot C_L^2)$$

For our aircraft:

$$C_{Do} = 0.01996 \quad \text{and} \quad k_1 = 0.3 \quad \text{Hence, } C_L \text{ for max. range is } 0.149$$

Using the definition of lift:

$$L = W = 0.5 \cdot \rho \cdot V^2 \cdot S \cdot C_L$$

With W = 0.9 · 66 000 · 9.81, V = M1.6 = 1.6 · 295 = 472 above 11 000 m, S = 170 sq. m, gives ρ = 0.2295. From ISA tables this density occurs at 14 000 m (46 000 ft) altitude, this is the initial supercruise height. This calculation involves the initial guess for the wing loading (i.e. 3808 N/sq. m). The equation above can be solved in terms of other values for wing loading to indicate the sensitivity of (W_{TO}/S) against initial optimum altitude:

Wing loading (N/sq. m)	3 000	4 000	5 000
Optimum altitude (m)	16 000	13 000	11 600
Optimum altitude (ft)	52 500	42 000	38 030

As fuel is used and the aircraft gets lighter the wing loading will reduce and the optimum cruise height will rise. On the return supercruise phase (and for the dash manoeuvres) when the aircraft is lighter the cruise height will be increased providing that the engine thrust is large enough to reach these altitudes.

Artificially fixing the supercruise height for the initial calculation at 14 000 m it is possible to determine the relationship of thrust to wing loading using the constraint equation above. The result is shown below (assuming $\beta = 0.9$):

Wing loading (N/sq. m)	2000	3000	4000	5000	6000	7000
Thrust loading (T_{SSL}/W_{TO})	0.905	0.656	0.548	0.496	0.473	0.465

(g) Initial climb to supercruise altitude

The required thrust to achieve the supercruise condition, as calculated above, must include sufficient climbing ability at the start of cruise. The minimum for this type of aircraft is 1000 ft/min (5.08 m/s). The thrust loading to give this rate of climb is calculated by the climb term $[(1/V) \cdot dh/dt]$ in the constraint equation, suitably adjusted to the take-off condition (i.e. multiplied by β/α). Hence,

$$\Delta(T_{SSL}/W_{TO}) = [(1/V) \cdot dh/dt] \cdot \beta/\alpha = (1/472) \cdot 5.08 \cdot (0.9/0.3) = 0.0323$$

(h) Dash at 50 000 ft altitude

This is similar to the supercruise case except that the starting mass will be lower due to the fuel used in the previous sector. We will assume $\beta = 0.8$. The calculation is performed with and without the climb requirement. The results are shown in Figure 8.10.

(i) Manoeuvres

There are three separate manoeuvres that have to be investigated (as described in section 8.3 as cases (a) to (c)). The constraint equation is used with the afterburning thrust ratio (1.5) for cases (b) and (c).

Case (a) is similar to the initial dash phase described above except that the aircraft weight is lower ($\beta = 0.69$). This will make it uncritical and therefore not worth investigating for the constraint analysis.

Case (b) is very critical as shown by the results plotted in Figure 8.10. This requirement overpowers all other constraints and will solely dictate the aircraft layout. For aircraft design this is an undesirable situation and calls into question the validity of this requirement. The specified climb rate of 200 ft/s (12 000 fpm) at the high altitude and high weight may be desirable for avoidance of threats but seems excessive in view of the stealth characteristics of the aircraft. It would be sensible to discuss this problem with the originators of the RFP to establish how 'firm' they are on retaining the requirement. Requirements often fall into two categories: 'demands' and 'wishes'. Part of the constraint analysis is concerned with distinguishing between these two types for the critical design requirements. To assist with the discussion it is worth showing the sensitivity of the climb requirement by performing the analysis for different values; in this case, for 100 and 150 ft/s. These extra cases are shown on Figure 8.10. The 100 ft/s case seems to offer the most 'balanced' design and still provide a respectable 6000 fpm climbing ability.

Case (c) is similar to case (a) but with the normal acceleration value (n) increased to 2, and with afterburning applied. As seen on the constraint diagram the case fits well with the other requirements.

8.7.1 Conclusion

The constraint analysis has shown that, in general, the aircraft requirements are well balanced. The exceptions to this optimism are concerned with the manoeuvre climb requirement and the airfield performance onto icy runways. Both of these present problems for the design. As discussed above, the climb requirement should be reduced to 100 ft/s. In the following work we will assume that this concession has been made by the customer. Operation from and onto icy runways is not avoidable so some extra retardation systems will have to be introduced or some other possibilities considered. Incorporating reversed thrust into the already complex, engine-nozzle system appears to be unfeasible. Braking parachutes will have to be used together with a reduction in the touchdown speed to lower the energy to be dissipated. The best solution would be the installation of an arrester-hook on the aircraft and some form of wire pick-up on the runway for those airstrips that are susceptible to icing. Such a concept is outside the remit for our design.

It should be remembered that constraint analysis is a very crude process. It is based on potentially inaccurate data that has been generated from the initial 'guesstimates' of mass, aerodynamic and propulsion values and characteristics. Nevertheless, it offers the first tests of the initial layout and provides a direction to first revision of the aircraft geometry.

8.8   Revised baseline layout

The most efficient aircraft layouts on the constraint diagram are those with lower values for thrust loading and higher values for wing loading. The original design point was set at a wing loading of 3826 N/sq. m and thrust loading of 0.6. From Figure 8.10, it is seen that this point violates the modified manoeuvre constraint. Moving to 4500 N/sq. m and 0.58 brings the design into the feasible region. This reduces the wing area by about 15 per cent and the engine by about 4 per cent. This should result in a reduction of the aircraft MTOM.

Using the detailed mass estimate calculated earlier (section 8.6.1) and assuming a saving of 2000 kg in empty mass (about 8 per cent) to reflect the new aircraft parameters above, provides an initial value of MTOM of **60 806 kg** (134 077 lb).

This makes the new wing area $= (60\,806 \times 9.81)/4500 = $ **133 sq. m** approx.
(i.e. 1425 sq. ft)

The static sea-level military thrust (both engines) $= (60\,806 \times 9.81) \times 0.58 = 346$ kN (77 782 lb)
SSL thrust per engine $=$ **173 kN (38 900 lb)**

It is now possible to modify the original aircraft general arrangement drawing and to make some detailed estimates for the aircraft mass, aerodynamic characteristics and engine performance.

8.8.1   General arrangement

The new general arrangement drawing of the aircraft forms the basis of the input to the detailed technical analysis for the next stage of the design process. The basic layout of the aircraft will not be changed from that devised previously but some of the principal dimensions will be different. We now realise the importance of reducing the aircraft maximum cross-sectional area and lengthening the 'fuselage'. This will be achieved by stretching the planform.

Table 8.3

LE sweep	TE sweep (forward)				
	30	25	20	15	10
60	15.18*	15.56	15.93	16.31	16.70
	17.53	17.10	16.70	16.31	15.93
65	13.98	14.27*	14.56	14.85	15.14
	19.03	18.63	18.27	17.91	17.57
70	12.66	12.87	13.08*	13.28	13.49
	21.03	20.67	20.34	20.03	19.72

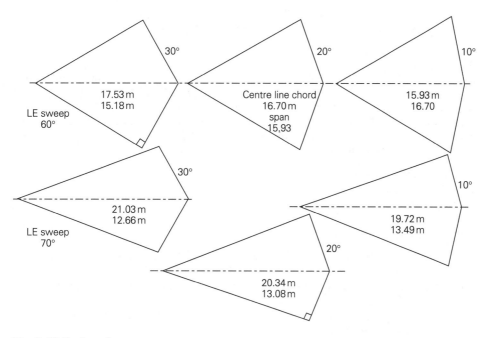

Fig. 8.12 Planform shapes

(a) Wing

For a diamond wing planform of a specified wing area (133 sq. m) it is possible to apply simple geometry to determine the span and centre line chord for various leading and trailing edge sweep angles (degrees) (Table 8.3).

In Table 8.3, the upper value is the wing span and the lower the centre line chord. Both values are in metres. The * values represent a wing tip angle of 90°. It is always wise to visualise geometric data. To appreciate the wing planform shapes the options in the table are drawn in Figure 8.12.

There are several considerations to take into account in making a choice of planform:

- To reduce wave drag a long centre line chord is desirable.
- Less TE sweep makes the trailing edge controls more effective.

- Less TE sweep pushes the centre of lift further aft. This could lead to aircraft balance and trim problems and demand larger control surfaces.
- A long centre line chord will give a deeper wing profile (for a given profile thickness ratio).
- A long centre line chord will reduce wing span and thereby reduce roll control but improve roll inertia.

Such concerns require a number of design compromises to be made. At this stage in the design process, the sensitivity of the control responsiveness and the aircraft balance issues are unknown. As our previous analysis has highlighted the need to reduce wave drag, our choice will be towards this aspect. We will select the 70° sweep with the 90° wing tip angle. Rounding the exact values shown above gives:

Wing span = 13 m,   Centre line chord = 20.5 m,   Wing area = 133.2 sq. m
Wing span = 42.6 ft,   Centre line chord = 67.2 ft,   Wing area = 1432 sq. ft

(b) Weapons
In order to arrange the fuselage shape it is necessary to identify the requirements for weapon storage. All the weapons are carried internally to reduce drag and radar returns. The design brief defined the type and combination of weapons to be carried. There are two categories of weapon listed: air-to-surface munitions and an air-to-air defence missile. The latter is the only form of self-defence on the aircraft. There are five different types of munitions specified:

- general purpose (GP) guided bombs,
- cluster bomb units (CBU),
- direct attack penetrators (JDAM),
- stand-off weapons (JSOW),
- small smart bombs (SSB).

The air-to-air missile is the AIM-120/AMRAAM which is commonly carried on other US military aircraft.

Descriptions and dimensions for the weapons are easily found in the aeronautical press[6] and web sites.

Most of these weapons are already used on other aircraft but some are externally mounted, therefore some detailed modifications for internal storage will be required.

The largest weapon in the list (Table 8.4) is the GBU. This will define the required weapon bay dimensions. The internal measurements will depend on the choice between four abreast or two abreast layouts (m/ft):

Layout	Length	Width	Depth
4 abreast	4.7/15.4	3.1/10.2	0.76/2.5
2 abreast	9.5/31.1	1.6/5.25	0.76/2.5

(c) Layout
Making some assumptions with regard to the engine size and installation, it is now possible to draw the revised baseline general arrangement (GA). This is shown in Figure 8.13.

To avoid the possibility of unstable flow conditions at the apex of the delta wing planform, a fuselage extension has been added (agreeing with the assumed length increase assumed in section 8.6.2). This will provide a separation of the airflow at the nose of the aircraft, it will add length which will reduce wave drag, and it will

Table 8.4

Weapon name	Guidance	Number	Size, length × dia. (m/ft)	Configuration
GBU-27	Laser	4	4.7/15.4 × 0.76/2.5	2 × 2
Mk-84 LDGP	None	4	3.6/11.8 × 0.45/1.5	2 × 2
JDAM	GPS	4	3.0/10.0 × 0.50/1.6	2 × 2
AGM-154 JSOW	Internal	4	4.3/14.0 × 0.33/1.1	2 × 2
SSB	N/A	16	2.0/6.50 × 0.15/0.5	4 × 4

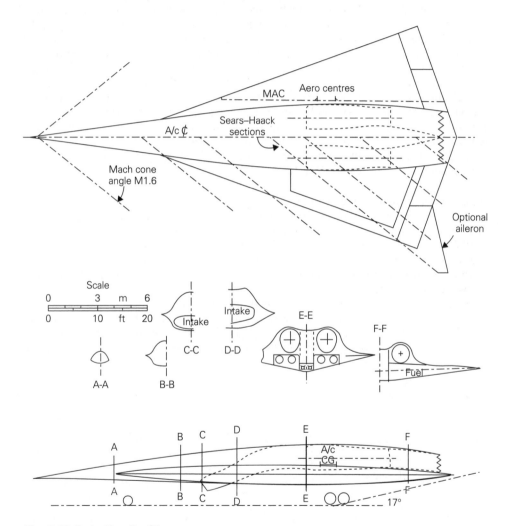

Fig. 8.13 Revised baseline GA

provide a useful storage volume to house sensitive sensors and flight instruments. The increased centre line chord provides sufficient volume to accommodate the weapon bay in a four-across configuration and part of the engine depth. This produces a blended body shape for the aircraft. For stealth and to avoid flow problems at high angles of attack, the intakes have been positioned on the underside of the wing profile. Without details of the aircraft centre of gravity position but knowing that the centre of lift is slightly forward of the 50 per cent MAC in the low-speed flight cases, the landing gear has been positioned relative to the centre of area of the wing planform. With the main units in this position there is sufficient tail-down angle to allow for tail-down rotation on take-off. The position of the nose unit may present intake ingestion problems that may demand wheel-debris shielding. Moving the nose units behind the intakes does not seem feasible.

Although the layout looks practicable, there are three areas of concern. The first relates to the reduced roll control that has arisen due to the smaller wing-span and area. To overcome this potential deficiency it may be necessary to modify the wing tip to extend the aileron surface. This is shown on the new aircraft GA as a surface with a 60° degree LE sweep and 20° TE sweepback. The second potential problem area involves the complication of geometry at the engine nozzle/wing trailing edge. The nozzle will need to have some vectoring capability to provide the aircraft with pitch control (particularly at high angles of attack). How this will be provided, without excessive complication to the wing structure and flow conditions at the wing intersection, has still to be realised. Finally, the third problem area concerns the positioning of the intakes relative to the wing leading edge vortex flow, particularly at low speeds. This is a complicated layout problem which will require some detailed computational fluid dynamic investigation and, at later stages, wind tunnel testing.

(d) Cross-sectional area distributions

Before starting the detailed technical analysis tasks, it is possible to use the aircraft GA to assess the volume distribution for wave drag evaluation and to determine the available fuel tank capacity relative to the estimated fuel load.

Some of the normal (90°) sections are shown on the GA and can be used to determine the cross-sectional areas at each station. These are plotted in Figure 8.14 and represent the values for the Mach-one case.

For higher Mach number cases, the sectional areas must be calculated as the surface area on the Mach cone intersection projected forward. For our aircraft, we will determine the distribution at Mach 1.6. In this case, the Mach cone semi-angle $\alpha = (90 - 51.3) = 38.7°$. We can approximate the cone section areas by applying the

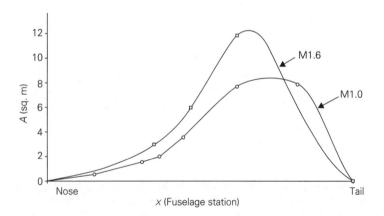

Fig. 8.14 Sears–Haack area distribution

ratio of the 'normal' section to the cone surface area:

$$\pi r l / \pi r^2 = l/r \quad \text{and} \quad l = r/\sin\alpha$$

where r is the base radius of the cone and l is the sloping length. For a cone angle of $38.7°$, $l/r = 1.6$.

Using this ratio to adjust the M1.0 areas gives the curve also shown in Figure 8.14. The actual area distribution, although smooth, shows a skew profile resulting from the arrow-wing planform. It is impossible to correct this without making impractical changes to the internal layout. Our current layout is not able to achieve the perfect Seers–Haack distribution but is regarded as relatively good compared to other aircraft. A distribution factor (E_{WD}) of between 1.2 and 1.4 seems reasonable.

(e) Fuel volume

As we have not yet performed a detailed mass prediction on our new layout we do not know the exact fuel load to be carried. We will have to use the value from previous calculations (i.e. a fuel load of 30 360 kg (66 944 lb)). Assuming a specific gravity of 0.8 for the fuel gives a required volume of (30 360/800 =) 38 cubic metres, or 10 300 US gal (or 66 944/6.8 = 9845 US gal if the specified JP8 fuel is used). From the aircraft GA, the central fuel tank = $2.0 \times 1.0 \times 7.5 = 15\,\text{m}^3$ and each wing tank = $7 \times 2.5 \times 1.0 = 17.5\,\text{m}^3$, making a total volume of $50\,\text{m}^3$. Allowing 10 per cent for structural intrusion and internal systems still leaves $45\,\text{m}^3$. This is more than adequate for the estimated requirement which we suspect is slightly high.

8.8.2 Mass evaluation

With the aircraft geometry determined in the general arrangement drawing it is now possible to make a detailed assessment of the aircraft mass. Using empirical formulae in design textbooks slightly modified to suit the particular features of our aircraft, each aircraft component can be evaluated. The results are shown in Table 8.5.

The blended profile of our aircraft layout makes it difficult to distinguish between the wing and body mass components. The total mass of these components roughly equals that expected for the wing alone of a traditional design. The compact structural arrangement of the blended body is likely to give substantial savings in structural mass so the result above can be accepted until more detailed analysis can be attempted. As mentioned in the initial mass estimation (8.6.1), the engine mass on our aircraft is unusually large. This can be explained by the need for the supercruise/dash speed and the high altitude performance. Extra allowance has been given to the avionics and cockpit systems mass to account for the sophisticated nature of the aircraft operation. Anticipating a lower MTOM from this analysis, the fuel mass has been reduced from the earlier value but still gives about a 48 per cent fuel mass ratio. The overall result shows the new design MTOM to be substantially lower than previously anticipated. This should prompt a review of the aircraft geometry and a further mass iteration. Without the confidence of the detailed aerodynamic and performance analysis this might be presumptuous, so we will continue the design process without altering our revised baseline details and accepting the above mass values.

8.8.3 Aircraft balance

Although the aircraft control limits have not yet been determined, it is still worthwhile to check if the aircraft configuration provides a sensible location for the centre of gravity excursions. Using the aircraft GA, Table 8.5 can be extended by adding the

Table 8.5

Component	lb	kg	% MTO	Arm (m)
Wing	6 157	2 800	5.4	18.0
Control surfaces	832	378	0.7	23.5
Body	5 210	2 368	4.6	16.5
Main gear	2 835	1 289		19.7
Nose gear	868	394	3.2 (u/c)	6.7
Intakes	2 633	1 197	2.3	13.5
$\sum$ STRUCTURE	18 535	8 406	16.2	
Dressed engine	15 600	7 091	13.7	
Installation	769	349	0.7	
Engine system	1 054	478	0.9	
$\sum$ PROPULSION	17 423	8 728	15.3	18.8
Fuel system and tanks	1 723	783	1.5	19.5
Aircraft systems	1 546	701	1.4	14.0
Avionics	2 370	1 077	2.1	9.5
Cockpit systems	1 440	653	1.3	10.0
Weapon systems	1 500	682	1.3	17.0
$\sum$ FIXED EQUIP.	8 579	3 891	7.6	
$\sum\sum$ EMPTY	44 537	21 025	39.1	
Crew and op items	1 100	500	1.0	10.0
Weapons	13 448	6 113	11.8	17.5
$\sum\sum$ ZERO FUEL	59 082	27 623	51.9	
Fuel	55 000	25 000	48.1	18.0 (central) 20.5 (wing)
Max take-off	114 082	51 739	100.0	

estimated positions of the component masses. With the aid of a spreadsheet, various combinations of aircraft loading can be assessed. The main results are shown on the excursion graph (Figure 8.15). Values of around 50 per cent MAC are acceptable for a supersonic delta wing and should prove feasible. The centre of gravity range for the flight cases is seen to be only about 6 per cent MAC (0.83 m or 2.7 ft). For the landing gear layout, the centre of gravity range (44 to 51 per cent MAC) should present no configurational difficulties.

8.8.4 Aerodynamic analysis

At this stage in the development of the project, the aerodynamic analysis must focus on a more accurate estimate of drag, an assessment of lift and lift-curve slope, and the determination of the aerodynamic centre of the wing. As the design matures it will be possible to refine these estimates using more sophisticated computational panel methods. As these require detailed geometric definitions of the full aircraft profile, which at this time is not fixed, it would be presumptuous to start such work now.

Drag

In the earlier part of the project, when the geometry of the aircraft was unknown, estimates of the aircraft subsonic drag were based on an equivalent skin friction coefficient.

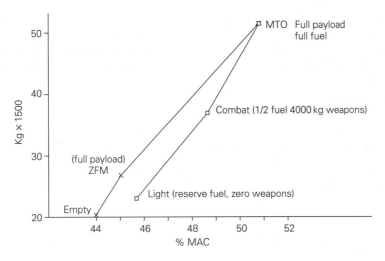

Fig. 8.15 Aircraft centre of gravity excursion plot

This is a crude method which is open to substantial inaccuracies as it cannot take into account any subtleties in the aircraft configuration. With more detail available about the aircraft layout, it is now possible to use the component build-up method to predict aircraft parasitic drag. Most design textbooks outline this process which combines the flat plate skin friction drag coefficient, a form factor to account for viscous effects, and an interference factor. Each component part of the aircraft is assessed separately, normalised to the aircraft reference area, and then summed to give the principal part of the zero-lift drag. Additional drag producing items (e.g. landing gear, flaps, external stores and fuel tanks, aerials and sensors, etc.) are analysed and their drag included, if appropriate to the flight case under investigation.

A word of warning and some advice

Aerodynamic analysis frequently involves the use of coefficients. These are convenient as it allows aerodynamicists to non-dimensionalise their parameters. For example, drag (which is a force) is divided by dynamic pressure ($q = 0.5 \times \rho \times V^2$) and by a reference area (e.g. $S =$ gross wing area) to give the drag coefficient (C_D). For a particular flight case, the value of a drag coefficient is useless without the knowledge of the accompanying reference area. As both of these numbers are always necessary, it is often better to quote drag in terms of drag area ($C_D \times S$). Drag area retains the definition of measurement units as it will need to be quoted in square feet, or square metres. This definition therefore allows a visualisation of the magnitude of the drag of components.

Providing that a consistent set of units is used, the aerodynamic coefficient will be the same value in each measuring system. Moving between the use of SI and 'British' units may cause errors unless care is taken with regard to the definition of the reference area units when calculating drag area.

For our blended-body layout, without the need for extra control surfaces, the component build-up involves only the wing surfaces and some allowance for the central body profile. Using a spreadsheet method, it is possible to determine parasitic drag for various combinations of altitude and speed. The following constants were used in the evaluation:

- ISA conditions
- Reference area = 133 sq. m (1430 sq. ft)
- Wing surfaces wetted area = 268 sq. m (2929 sq. ft)
- Body wetted area = 160 sq. m (1720 sq. ft)
- Wing reference length (MAC) = 13.8 m (45.2 ft)
- Body reference length (overall length) = 25.5 m (274 ft)
- Wing laminar flow = 10 per cent
- Body laminar flow = 1 per cent
- Interference factor = 1.0

As an example the values used to determine the drag coefficients (in SI units, for sea level and M0.2) are shown below:

- Reynolds number ($\times 10^6$) = 64.3 wing, 118.8 body
- Form factor = 1.0 wing, 1.07 body
- Skin friction coefficient (turbulent) = 0.00245 wing, 0.00225 body
- Skin friction coefficient (laminar) = 0.00165 wing, 0.00122 body
- Zero-lift drag coefficient = 0.00448 wing, 0.00263 body

Extra (miscellaneous) drag (base, up-sweep, leaks, protuberances drag) is assessed as $\Delta C_{do} = 0.00061$, and landing gear drag (when appropriate) as $\Delta C_{do} = 0.00376$ (including interference). As we shall see in the lift section the vortex lift increase on this configuration is sufficient to avoid the inclusion of flaps and associated drag and system complications. Adding the extra drag to that for the wing and body above gives:

$$C_{DO} = (0.00448 + 0.00263 + 0.00061) = 0.00772 \text{ (clean), and } 0.01148 \text{ (with u/c down)}$$

The results for other flight cases are shown in Figure 8.16.

Using methods described in reference 4, the critical Mach number for our wing is calculated as M0.849. Just prior to and beyond this speed, supersonic flow over the aircraft creates a substantial increase in drag (wave drag).

As described earlier (section 8.6.2) reducing wave drag is a vital objective in this design. It is important to make changes to the aircraft layout to bring this about. The revised GA, even with the smaller wing, was analysed and found to generate an unacceptable wave drag. To reduce the drag the wing thickness was reduced to 6 per cent. This was acceptable as the fuel volume was found previously to be more than adequate. The Sears–Haack volume distribution graph for the thinner wing design and with an extension to the nozzle length is shown in Figure 8.17 together with the ideal Sears–Haack distribution.

The cross-sectional areas shown in Figure 8.17 have been calculated around a Mach cone angle of about 38°. This corresponds to the cruise speed of M1.6. To improve the shape of the distribution some area ruling can be applied to the fuselage profile around the forward wing junction. The effect of such a modification to the shape of the aircraft is shown in Figure 8.18. The resulting area distribution can be seen in Figure 8.19.

The advantage of the blended-body configuration together with area ruling has created a volume distribution that is as close to the ideal 'Sears–Haack' profile as can be expected. It should be possible to have confidence in using an efficiency factor E_{WD} of

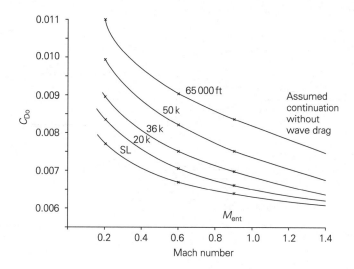

Fig. 8.16 Subsonic zero-lift drag coefficient

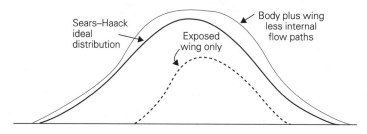

Fig. 8.17 Revised Sears–Haack distribution

1.4 for our aircraft. The wave drag area is calculated using the formula below, with the inputs shown:

$$D/q = E_{WD}(1 - 0.386(M - 1.2))^{0.57}(1 - (\pi \times LE\ sweep^{0.77}/100))$$
$$\times (4.5\pi (A_{max}/length)^2)$$
$$C_{Dwave} = (D/q)/S$$

where $E_{WD} = 1.4$
LE sweep $= 70°$
$A_{max} = 8.6$ sq. m (92.4 sq. ft)
length $= 25.5$ m (83.6 ft)

Giving, for $S = 133$ sq. m (1430 sq. ft):

$$C_{Dwave} = 0.01691\ at\ M = 1.2, \quad = 0.01646\ at\ M = 1.4$$
$$= 0.01624\ at\ M = 1.6, \quad = 0.01607\ at\ M = 1.8$$

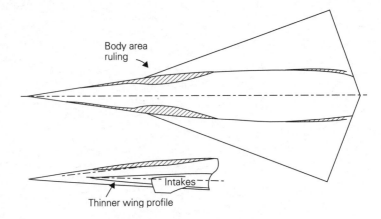

Fig. 8.18 Area ruling

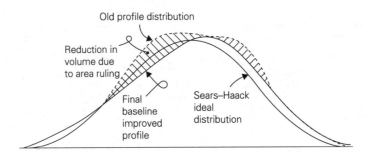

Fig. 8.19 Final Sears–Haack distribution

Adding these to the zero-lift values and sketching the transonic region allow us to draw the zero-lift drag versus speed graph, Figure 8.20.

This shows that at the initial dash condition (M1.6, 50 000 ft), the total aircraft $C_{Do} = 0.0205$. The wave drag contributes nearly 80 per cent of the drag. This illustrates that the significance of reducing wave drag is obvious. It can be seen from the equation for wave drag that it is inversely proportional to aircraft length squared. As we have anticipated a potential problem area in the flow conditions at the rear of the aircraft, a future modification may be required to extend the nozzle region (as shown in Figure 8.21).

This change, together with a small stretch to the aircraft nose, could increase the length from 25.5 m (85 ft) to 27.5 m (90 ft). This increase to the aircraft length would reduce wave drag by 14 per cent. A very small penalty would be incurred in the parasitic drag but the overall effect would be to lower the zero-lift drag coefficient to 0.0183. Although these changes will not be taken into account at this time, the calculations show how sensitive the drag estimate is to aircraft configuration.

Lift and lift-curve slope

To determine the drag due to lift requires the estimation of the induced drag factor (k_1). A vortex-lattice method is the most appropriate way to estimate the induced drag factor but this is too time consuming and involved to be considered at this time.

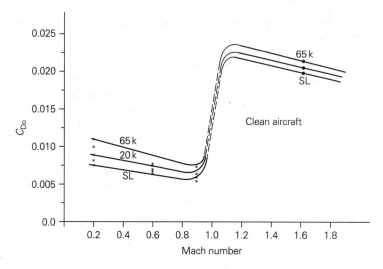

Fig. 8.20 Zero-lift drag versus Mach number

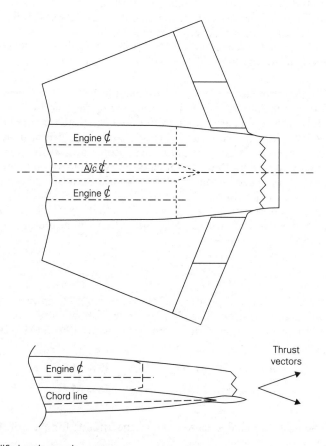

Fig. 8.21 Modified engine nozzle geometry

Simple methods, however, may lead to substantial inaccuracy. To add confidence to the estimation two different methods will be tried.

The Oswald span efficiency method, with $A = 1.27$ and LE sweep $= 70°$ gives a value of $e = 0.493$. Using the formulae for $k_1 (=1/\pi Ae)$ gives $k_1 = 0.508$. This value seems high when compared to conventional trapezoidal, long span wings but for our low aspect ratio, delta planform, the spanwise air loading is far from optimum. McCormick[10] and other aerodynamic textbooks provide methods of predicting (k_1) that are specific to supersonic delta wings. As the planform will produce zero leading edge suction $k_1 = (1/C_{L\alpha})$, where $C_{L\alpha}$ is the wing lift curve slope.

For our wing, the two-dimensional lift curve slope $= 4/(M^2 - 1)^{0.5}$
For LE sweep of $70°$ at M1.6, the method in reference 10 determines the finite wing
$C_{L\alpha} = (\mathbf{0.6}C_{L\alpha 2D}) = 0.6 \times 4/(1.6^2 - 1)^{0.5} = 1.92$
This gives $(k_1) = 0.521$.

There are several other methods that could be used to determine (k_1) but as these give approximately the same value an average will be used (i.e. 0.514).

The combination of zero-lift and induced drag provide the aircraft drag equation.

At the initial dash phase, $C_D = 0.0205 + 0.514 C_L^2$
With the aircraft at 80 per cent of the take-off mass, $C_L = 0.146$
This gives the aircraft drag coefficient $= 0.03146$
Therefore drag $= 0.5 \times 0.1864 \times 472^2 \times 133 \times 0.03146 = 86.88 \,\text{kN}$ (19 531 lb)
The aircraft lift/drag ratio $= 0.146/0.03146 = 4.64$

McCormick[10] also provides a method of predicting the lift characteristics for delta wings. The lift from the delta planform is augmented by the leading edge vortex flows (called vortex lift). The results are shown in Figure 8.22. Note that at higher angles of attack the lift-curve slope increases and this will reduce the lift-induced drag factor. We will assume an average value of 0.15 for the subsonic drag evaluations.

Raymer[4] presents some data that recommends a maximum angle of attack of about $35°$. To avoid flying too close to this condition we will limit our design to $30°$ which gives a maximum lift coefficient of 1.4. At these high angles of attack the aircraft drag is roughly proportional to $(\sin \alpha)$ times the lift. This high drag will be useful to offset the horizontal component from the vectored thrust that will be required to control the high pitch attitude.

Centre of lift

Nicolai[12] provides data on the position of the aerodynamic centre for a symmetric profile, delta wing configuration. The chart gives the following results relative to the distance behind the leading edge of the mean aerodynamic chord (MAC):

$$M0.2 = 35\% \text{ MAC}, \quad M0.8 = 39\% \text{ MAC},$$
$$M1.0 = 50\% \text{ MAC}, \quad M1.6 = 48\% \text{ MAC}$$

This data will be useful to enable us to balance the aircraft for each of the flight and loading (weight) cases.

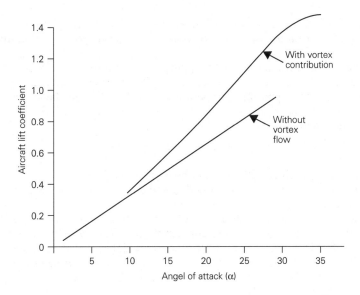

Fig. 8.22 Aircraft lift coefficient versus angle of attack

8.8.5 Propulsion

Before aircraft performance can be estimated it is essential to have data on the available thrust and fuel flow from the engine. In this design project, the AIAA provided some generalised performance data for a low bypass supersonic cruise turbofan engine. The quoted static sea-level thrust is 26 356 lb per engine.

Our constraint analysis indicated a $(T/W)_{SSL}$ of 0.58. For our predicted MTOM (51 739 kg: 114 082 lb) this relates to two 33 080 lb thrust engines. We therefore need to scale the given data by a factor of 1.25. This represents about the maximum feasible limit on scaling engine data. As the engine is a current design and as we are scaling it into a larger engine, it should be possible to envisage a new engine that is more fuel efficient than this design. Mattingly[11] provides programs that allow the design of a new engine. This method could be used to optimise an engine specifically suited to our specification. Although this has not been done in the present study, this type of aircraft-engine design integration can make an interesting team project as it involves design compromises that need to be made between airframe and engine requirements.

Until the available engine data has been assessed and some aircraft mission analysis conducted, it is not appropriate to include any factors to account for a new engine configuration. The data supplied by AIAA although corrected for installation effects (including intake and nozzle flow losses and off-takes) was in the generalised (non-dimensioned) format typical of data from engine manufacturers. Both the net thrust and the fuel flow values were divided by the pressure ratio (δ_t) which relates the air pressure at the engine fan face to the ambient air pressure at sea level. This is calculated as the product of the local total, or stagnation, pressure ratio (P_t/P) and the ambient, far-field ratio (P/P_{SL}). Both of these are functions of air properties as shown below:

$$(P_t/P) = 1 + ((\gamma - 1)M^2/2)^n$$

where $M =$ flight Mach number and $n = (\gamma/(\gamma - 1))$.

For air at ISA conditions $\gamma = 1.4$, giving $[(\gamma - 1)/2] = 0.2$ and thereby, $n = 3.5$. The variation of (P/P_{SL}) (commonly denoted as (δ)) with altitude can be found in most ISA tables.

The value of fuel flow provided by engine manufacturers is also divided by total temperature ratio (θ_t) raised to the power of 0.6. Where (θ_t) is calculated as the product of the local total, or stagnation, temperature ratio (T_t/T) and the ambient, far-field ratio (T/T_{SL}).

These are defined below:

$$(T_t/T) = 1 + ((\gamma - 1)M^2/2)$$

where $M =$ flight Mach number and as above $[(\gamma - 1)/2] = 0.2$.

The variation of (T/T_{SL}) (commonly denoted as (θ)) with altitude can be found in most ISA tables.

Using the functions above with the AIAA engine data it is possible to determine the installed net thrust and specific fuel consumption ($=$ fuel flow/thrust) of the engine against aircraft Mach number and altitude. This data, scaled to a T_{SSL} thrust of 33 080 lb, is shown cross-plotted in Figures 8.23a to 8.23c. These graphs are drawn in 'British' units as this is how the original data was given (note: 1 lb = 4.448 N and sfc values are the same in N/N.hr).

The AIAA engine data is based on an axi-symmetric, translating, centrebody intake design with an ejector nozzle. Although the nozzle design on our aircraft is likely to be more complex, it is not expected to be any less efficient. Our intake should be designed to provide lower losses at the operating conditions. This will involve the use of a moveable ramp and tip at the front of the intake duct, to better match the low- and high-speed requirements. This may provide up to about 10 per cent improvement but with the mechanical complexity it will be less stealthy. Will the propulsion advantage outweigh the disadvantage in stealth (a classic design decision)?

8.9   Performance estimations

With the detailed analysis of the aircraft mass, aerodynamic coefficients and engine characteristics, it is now possible to assess the aircraft performance in all flight conditions. This involves analysis of three different operational modes:

1. Manoeuvring performance
2. Mission analysis
3. Field performance

8.9.1   Manoeuvre performance

(Note: more detailed explanation of the methods used in this section can be found in design textbooks, e.g. reference 4 or 7. In order to avoid confusion on the various graphs in this section they have been drawn in 'British' units only.)

Assessment of the manoeuvring capabilities of military fighting aircraft is an essential part of the performance analysis. The methods used for this work are based on the calculation of the specific excess power (SEP) available in various flight conditions (speed, height, aircraft weight and load factor). SEP has the units of rate of climb (ft/s or m/s) as defined in Manoeuvring below.

From the previous sections on propulsion, aerodynamics and mass we have sufficient information to calculate SEP at any point in the aircraft flight envelope.

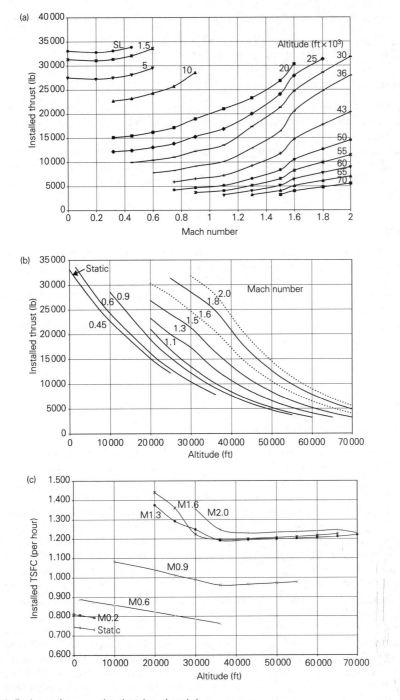

Fig. 8.23 Engine performance data (graphs a, b and c)

The original problem definition (section 8.3) defined four manoeuvrability criteria:

(a) SEP (1g) military thrust (dry), 1.6 M at 50 000 ft = 0 ft/s.
(b) SEP (1g) maximum thrust (wet), 1.6 M at 50 000 ft = 200 ft/s.
(c) SEP (2g) maximum thrust (wet), 1.6 M at 50 000 ft = 0 ft/s.
(d) Maximum instantaneous turn rate, 0.9 M at 15 000 ft = 8.0°/s.

Each of these criteria must be assessed at the aircraft manoeuvre weight (defined as 50 per cent internal fuel with two AIM-120 and four 2000 lb JDAM) = 81 797 lb (37 100 kg).

The manoeuvring and turning performance estimates are based on the evaluation of the available specific excess power. However, the flight condition to be considered in manoeuvre and turning cases are different so the analysis will be done in two separate sections.

Manoeuvring

The best way to approach the manoeuvring analysis is by generating data matrices of altitude and aircraft speed. These provide the values for aircraft drag and engine thrust at the aircraft weight and load factor to be considered. Computer programs or spreadsheet applications are the best way to perform the calculations repeatedly for different flight cases.

The specific excess power (SEP) is calculated at each point (height and speed) of the data matrix by the formula:

$$SEP = [(T - D)/W]V$$

T = thrust
D = drag
W = aircraft weight
V = aircraft forward speed

For the aircraft at the manoeuvre weight with 1g loading and dry thrust, the SEP values are shown plotted in two different formats in Figures 8.24a and 8.24b.

These data maps show that at M1.6 and 50 000 ft, the aircraft exceeds the zero SEP requirement (see (a) above). Figure 8.24b shows how the introduction of wave drag affects the SEP. Similarly, a set of curves can be drawn for the wet (afterburning) thrust cases for 1g and 2g loading. These are not shown in full but the SEP versus height curves for an aircraft speed of M1.6 have been reproduced on Figure 8.25. This shows that the zero and 200 ft/s SEP requirements ((b) and (c) above) have been exceeded. This, more detailed analysis, seems to have shown that the original problem with the 'wet/200 fps' requirement is now overcome. This is a good example of not rushing to make changes too early in the design process because the predicted SEP is very sensitive to changes in the drag and thrust values.

The process of evaluating SEP developed for the manoeuvre assessments can also be used to specify the optimum climb profiles for the aircraft. The most critical case is the initial climb to the cruise height (assumed to be 50 000 ft) following take-off at maximum weight. Using an assumed average weight in the climb of 0.9 MTOW and a load factor of one, SEP maps for dry and wet thrusts can be drawn. These can be cross-plotted to provide SEP contours as shown on Figure 8.26.

It is possible to draw lines of constant energy height (i.e. potential and kinetic energy = altitude + aircraft speed squared divided by 2g) onto this graph. The combination of the energy height and SEP contours are used to identify the quickest time

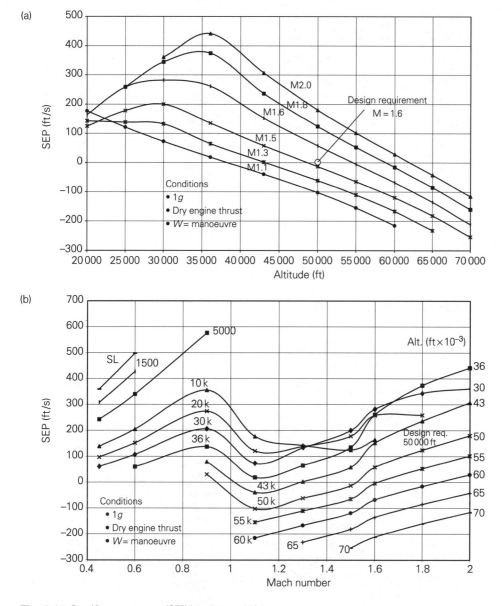

Fig. 8.24 Specific excess power (SEP) (graphs a and b)

to altitude. This is achieved by following a line drawn through the 1*g* SEP lines, perpendicular to the energy height contours. This is shown on Figure 8.26 as line A. The problem with this profile is that, as the aircraft is relatively underpowered in the dry thrust condition, the line extends past the zero SEP contour (i.e. obviously an unfeasible criteria). This difficulty arises due to the penalty imposed on the SEP by the wave drag at transonic speeds. The classical way of overcoming this problem is to perform a zoom-climb flight profile. In this, the pilot climbs the aircraft until the climb rate

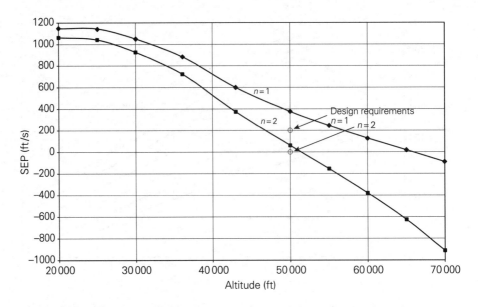

Fig. 8.25 SEP for M1.6 versus altitude

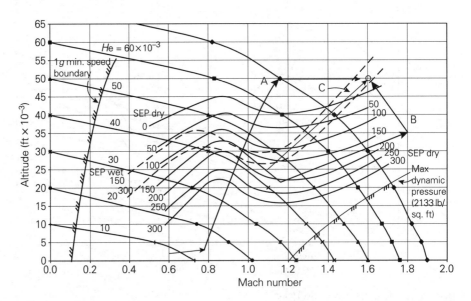

Fig. 8.26 Energy height plot

deteriorates then dives (through the initial sound barrier) and then at an increased speed and high-energy returns to the climb condition. The dive profile follows the constant energy height contour. With dry thrust, a possible zoom-climb profile could follow that shown as line B. With afterburning the climb could be quicker if profile C was used. Profile B is shown to overshoot the final cruise speed to make the final section perpendicular to the SEP lines (i.e. transferring kinetic energy for potential energy).

Detailed evaluation of the climb performance would involve a step-by-step calculation following the height and speed profile defined in Figure 8.26. This has not been done in this study.

Turning

Turning capability is relatively easy to assess providing that a database of aircraft drag and engine thrust against altitude and aircraft speed is available. The turn diagram is drawn for a specified aircraft weight and altitude ($W_{man.}$ and 15 000 ft in our case). As shown below, several significant manoeuvring parameters can be determined from the turn diagram. This makes it a useful device for comparing the effectiveness of different design configurations in a quantifiable way. For example, in trade-off studies, different variants of a baseline design can be assessed. Also, using this diagram at the conclusion of the project, the new aircraft can be directly compared to known competitor aircraft or other threats.

The analysis starts by drawing the generalised turn diagram. This is a graph of turn rate against aircraft speed. The formula relating these two parameters is:

$$\text{Turn rate } (^\circ/\text{s}) = 57.3 \left[g(n^2 - 1)^{0.5}/V \right]$$

where g = gravitational acceleration (32.2 ft/s^2, or 9.81 m/s^2)
$\quad\quad\quad n$ = manoeuvre load factor = (L/W)
$\quad\quad\quad V$ = aircraft speed

Note, the turn rate formula is not specific to a particular aircraft design.

From the formula above a series of curves can be drawn for each value of load factor (n), as shown in Figure 8.27.

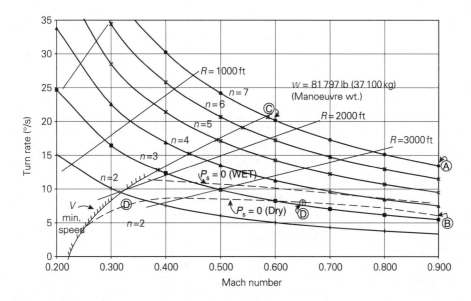

Fig. 8.27 Turn performance graph

As the radius of the turn equals the turn rate divided by the aircraft speed, it is possible to construct a set of straight lines on the graph that represent specific turn radii.

The turn diagram is made specific to a particular aircraft by introducing three boundaries:

1. Maximum positive structural normal acceleration factor (n_{max}). For our design, this is set in the design requirements at $+7g$.
2. Maximum structural dynamic pressure (q_{max}). The speed at which this limit is reached at various altitudes is determined from the equation: $q = 0.5\rho V^2$. For the specified requirement of 2133 lb/sq. ft the calculated values are shown in Table 8.6.
3. The aircraft minimum speed boundary at the flight condition (e.g. $W_{man.}$ and 15 000 ft). This can be determined for each load factor (n) from:

$$V_{min} = Wn/(0.5\rho S C_{Lmax})$$

Assuming that for our aircraft $C_{Lmax} = 1.4$ (limited by angle of attack) (see Table 8.7).

The maximum g and minimum speed boundaries are plotted on Figure 8.27 but the dynamic pressure boundary falls outside the scope of the graph. All that is now needed are the contours of the zero SEP (0 ft/s) for the dry and wet engine thrust.

This is done by plotting the full SEP contours (as shown on Figures 8.28a and b) and transferring the (0 ft/s) intersection speeds onto the appropriate (n) curves on Figure 8.27. Several points on the turn diagram are of interest:

(A) The maximum instantaneous turn rate for M0.9 is limited by the n_{max} line at a value of 13.4°/s (the associated turn radius is 4069 ft). The turn rate is in excess of the requirement of 8.0°/s.

Table 8.6

Altitude (ft)	V (ft/s)	Mach no.
SL	1340	1.2
5 000	1443	1.29
10 000	1559	1.40
20 000	1835	1.64
30 000	2190	1.96

Table 8.7

n	V_{min} (ft/s)	Mach no.
1	234	0.221
2	331	0.313
3	405	0.383
4	467	0.442
5	523	0.494
6	573	0.524
7	618	0.585

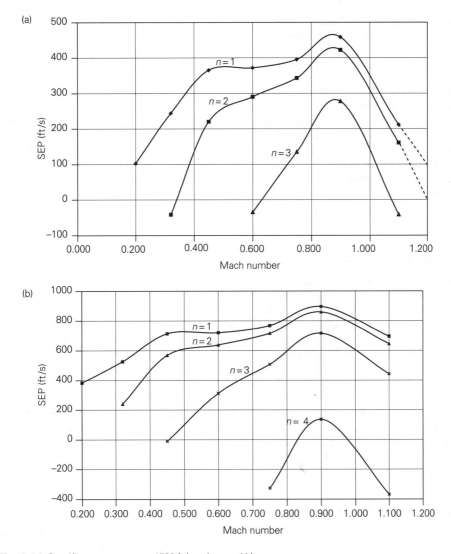

Fig. 8.28 Specific excess power at 1500 ft (graphs a and b)

(B) The sustained turn rate (dry) at M0.9 is limited by the zero SEP boundary at 5.4°/s. This gives a radius of turn of 10 097 ft. The wet thrust, zero SEP boundary gives a turn rate of 8°/s (6815 ft radius).

(C) The intersection of the minimum speed and n_{max} boundaries defines the corner point. This gives the highest (instantaneous) turn rate for the aircraft* of 21.0°/s at a corner speed of 365.7 kts. The turn radius is only 1686 ft which explains why fighter pilots attempt to get to the corner point in dog fights!

(D) The peak value on the zero SEP curves give the highest sustained turn rate. For the dry engine the rate is 8°/s at 407 kts (note how flat the sustained rate against speed is on this aircraft). The corresponding values for the wet thrust are 12°/s at M0.35 (4921 ft radius). This curve is also very flat.

(E) Although point D above gives the max. sustained turn rate, the tangent of a radial from the origin to the zero SEP curve gives the smallest sustained turn radius. In the dry thrust case, the values are 7.5°/s at 200 kts with a radius of 2582 ft. The wet thrust intersection coincides with the minimum speed boundary.

* The minimum speed boundary in this calculation assumes that the high angle of attack required to achieve the max. C_L value is controllable. It is likely that in our design, this may only be possible with a contribution from vectored thrust. The component of force from the thrust vectoring has not been included in the calculations because this would require more aircraft and propulsion details than are available at this stage. As we are relying on assistance from thrust vectoring for landing control, it may be possible to design the system to provide an integrated aerodynamic and propulsion control system in the turn manoeuvre without much additional complexity. We have easily met the instantaneous turn requirement so we will assume that the minimum speed boundary is achievable and not critical.

Recommendations

All of the specified manoeuvre and turn requirements have been easily met with the current design but a word of caution is appropriate. As the value of SEP at a particular flight condition is dependent on the difference between two relatively large numbers (thrust and drag), small percentage changes in either will result in large variations in SEP. At this stage in the design process, when only crude estimates have been made about aerodynamic and propulsion characteristics, this must concern us. For example, when considering flight at high manoeuvre load factors, the lift-induced drag becomes a significant component of drag. As this is dependent on the estimation of the induced drag factor, which is difficult to predict accurately for our planform, there could be uncertainty in the 'high g' performance. Also, the engine performance is affected by the detail layout and control mechanisms in the intake. A poor estimate of intake efficiency will significantly affect the net thrust available. For these reasons it is important to obtain better (higher confidence) estimates of these parameters in the next phase of the design process.

8.9.2 Mission analysis

There are four cruise stages to be assessed: outbound supercruise, outbound dash, return dash and return supercruise. Each of these stages is to be flown at M1.6. It is necessary to determine the optimum (based on minimum fuel burn) cruise height for each stage. From the engine data it is possible to extract the performance (net thrust and sfc) versus altitude for the cruise speed of M1.6, see Figure 8.29.

This provides the variation of sfc to be used to predict fuel burn by assuming that the thrust required equals the aircraft drag. From section 8.8.4, the aircraft drag polar at M1.6 is:

$$C_D = 0.0205 + 0.514 C_L^2$$

The dynamic pressure q ($=0.5\rho V^2$) in the stratosphere (above 36 089 ft) where the speed of sound is constant at 986 kts, can be determined for each height using the ISA formula for relative density multiplied by the sea-level value (0.002378 slugs/cu. ft) as shown below:

$$q = 0.5(0.2971e^{-x})0.002378(1.6 \times 986)^2$$

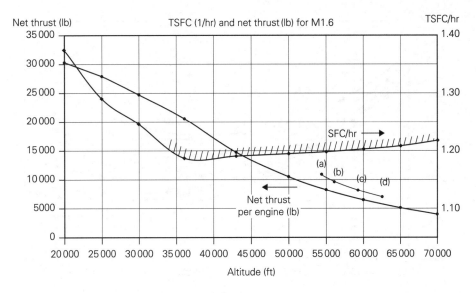

Fig. 8.29 Engine performance at M1.6

where $x = (H - 36089)/20806.7$
        $H = $ altitude (ft)

For our aircraft, the reference area is 1430 sq. ft (133 sq. m).

As we are unaware of the fuel burnt in each segment at this stage in the design process, it will be necessary to make some assumptions regarding the weight of the aircraft at the start of each stage, as shown below:

(a) Outbound supercruise = 0.9 MTOW
(b) Outbound dash = 0.8 MTOW
(c) Return dash = 0.7 MTOM
(d) Return supercruise = 0.6 MTOW

And if the return stages follow the release of munitions:

(e) Return dash = 0.7 MTOM – 8000 lb
(f) Return supercruise = 0.6 MTOW – 8000 lb

where, from section 8.8.2, MTOM = 114 082 lb (51 739 kg).

The aircraft weight defines the C_L which in turn defines the C_D from which the aircraft drag is calculated. This is multiplied by the engine sfc to obtain the fuel used per hour. This procedure is easily performed using a spreadsheet method.

The results are shown in Figure 8.30.

This clearly shows an optimum altitude for each stage. The optimum heights are cross-plotted against aircraft weight in Figure 8.31. The associated fuel consumption is also plotted on this graph.

At this point it is possible to use the fuel consumption results to determine the overall fuel burnt on the mission (assuming that the fuel consumption in each stage is the average between the start and end values). The time spent on each stage is the stage distance divided by the aircraft speed. As the speed is constant (933.5 kt), the

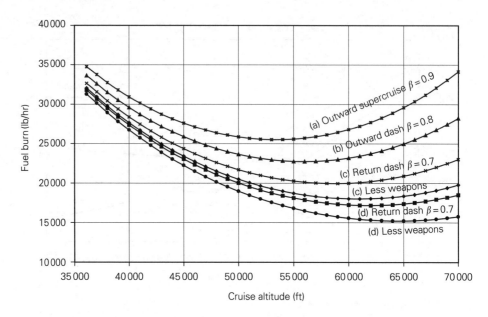

Fig. 8.30 Aircraft fuel burn versus altitude

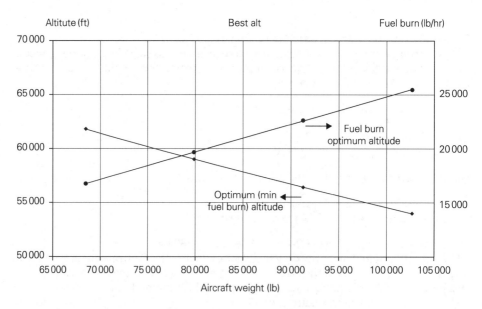

Fig. 8.31 Optimum cruise versus aircraft weight

supercruise stages of 1000 nm take 1.07 hr and the dash stages of 750 nm take 0.80 hr. The analysis is shown in Table 8.8.

This is much larger than originally estimated due to the lower lift to drag ratio (4.87 compared to 5.56 assumed earlier). This would mean that the aircraft MTOW

Table 8.8

	Fuel burn per hour				
Stage	Start	End	Average	Time	Fuel used (lb)
(a)	25 521	22 742	24 131	1.07	25 820
(b)	22 742	19 968	21 355	0.80	17 148
(c)	19 968	17 178	18 573	0.80	(14 914)
(d)	17 178	14 000*	15 589	1.07	(16 680)
(e)	18 012	15 219	16 620	0.80	13 346
(f)	15 219	13 000*	14 109	1.07	15 097
					71 411 lb

* Guessed values.

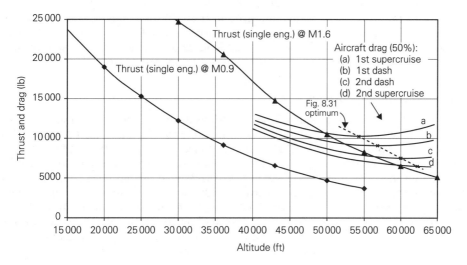

Fig. 8.32 Engine thrust at M0.9 versus altitude

must be increased. As the above mission assumed that only 8000 lb of weapon load would be dropped and about 13 000 lb was used in the mass statement to account for different missions, we could substitute 5000 lb into the fuel load. This would mean that the aircraft weight would need to be raised by 11 400 lb. However, some of this penalty could be set against potential improvements in engine design as mentioned in section 8.8.5. (For example, if the sfc could be reduced from 1.2 to 1.1 the fuel load would reduce by 6000 lb.)

The cruise analysis predicted the drag. This can be compared to the available thrust which has been extracted from the engine data and shown on Figure 8.32.

The analysis shows that the engine needs to be slightly more powerful to fly at optimum (minimum fuel burn) altitudes. However, as the aircraft L/D ratio is lower than expected no change should be made to the engine until a more accurate estimation of aircraft drag is available.

Recommendations

Assuming a 5 per cent reduction in engine sfc is possible from a new design, it is suggested that a fuel load of 68 000 lb (30 840 kg) should be provided in the next review of the aircraft mass. The review should also reduce the weapons load to 9400 lb (4263 kg).

8.9.3 Field performance

There are four operational issues to be investigated in this performance section:

1. Normal take-off distances to the point at which the aircraft achieves lift-off.
2. Balanced field length and the decision speed, for single engine operation.
3. Approach speed.
4. Landing distances from aircraft touchdown.

The calculations for each of the cases above require an analysis of the forces on the aircraft (weight, lift, drag, thrust and ground friction). Our previous estimations of mass, aerodynamic and propulsion characteristics are sufficient to use as input to the analysis. The prediction of take-off and landing distances requires a step-by-step calculation which can be done using a spreadsheet application method.

Normal take-off distances

The take-off distance is the sum of the ground distance (s_G) and the rotation distance (s_R). The ground distance is that travelled along the runway up to the point at which the rotation speed is reached. The rotation distance is a nominal distance to account for the rotation of the aircraft to achieve the initial lift-off manoeuvre, prior to the climb from the runway. Take-off speed (V_{TO}) is defined as that reached at the point that the aircraft leaves the runway. To avoid inadvertent instabilities in the initial climb phase, this speed must be faster than that related to the lift coefficient at rotation in the take-off configuration. The allowance above stall on conventional aircraft is typically set at 15 to 20 per cent. As we are well away from the max. C_L angle in our aircraft, we can either no factor is applied or that the factor is small:

$$\text{Take-off speed } (V_{TO}) = [W_{TO}/(0.5\rho S C_{Lto})]^{0.5}$$

where for our aircraft:

$W_{TO} = 51\,739\,\text{kg} \,(114\,082\,\text{lb})$
$\rho = \text{ISA sea level air density}$
$S = \text{reference area} = 130\,\text{sq. m} \,(1340\,\text{sq. ft})$
$C_{Lto} = \text{maximum lift coefficient}^*$

*As the aircraft does not have any flaps, this is taken as the lift coefficient at the maximum aircraft tail-down attitude of 15°. From Figure 8.22 this is seen to be 0.52.
Using the values above gives:

$$V_{TO} = 36.3\,\text{m/s}\,(118\,\text{ft/s}, 70\,\text{kt})$$

The numerical integration of the ground distance covered is calculated in steps of aircraft speed from brake release (zero speed) to V_{TO}. Although the aircraft would accelerate during the rotation phase, which would reduce the calculated distance, we will concede this small advantage (inaccuracy) to make the calculation simpler. The formula to estimate ground distance is available in most textbooks and repeated below:

$$s_G = 0.5 \int (1/a) d(V^2)$$

where

$a = $ aircraft acceleration $= [T - D - \mu(W - L)]/M$

$T = $ take-off thrust. As there is only a small variation of thrust during the take-off speed change, we will assume that the thrust remains constant at the average energy speed (i.e. $0.707 V_{TO}$). From engine data, this relates to a thrust of 32 950 lb per engine.

$D = $ drag in the take-off configuration. This is calculated from the zero-lift drag coefficient estimated as 0.01148 in section 8.5.7, and the induced drag coefficient at subsonic speeds is assumed to be 0.15. With the wing at a 4° angle of attack on the ground, the lift coefficient (from Figure 8.22) is 0.15. Hence the aircraft drag coefficient is $0.01148 + (0.15 \times 0.15^2) = 0.01486$.

$\mu = $ the coefficient of ground friction without braking. Design textbooks suggest this is 0.04 for dry runways and 0.02 for icy ones.

$(W - L) = $ the ground reaction force. Where W is the aircraft take-off weight (114 082 lb, 507.44 kN) and L is the lift generated with the lift coefficient of 0.15 mentioned above.

$M = $ aircraft mass $= W/g$.

The ground distance, calculated by the step-by-step integration, is 583 ft for the dry runway and 563 ft for the icy one. In this case, the ice reduces ground friction and is therefore not critical except that the aircraft may be less directionally stable (see further comments in the landing section below).

The time spent in the rotation phase is assumed to be 3 seconds. Hence, at the take-off speed of 118 m/s, the distance covered during rotation (s_R) = 354 ft.

For normal take-off, at maximum take-off weight, the max. total take-off distance is 937 ft. Even if the usual 1.15 factor to account for pilot and atmospheric variability is applied to this figure, it is still within the 8000 ft specified in the design brief. Therefore, the all-engines take-off distance is shown to be not critical.

Balanced field length

If an engine fails during the take-off run, the pilot must make a decision either to continue the take-off with only one engine working, or to abort and bring the aircraft to rest further down the runway. If the failure occurs late in the take-off run he would naturally continue and vice versa if it happened earlier. The aircraft speed at which it is better to continue the take-off is called the decision speed. The pilot will be aware of this speed from the aircraft flight manual before starting the take-off manoeuvre. To determine this speed, it is necessary to calculate separately (for each of the possible speeds at which an engine might fail) the distances required to effect an 'accelerate-go' and an 'accelerate-stop' manoeuvre.

For the accelerate-go case, the calculation includes 1 second of travel after the engine failure to recognise the fault and to take the necessary actions. After this period, the failed engine is assumed to be shut down and a 'normal' take-off performed with only the remaining engine producing thrust. During this time, no changes to the aircraft configuration are allowed. An increase in drag is applied to account for the drag from the failed engine and the trim forces from the control surfaces required to stabilise the asymmetric flight condition.

For the accelerate-stop case, again a 1 second delay is applied before any action is taken. After this time, a 3 second allowance is given to account for the application of brakes and the deployment of other drag devices (e.g. air brakes, reverse thrust, drag chutes). As our engine is relatively complex due to the reheat and vector thrust mechanisms, it is unlikely that thrust reversal will be available. For reasons of stealth and aerodynamic efficiency, the smooth wing profiles will not be disturbed by the installation of air brakes. Hence, the deployment of braking parachutes seems to be the preferred method of providing extra retardation at high speeds. Reference 12 provides a value for the drag area of parachutes. Using the figure of 1.4 times the canopy maximum area gives a ΔC_D of 0.076 for two 2 m (7 ft) diameter drag chutes.

The results of the calculation using the previous aircraft characteristics and the operational assumptions above for the accelerate-go and accelerate-stop cases, for dry and icy conditions, are shown in Figure 8.33.

The intersection of the lines for the go and stop cases define the decision speed and the balanced field length. These distances are again substantially less than the required 8000 ft specified. In fact, the high thrust to weight ratio of the aircraft means that, if necessary, the take-off could be achieved with only one engine operating from the start (this is not a common feature on most aircraft).

Calculations show that single-engine take-off can be achieved in 1688 ft for a dry runway and 1596 ft for the icy condition.

Approach speed

The approach speed is dependent on the value of the maximum C_L in the approach condition and the maximum aircraft landing weight. Using the high angle of attack

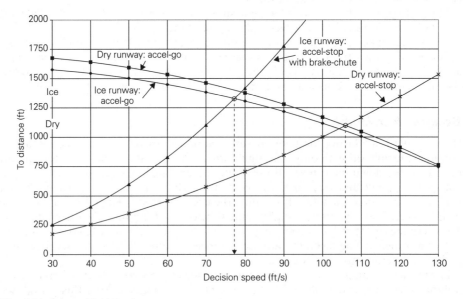

Fig. 8.33 Balanced field length curves

on approach as described in section 8.7 and the lift data in Figure 8.22 at an assumed angle of attack of 30°, provides a C_{Lland} of 1.4. The maximum landing weight is set by the operational requirements of the aircraft. If it is necessary to allow for a landing immediately following take-off (e.g. emergency due to system or engine failure) the landing weight could be up to 95 per cent of the take-off weight. If it was possible to burn or dump fuel before landing then a lower landing weight could be set. To avoid penalising the aircraft for the exceptional emergency case we will assume the more conventional landing weight of MTOW less 50 per cent of fuel. For our aircraft, this definition makes the landing weight:

$$W_{land} = 114\,082 - (0.5 \times 55\,000) = 86\,582\,\text{lb} \ (39\,266\,\text{kg})$$

For many conventional aircraft, the minimum approach speed is set at 1.3 times stall speed. As our aircraft must be fully automated for landing (due to the poor pilot visibility) and will have precision positioning systems we can assume this safety factor to be reduced to 1.2. In this landing case (as compared to the take-off), the aircraft is flying close to its maximum C_L so a factor is still appropriate.

Therefore:

$$V_{land} = 1.2[86\,582/(0.5 \times 0.002377 \times 1340 \times 1.4)]^{0.5}$$
$$= 236.5\,\text{ft/s} \ (72.1\,\text{m/s}, 140\,\text{kts})$$

This seems reasonable compared to estimates of the approach speeds for similar military aircraft (F-14 = 134, F-117 = 144, Su-33 = 194!, B-2 = 140, B-52 = 140 kts). However, the analysis for our aircraft was based on assumptions for the landing weight and C_{Lmax} for each aircraft which may be in error, so a sensitivity study was undertaken. The result is shown in Figure 8.34.

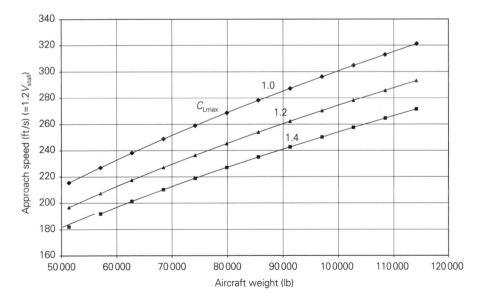

Fig. 8.34 Approach speed versus aircraft weight

Landing distance

Landing distance is computed in a similar method to that for take-off except that thrust is set to zero. To stabilise the aircraft on the ground and to apply maximum braking, a free-roll on touchdown of 3 seconds is assumed. In conventional landing procedures, the touch-down speed is lower than the approach speed due to the drag produced in the flare phase. In our design the high angle of attack on approach will be reduced prior to landing to avoid scraping the rear fuselage. This may suggest that the touchdown speed will be higher than the approach speed. However, to simplify the calculation we will assume that the touchdown speed is equal to the approach speed.

The detailed landing calculation shows that, at the landing weight assumed above, the unfactored distance is 2535 ft (773 m) on a dry runway. On an icy runway, the distance increases to 9273 ft (2828 m). This is beyond the available runway length of 8000 ft specified in the project brief. It will therefore be necessary to use brake chutes to reduce the distance. Braking parachutes are particularly useful devices as they are most effective at higher speeds when the aircraft brakes are less powerful (due to the unwanted lift reducing the ground reaction force). Using the two 7 ft diameter chutes described previously, the landing distance on an icy runway is reduced to 7047 ft (2150 m). This brings the distance within the available length. In fact the aircraft would be able to land at 95 per cent MTOW within the 8000 ft allowance. Figure 8.35 shows the variation of unfactored landing distance against landing weight.

Although the results above look acceptable, it must be remembered that the landing manoeuvre may not be as precise as we have assumed in the analysis. For example, the approach speed may be higher than expected or the aircraft may overshoot the runway threshold due to gust disturbance just prior to touchdown. To guard against such possibilities it is common practice to apply a factor to the calculated landing distance. Typically, this is set at 1.67. Applying this to the dry distance of 2535 ft and the icy distance of 7074 ft increases them to 4233 ft and 11 768 ft. The normal, dry runway landing is still acceptable but clearly the icy one is still much too long.

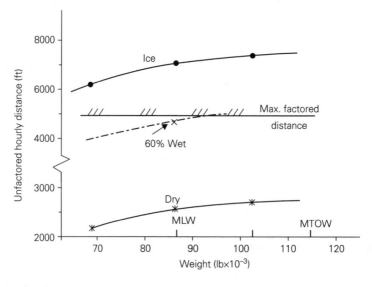

Fig. 8.35 Landing distance versus aircraft weight

As military airfields are fully serviced, it is not unreasonable to expect that in icy conditions the runway surface will be treated to dissolve the ice (as on highways). Recalculating the landing distances using the accepted runway friction coefficient for wet surfaces (0.3) over the last 60 per cent of the runway length, instead of that for ice (0.1), reduces the landing distance to 4715 ft (unfactored) and 7874 ft (factored). This is within the allowable runway length. Treating the runway to avoid ice contamination will also avoid potential directional instabilities and skidding problems.

8.10 Cost estimations

Estimating the costs of future aircraft has always been seen as an inexact science. Evidence from previous design programmes show that even the seasoned professionals in industry do not have a good track record at making such estimates. For students, and even faculty, in an academic environment it is impossible to predict the *absolute* costs associated with a new project. Too many of the factors that are needed are only available within a commercial organisation. Such factors relate to the accountancy practices used, the organisation of the company (or more likely the consortia of companies that are formed to share the design and manufacturing tasks), the interrelationship between government and industry, and many more non-technical issues.

For military projects, the need to incorporate modern and advanced technologies is paramount. The timescales involved in the development of such technologies often overlaps the aircraft development period. This leads to uncertainties in the costs incurred. For our project there are at least six technological areas (e.g. stealth, propulsion, aerodynamic design, structures and materials, and systems) which need to be matured before an exact cost can be assumed. Notwithstanding these difficulties, it is often financial parameters that are used to choose between different design options. It is therefore essential to be able to determine *relative* costs to create a framework for such decision making and to be able to compare our design with competitor aircraft.

Fortunately, historical data shows that many of the cost parameters are related to aircraft design variables (e.g. aircraft empty weight, installed engine thrust, number of engines, aircraft operational speed, and the overall system complexity). Other factors are related directly to manufacturing variables (e.g. labour rates, number of aircraft produced and the production rate). Due to the variability of the value of a currency with time, it is always essential to 'normalise' the quoted cost numbers to a specific date (year). This means that inflation rates for the currency must be applied to any data used. Cost estimates must always state the year to which they are indexed.

Several aircraft design textbooks provide details of cost estimation methods but in this study the method published by the Society of Allied Weight Engineers (SAWE)[13] is used. This paper describes fully all the details required to estimate the significant cost values at the preliminary design stage. It also provides a spreadsheet method and example. The method is based on regression of historical data from aircraft of specific types. As new designs will be more technically complex than older aircraft it is necessary to apply factors to account for the increase in costs associated with these new features. Our aircraft has many new technical features including new structural materials and construction processes, a sophisticated flight and weapon control system, vectoring engine nozzles, efficient high altitude and fast flight, and advanced stealth features. Each of the technical factors in the SAWE method will need to be set at high

Table 8.9

Phase	Development	Production
Engineering cost	5 738	4 678
Development support	1 174	
Flight testing	943	
Tooling labour	1 629	1 623
Manufacturing labour	2 369	12 030
Quality and security	313	1 564
Materials and equipment	265	6 615
Propulsion systems	173	4 336
Avionics	189	4 735
Total programme	12 792	35 585
Acquisition cost per aircraft ($M)		232.6
Recurring flyaway cost per aircraft ($M)		178.0
Recurring cost/lb empty weight ($)		3995.0

values to match these innovations. Details of the factors used in the analysis are shown below:

	Factor
1. Advanced technology features (ATF)	2.0
2. Flight test requirements to prove ATF	1.3
3. Application of advanced materials	1.5
4. Incorporation of stealth technologies	1.3
5. Cost burden of project security	1.3

Each of these factors is equal to, or higher than, the advanced fighter example used in the report.

Applying the method to our aircraft, with the factors above, and assuming a production run of 200 aircraft, gives the following cost breakdown ($M, FY2000) (see Table 8.9).

Clearly, the recurrent flyaway cost exceeds the $150M mentioned in the design brief. There are several strategies that can be used to reduce the cost to the specified target:

- To accept a reduction in the capability of the design. This is probably the worst of the options for the military to take. It is unlikely to be acceptable unless the operational requirements placed on the Defence Department by the government are altered.
- To reduce the number of aircraft to be produced to match the available budget. If all the overhead costs could be held proportional, this would mean that only 168 aircraft could be afforded.
- To produce more aircraft than is needed for the US military by supplying aircraft to friendly (NATO) countries. This may not be feasible for political and national security reasons. However, many modern military programmes (including the Eurofighter and the JSF aircraft) are produced by international consortia. To investigate the effect on costs of increasing the production volume, the cost method used above was applied to the production of 500 and 1000 aircraft. As the development overhead is

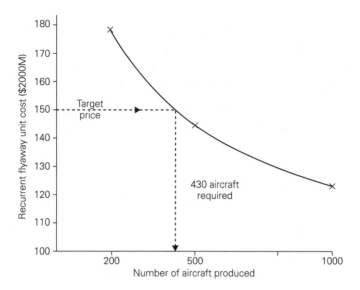

Fig. 8.36 Aircraft recurrent cost versus production run

shared by the increased number of aircraft produced the flyaway cost is substantially reduced, providing that additional costs due to the collaboration can be avoided. Figure 8.36 shows the results of the investigation. From this graph, it would be possible to reach the recurrent unit cost target of $150M if 430 or more aircraft are manufactured (and sold!).

In an attempt to judge the accuracy of the cost method, details of the F-22 aircraft were input and analysed. This showed that, at FY2000 prices, the aircraft would cost about $141M. Investigating published data from the US National Audit Office and other government reports suggests that the aircraft actual unit cost is about $94M. This suggests that the published figures have been misinterpreted, the costs may have been inaccurately extrapolated to FY2000, or that part of the development cost could have being transferred to a different accounting record. Alternatively, the method may simply overestimate the cost of the F-22 aircraft. The price does seem to be high relative to our aircraft which is larger and more capable than the F-22. This leaves the accuracy of the method under suspicion but does provide us with a 'ballpark' figure to use in subsequent trade-off studies. The value of weight saving ($/lb), as defined in reference 13 and shown above, reduces to 3221 for 500 and 2784 for 1000 aircraft. This type of data will be very useful in subsequent trade-off work as it links cost changes to aircraft weight.

Estimation of aircraft life cycle costs (LCC) for the aircraft are considered to be much too speculative at this stage in the design process, so this calculation has not been attempted.

8.11 Trade-off studies

As we now have developed all the necessary techniques to analyse the aircraft configuration, we can investigate if the aircraft characteristics are the best choice for our purposes. This is done by sequentially making small changes to the aircraft parameters and comparing the results to the baseline values. These investigations are called 'trade-off studies'. They may take different forms depending on the purpose of the investigation. For example, to determine the best choice of wing and thrust loading, to identify any constraint that is imposing a critical design penalty on the aircraft, to test the sensitivity of assumptions that have had to be made to complete the performance analysis, and to make a more informed selection of geometric and other characteristics. In some reports and textbooks, such investigations may be referred to as 'parametric studies' or 'sensitivity analyses'. Examples of such studies are given in references 4 and 14.

The list of possible trade-off studies that can be undertaken on a project is obviously large. The selection of which to choose is dependent on the type of aircraft and the purpose of the study. Here are some suggestions relating to our aircraft:

- To review the selection of aircraft wing loading and associated thrust loading. This choice was made previously in the constraint analysis using very crude assumptions.
- To understand, with more accurate analysis, the influence of each of the design constraints and to recommend changes to these if appropriate.
- To investigate the trade-offs between aircraft parameters (e.g. wing aspect ratio, thickness, sweepback, etc.) and aircraft weight or performance. These parameters were previously chosen to be similar to existing layouts. This type of trade-off will provide a more rational basis for the values selected and provide a more efficient configuration.
- Test the sensitivity of the assumptions made in the aerodynamic and propulsion analyses (e.g. drag and lift assessments, engine performance). These results will allow us to focus subsequent work on improving the estimation of those characteristics that are seen to be most critical to the design.
- To investigate the influence of known critical design drivers. For example, in our design the engine specific fuel consumption translates to the fuel mass and then to the aircraft performance. Making changes to the engine design to improve sfc will affect several other design parameters (e.g. drag and weight). There must be an optimum choice of engine configuration to minimise aircraft weight and cost.

As the aircraft system and weapon cost are fixed by the design specification, the main variables contributing to aircraft cost are the aircraft empty weight and engine size (thrust). The cost estimation has provided a value for the value of weight saving ($ per pound) and the price of engines. It is therefore possible to translate changes in aircraft weight and thrust directly to aircraft cost.

Many of the choices made in the trade-off studies require a definition of the objective (or goal). In some cases, this may be stated simple as 'minimum wing weight', 'minimum fuel used', 'minimum aircraft price'. Sometimes a combination of parameters is used (e.g. weight and size, or structure weight and fuel weight). The ability to use the cost trade-off value in such cases will be very useful.

The difficulty of using trade-off studies lies in the assumptions used in their analysis. It would be very time consuming to have to individually analyse the various combinations of configurations in the detail that has been used to study the baseline design in the previous sections of this chapter. Trade-off studies at this point in the design process do not make substantial changes in the basic aircraft layout. They concentrate on relatively small modifications (e.g. 5, 10, of 20 per cent variations), therefore some of the aircraft parameters may not change significantly in the pursuit of the overall answers. Recognising such parameters allows us to hold them constant, or make them change relative to some other variable, and thereby reduce the amount of work. (For example, the aircraft wetted area that is used in the drag calculation can be somehow related to changes in wing area.) Choosing the assumptions to make at the start of the trade-off studies is the most difficult part of the process.

As trade-off work involves the repeated calculation of similar types of analysis, it is appropriate to use some form of computer assistance. This may be in the form of specifically written computer programs or the use of spreadsheet application software. In this way, and by making suitable assumptions as mentioned above, small variations in aircraft parameters can be quickly assessed and graphs produced to illustrate the trends. The use of such methods must be tailored to the specific aircraft configuration and the type of study to be followed. Unless one is fully conversant in the use of commercial programs and aware of their limitations (i.e. their validity to the problem), it is unwise to simply 'turn the handle' to get results to specific types of study.

It is not possible within the limits of this chapter to perform any trade studies in sufficient detail. However, there are plenty of opportunities for students who have followed the development of the aircraft this far to continue with their own investigations. The question that is still unanswered in this chapter is 'what is the best (not optimum) configuration for this aircraft?' This leaves plenty of scope for coursework!

8.12   Design review

From the analysis above, we have shown that the aircraft meets all of the design requirements apart from the specified range. As the aircraft will be analysed in more depth with respect to aerodynamic (drag) and propulsion (sfc) characteristics in the following phases of the design process, it would be unwise to make any substantial changes to the configuration at this time. The suggestion to increase the aircraft length by extending the engine nozzles made previously will reduce wave drag and this may rectify the range deficiency.

It is now appropriate to redraw the aircraft general arrangement to include the minor alterations suggested in the previous design process. This drawing together with a more detailed internal arrangement drawing and an initial specification of the structural framework can be seen in Figures 8.37, 8.38 and 8.39.

At this stage in the design process, it is advisable to compile a detailed description of the aircraft so that the work that follows (often by different specialists) will have a common basis. The section below is typical of the detail that should be included in such a description.

8.12.1   Final baseline aircraft description

Aircraft description

Aircraft type: Two-seat, high altitude, supersonic, low-observable, deep
                   interdiction aircraft.
Design features: Mid-wing, diamond planform, blended body, tailless,
                   twin-engine layout. All weapons stored internally in a central
                   bomb bay below the engine and equipment compartments.
                   Side-by-side, high mounted, low-bypass engines with 2D
                   variable geometry, under-wing intakes positioned close to the

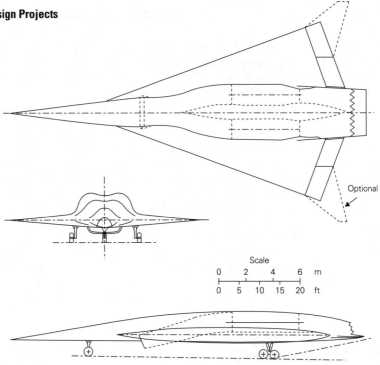

Fig. 8.37 Final baseline aircraft GA

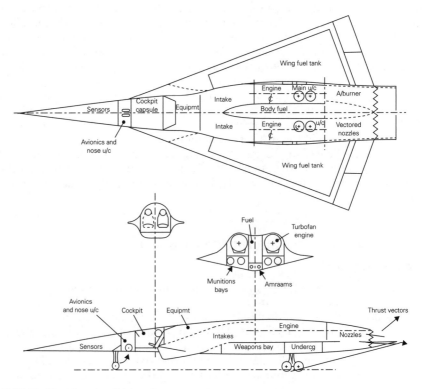

Fig. 8.38 Final baseline aircraft internal arrangement

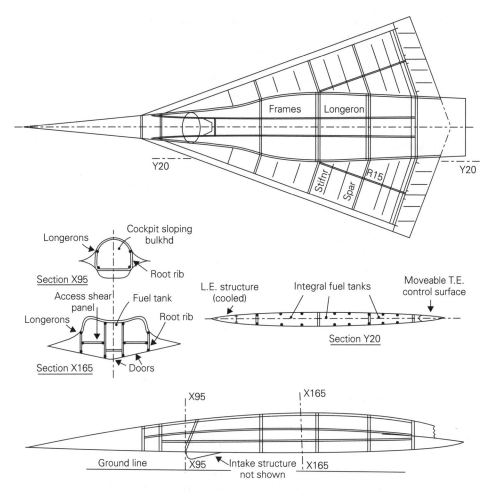

Fig. 8.39 Final baseline aircraft structural framework

wing leading edge. Afterburning and vectoring rectangular
nozzle positioned to the rear of the wing trailing edge.
Mid-fuselage, side-by-side, twin pilot cockpit with limited
external view. Access to the cockpit is through the forward bomb
bay bulkhead. Artificial pilot vision and automatic flight control
system. Cockpit capsule-escape system. Conventional tricycle
retractable landing gear.

Stealth features: Very low radar cross-sectional area, achieved by the blended
profile with aligned external geometry and structure, and the
application of radar absorbent materials and structure. Structure
cooling to reduce kinetic heating. Shielded and intercooled engine
exhaust flow. Polymer coatings to reduce infrared signature, and
sound-profiling to reduce the sonic boom.

Structure: Integrated wing and body internal and profiled structural
framework. Extensive use of composite structural materials
with RAM and RAS applied to reduce observable signature.
Design limits $+7/-3g$, $V_D = M2.0$ and max. dynamic
pressure $= 2133$ lb/sq. ft (equivalent to 800 kt at SL).

Weapons: Common racking for combination weapon loads as defined below:
(4) Mk-84 LDGP + AIM-120
(4) GBU-27 + AIM-120
(4) 2000 lb JDAM + AIM-120
(4) AGM-154 JSOW + AIM-120
(16) 250 lb small smart bomb

Aircraft data

Dimensions:			
	Overall length	87.0 ft	26.5 m
	Overall span	42.6 ft	13.0 m
	Overall span (option)	52.5 ft	16.0 m
	Overall height	11.5 ft	3.5 m
	Wing aspect ratio	1.27	
	Wing taper ratio	0	
	Wing LE sweep	70°	
	U/C wheelbase	42.3 ft	13.5 m
	U/C track	14.8 ft	4.5 m

Areas:			
	Wing planform (ref)	1430 ft^2	133 m^2
	Exposed wing	700 ft^2	65.0 m^2
	Total wetted	2472 ft^2	230 m^2
	Max. cross-section	91.5 ft^2	8.6 m^2
	Elevators	56.0 ft^2	5.2 m^2
	Ailerons (normal)	56.0 ft^2	5.2 m^2
	Ailerons (option)	134 ft^2	12.5 m^2

Weight (mass)			
	Max. TO (design)	114 082 lb	51 739 kg
	Empty	44 537 lb	21 025 kg
	Manoeuvre	81 797 lb	37 100 kg
	Landing (90% MTO)	102 674 lb	46 565 kg
	Fuel load	55 000 lb	25 000 kg
	Fuel (US gals)	10 300	
	Weapons (max.)	13 448 lb	6113 kg

Loadings:			
	Wing loading (max.)	80 lb/sq. ft	3815 N/sq. m
	Thrust/Weight (TO)	0.58	Dry
	Thrust/Weight (combat)	0.18	Dry at 50 000 ft

Engines (each):		
	Thrust (ssl, dry)	33 080 lb
	SFC dry (TO)	0.85
	SFC dry (cruise)	1.20
	Weight: mass	7800 lb
	Bypass ratio	0.6

Aerodynamics:	Supersonic cruise	$C_{Do} = 0.0205$
		$C_L = 0.146$
		$C_D = 0.0315$
		$L/D = 4.65$

Subsonic (clean) $C_{Do} = 0.0077$
    (TO and land) $C_{Do} = 0.0115$
    (approach) $C_L = 1.4$ (HAA)

Performance: *Mission*:
    Cruise speed   M1.6
    Cruise height   54 to 63 000 ft
    Range   3500 nm
    Manoeuvre (SEP at M1.6 at 50 000 ft):
    1g dry thrust   60 ft/s
    1g afterburning   370 ft/s
    2g afterburning   50 ft/s
    Turning (M0.9 at 15 000 ft):
    Instantaneous   13.4°/s (4069 ft radius)
    Sustained   5.4°/s (10 097 ft radius)
    Sustained (A/B)   8.0°/s (6815 ft radius)
    Max. instantaneous   21.0°/s at 366 kt (1686 ft radius)
    Field:
    TO run @ MTO   1009 ft dry (unfactored)
    Speed V_2 70 kt
    Balanced field   1150 ft normal, 2320 ice (unfactored)
    Approach speed   140 kt
    Landing roll (dry)   2530 ft (unfactored)
       (ice)   7000 ft (unfactored)
       (wet)   4700 ft (unfactored)
Recurrent flyaway unit cost (FY2000) $178 M for 200 production
          $150 M for 430 production
          $123 M for 1000 production

8.12.2 Future considerations

Although the aircraft layout appears feasible, there are a number of outstanding issues that must be resolved. The main concern relates to the directional control and stability of the configuration. There are several examples of this type of tailless aircraft flying to give confidence but these do not fly as fast and are not expected to approach and turn at such a high angle of attack. The design of the vectored thrust and the flight control systems are intrinsically integrated into such analysis. Other issues relate to the technologies required for stealth. These include the performance of new radar absorbent materials and structure, the dissipation of the sonic boom, infrared reduction of the nozzle area and of the kinetic heating of the structure. Finally, little has been done so far to define the systems integration on the aircraft. Many of the expected improvements in the aircraft capability are linked to the system design. Hence, there are plenty of opportunities for further individual studies.

Obviously, this aircraft is a very sophisticated weapon system that relies on the total integration of many diverse technology developments. Working in such a design environment, it is difficult to accurately determine the effect of the technologies on aircraft mass, supportability, efficiency, costs and timescales. These problems are common in advanced aircraft design and add to the interest and fascination of the work.

Before starting the subsequent detailed design stages, it is worth considering future developments for the aircraft. This may colour future decisions on the layout and capabilities of the aircraft. Most existing aircraft have not been limited to their initial design specification. They have been developed from the original concept by extending their capabilities (speed, manoeuvre, range, payload, weapons, etc.) or by adapting them for other roles. If such developments could be anticipated, the aircraft would be more versatile and easier to modify later in its operational life span. For example:

- What new weapons might the aircraft need to carry?
- Could an uninhabited version be envisaged?
- Would more self-defence be necessary in the future?
- Can future threats be anticipated?
- Will stealth features need to be reconsidered as offensive systems are improved?
- Could a shorter range version utilise the reduction in fuel load by increasing payload? And what would this mean in respect to weapon storage and delivery systems?
- Would flight refuelling be a useful extra feature? And how would this affect the aircraft payload range and stealth issues?

The application of a SWOT (strength, weakness, opportunity and threat) analysis of the aircraft and the operational environment in which it functions provide a more procedural method of arriving at the sort of questions listed above.

8.13 Study review

This study has provided an example of the design of an advanced-technology, military aircraft. It has demonstrated some of the methods and techniques needed to analyse modern high performance, stealthy configurations. The principal design driver for the project has been a combination of stealth technology and efficient supersonic aerodynamic performance. Balancing the demands of two such significant requirements together with combat effectiveness is common in the design of military aircraft. Layout considerations that provide the necessary stealth characteristics and the minimisation of wave drag have led to the unusual profiling of the aircraft. The analysis of the manoeuvring performance has provided a good example of the 'energy height' and 'specific excess power' methodologies. The assumptions made with respect to low observables, minimised wave drag and thrust vectoring in the project may be regarded as somewhat optimistic and unachievable in the 'real' world but they are intended to offer a forward vision for combat aircraft design. Although incomplete, the study provides a useful starting point for several continuation projects that would be suitable for under-post-graduate coursework.

The author acknowledges the work of previous students on this project[15,16] in completing their submissions to the AIAA design competition. Their groundwork in data collection and the development of analytical methods has assisted in the writing of this chapter.

References

1 www.aiaa.org.
2 www.aiaa.org/education/undergradaircraft.pdf.
3 Jenn, D., *Radar and Laser Cross Section Engineering*, AIAA Education Series, 1995, ISBN 1-56347-105-1.
4 Raymer, D. P., *Aircraft Design: A Conceptual Approach*, AIAA Education Series, 1999, ISBN 1-56347-281-0.
5 Brandt, S. A. *et al.*, *Introduction to Aeronautics: A Design Perspective*, AIAA Education Series, 1997, ISBN 1-56347-250-3.

6 *Aviation Week Source Book*, published annually in January.

7 Eshelby, M. E., *Aircraft Performance – Theory and Practice*, Butterworth-Heinemann and AIAA Education Series, 2000, ISBN 1-56347-250-3 and 1-56347-398-4.

8 *AIAA Aerospace Design Engineers Guide*, 1998, ISBN 1-56347-283-X 1.

9 ESTOL aircraft landing profile for the X31 demonstrator (www.aviationnow.com).

10 McCormick, B. W., *Aerodynamics, Aeronautics and Flight Mechanics*, Wiley and Sons, 1979, ISBN 0-471-03032-5.

11 Mattingly, J. D., *Aircraft Engine Design*, AIAA Education Series, 1987, ISBN 0-930403-23-1.

12 Nicolai, L. M., *Fundamentals of Aircraft Design*, METS Inc., San Jose, California 95120, USA, 1984.

13 Society of Allied Weight Engineers Inc., J. Wayne Burns, 'Aircraft cost estimation methodology and value of a pound derivation for preliminary design development applications', SAWE Paper No. 2228 Cat. No. 29, May 1997.

14 Jenkinson, Simpkin and Rhodes, *Civil Jet Aircraft Design*, AIAA Education Series and Butterworth-Heinemann, 1999, ISBN 1-56347-350-X and 0-340-74152-X.

15 Southampton University, UK, 'AIAA undergraduate team design competition – Group report', June 2002.

16 Loughborough University, UK, 'AIAA undergraduate team design competition – Group report', June 2002.

9

Project study: high-altitude, long-endurance (HALE) uninhabited aerial surveillance vehicle (UASV)

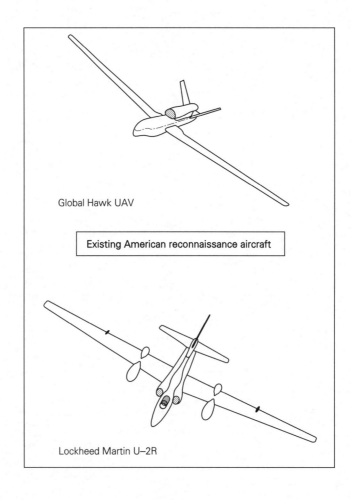

Global Hawk UAV

Existing American reconnaissance aircraft

Lockheed Martin U–2R

9.1   Introduction

This project study was stimulated by a technical paper[1] presented by European defence staff in 2000. The study has been used in this book to highlight the unique problems associated with operations at high altitudes and for long periods. A novel configuration has been investigated to allow comparison with conventional designs. The system and communication problems associated with remote and autonomous control of the uninhabited vehicle has not been addressed as this is outside the scope of the initial/conceptual aircraft design task. Such an investigation could form a suitable topic for system/electronic/electrical/avionics design courses.

9.2   Project brief

Aerial reconnaissance has always been an essential feature of military intelligence. The first use of aircraft in a military context was as artillery spotter planes at the start of World War I. At this time, airships were used for reconnaissance but they soon became too vulnerable to ground fire. High-altitude surveillance was perfected during the start of the Cold War. At this time, anti-aircraft munitions were unable to reach high flying aircraft. The incident in which a US pilot (Gary Powers) flying a U2 'spy plane' at 65 000 feet was shot down over Russia curtailed such operations over hostile territory. The exploitation of the pilot by his captives and the ensuing political and diplomatic consequences has given rise to the requirement for unmanned flights in dangerous missions. Surveillance has subsequently been more safely undertaken by sophisticated satellite systems.

Following the end of the Cold War, many national airforces have been deployed in international peacekeeping roles for the UN and other bodies. Part of such activities involves the monitoring of 'no fly' and demilitarised zones. This requires continuous (day and night), all-weather surveillance over large areas. Although, in such operations, there is only a small chance of a threat to the aircraft, the political consequences of dealing with unfriendly governments holding a pilot as hostage (like Gary Powers and Gulf War prisoners) sets a requirement for unmanned autonomous operations. There are few such aircraft in existence. Many airforces have piloted surveillance aircraft but these are used for tactical military support (e.g. target designation and damage assessment). They are mostly operated over short range, at modest altitude and for short duration.

An aircraft possessing the capability to monitor for long periods and operate from remote bases would also be appropriate for some civil or quasi-civil operations. Such roles may include:

- Maritime patrol
- Drug law enforcement
- Remote high-value facility protection
- Civil disorder
- Border control and police surveillance
- Traffic intelligence
- Environmental protection
- Disaster management

As a research vehicle for environmental studies, such an aircraft could be used to monitor and report on atmospheric/climatic conditions, weather intelligence (providing near-earth observations) to supplement satellite data. The ability to fly for long endurance at high altitudes could also be used to provide communication links where radio or satellite facilities are unavailable or inadequate.

9.2.1 Aircraft requirements

The design requirements for the aircraft are based on data from a design/research paper presented at a conference in 2000.[1] They are relatively straightforward:

- payload (reconnaissance systems) 800 kg (1760 lb),
- ability to easily reconfigure the systems on the aircraft,
- ferry range, from the home airfield = 6000 nm,
- operational radius from advanced base = 500 nm,
- patrol duration 24 hours,
- all weather, day/night capability,
- short, rough field (unspecified) capability without the need for specialised ground launch or recovery systems (e.g. catapult/rocket launch, arresting wires),
- quick operational readiness,
- quick turnaround between missions,
- cost efficient (not necessarily minimum first cost) system,
- safety systems to avoid or reduce co-lateral damage in the event of an aircraft failure,
- structural loading $+2.5/-1.25g$.

For comparison, the US Global Hawk HALE-UASV is reported to be capable of flying 1200 nm, spending 24 hours on patrol at 60 000 ft (about 18 km) and returning to base. This is a reputed 32 hour mission!

9.3 Problem definition

We all know that if we stand on the shoreline and look out to sea, the horizon is only about five kilometres away. If we climb up to the top of the cliff, we can see much further out to sea. From a study of observation geometry, we understand that the maximum observation range is a function of height above sea level and the radius of the earth (Figure 9.1). The earth is not a pure sphere. It is flattened at the polar regions relative to the equator. The exact values[2] for the earth radii are 6356.9 km at the poles and 6378.4 km at the equator. In the analysis below, the often-quoted mean value of 6378.1 km (3444 nm) is assumed.

When we are flying, if we look down below the horizon the observation range becomes shorter and the view is clearer. This improved vision is partly due to reduced atmospheric contamination and ground clutter that adversely affect longer viewing distances. Simple trigonometry can be used to determine the observation range at various altitudes and for different downward-viewing angles (known as slant angles). These results are shown in Figure 9.2.

At an altitude of 20 km (about 65 000 ft), the horizon (zero slant angle) is about 500 km (270 nm) away. Looking down 5°, the distance to the ground reduces to about 200 km (108 nm). The corresponding figures for 10 km altitude are 350 km (190 nm) and 100 km (54 nm) respectively. If the aircraft reconnaissance equipment is powerful enough, it is obviously an advantage to operate at high altitude to observe a greater

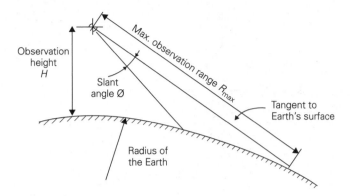

Fig. 9.1 Observation geometry

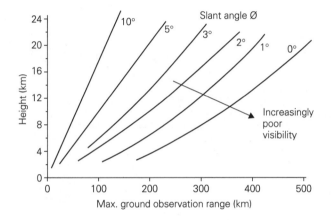

Fig. 9.2 Observation range

surface area. Published photographs of ground surveillance from satellites show that such reconnaissance equipment is feasible and available. Operating at high altitudes means that a greater ground area can be observed. Therefore, fewer missions or aircraft are required to sweep a territory. High-altitude operation also provides a stealthier mission which reduces aircraft vulnerability. In the situation where the territory to be observed is 'hostile', the aircraft could stand off from a border and still be capable of observing over the border for a considerable distance (Figure 9.3).

The aircraft is required to be capable of operating in all weather conditions. In this respect, one of the most important factors to be considered is the strength of winds. The average wind speed on a statistical basis varies with altitude as shown in Figure 9.4. Flying in constant wind has little influence on the aerodynamic characteristics of the aircraft as the forces on the aircraft are related to the relative/local air movements. It does, however, affect the perceived speed over the ground. This may influence the performance of the search equipment. Winds also affect the aircraft climb performance relative to the ground. In any event, it is desirable to operate in the calmest atmospheric conditions available. Figure 9.4 shows that average wind speeds are less at 18 to 20 km altitude. This region is selected as the best patrol altitude.

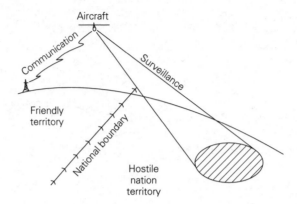

Fig. 9.3 Stand-off surveillance

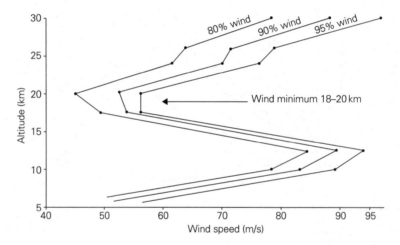

Fig. 9.4 Wind speeds[1]

Winds are seldom present in a stable direction or at a constant speed. This variation leads to the generation of vertical wind loading on the aircraft (gust effects). Although gusts are less pronounced at higher altitudes they must still be considered. Such gust activity leads to dynamic changes in aircraft lift. Aircraft with a low wing loading or with a high value of 'lift-curve slope' are more susceptible to gust disturbance. Aircraft designed for high altitude operation are likely to require both of these aerodynamic characteristics. The influence of gusts on the aircraft flight behaviour will need to be carefully assessed especially when operating at low altitudes (e.g. take-off and landing). Figure 9.4 indicates that at low altitude, average wind speed can be almost double that at the operating height. To mitigate the effects of gusts on the aircraft dynamic behaviour, it will be necessary to equip the aircraft with powerful gust load alleviation and some form of auto-stabilising systems. If the surveillance equipment is affected by gust activity it will be necessary to fit the aircraft with ride stabilising control systems.

Flying at high altitude is affected by an increase in stall speed due to the reduction in air density. There is also a reduction in the speed of sound due to the reduced air temperature and this will reduce the speed at which aerodynamic buffet will be felt. These two effects combine to severely reduce the available aircraft speed range (i.e. minimum to maximum operating speeds). This consequently increases the pilot/system workload to avoid the stall or buffet regions. The U2 aircraft was notoriously difficult to handle in this context. At a later stage in the development of the aircraft a detailed analysis of this problem will need to be undertaken to establish the best operating speed and the consequences on the design of the aircraft flight control system. At this early stage in the project, these issues are not considered in any more detail.

9.4 Initial design considerations

As the observation area increases with operational height and the vulnerability of the system is reduced, a satellite-based system may be regarded as a natural choice. However, such systems are known to be expensive and are inflexible in operation. For example, changing the orbit of a satellite to observe a new territory is often impossible and always time-consuming, complex and expensive. Together with these difficulties, it is often impossible to change the reconnaissance equipment to match the new operational requirements. For these reasons, a satellite-based system is not considered suitable for the current brief.

A lighter-than-air aircraft (airship or balloon) could achieve the long endurance specified for the system but these are, in general, large and slow-moving vehicles. This makes them unsuitable for quick response. They are also vulnerable to hostile action and incapable for all-weather operations. Such aircraft are therefore considered unsuitable options for this project.

A specifically designed aircraft would seem to offer the best option to meet the design brief. The design could be relatively cheap, very flexible in operation, quickly refitted to suit different missions and be able to be sent rapidly to new areas of strategic significance. Such aircraft could be uninhabited to avoid pilot/observer vulnerability in potentially hostile areas. The design could also be designed to house a pilot for operations that were deemed to be safe and benefit from human observation and on-board control.

9.5 Information retrieval

Reviewing significant reconnaissance aircraft, found in the published literature,[3,4] produces the following information. Most air forces in the world have aircraft in their fleets that are used for reconnaissance. Many of these are converted civil transport aircraft (mainly turboprop regional airliners). The list below is representative of such types:

Embraer Brazilia
De Havilland Dash 8
BAE(Hindustan) HS748
Boeing 707, 747
Fokker 50
Ilyushin 76, 18, 86
Tupolev 134
Saab 34
Pilatus BN2T
Fairchild Metro

None of these aircraft would be suitable for the specified missions as they are incapable of flying at high altitude and would not be able to achieve the specified range or endurance.

Three special-purpose aircraft were found in the review and are described below.

9.5.1 Lockheed Martin U-2S

This is a re-engined and newly equipped/upgraded version of the aircraft type that was piloted by Gary Powers and shot down on that fateful mission over Russia in the 1960s. The new version was first delivered into US service in 1994. The more fuel-efficient new engine has increased the aircraft range, endurance and operating capability. The high aspect ratio, large area wing is of traditional design and construction. The slender fuselage accommodates a single pilot and carries conventional tail surfaces and rear air brakes. Below are some of the aircraft details:

Wing span	31.4 m	111 ft
Overall length	19.2 m	63 ft
Wing area (ref. gross)	*sq. m	*sq. ft
Wing aspect ratio	*	
Empty mass	8074 kg	17 800 lb
TO mass	18 145 kg	40 010 lb
Cruise speed	192 m/s	373 to 470 kt
Ceiling (operational)	22.4 km	70 to 73 400 ft
Ceiling (absolute)	27.4 km	90 000 ft
Range	3800 nm	
Duration	12–15 hours	
Structural limit	2.5g	
Engine: F188-GE-101 turbofan		
Max. thrust (total)	84.5 kN	19 000 lb

(*As some of the details of the new U-2S aircraft are unavailable, the table above does include figures from the earlier model.)

New sensors installed in the updated equipment include electro-optical multi-spectral cameras and advanced synthetic aperture radar systems. Equipment is housed in a forward interchangeable nose bay, behind new fuselage hatches and in wing mounted pods.

9.5.2 Grob Strato 2C

This German aircraft first flew in 1995 (as a proof of concept vehicle). To maintain sufficient power up to an operating height, reputed to be 24 km (78 700 ft), it has two compound piston engines incorporating a two-stage turbocharger with intercooler. The two engines drive large five bladed pusher propellers mounted aft of the wing trailing edge. The engines are rated at 300 kW (400 hp) each. The large area, high mounted, straight, laminar flow wing has a high aspect ratio. The conventional pressurised fuselage carries tail surfaces, an outrigger mounted main undercarriage and nose unit. The accommodation comprises two flight crew and two observers, work stations, rest areas and associated galley and toilet provision. Details are shown below:

Wing aspect ratio	21.3	
Empty mass	6650 kg	14 663 lb
TO mass	13 350 kg	29 440 lb
Payload	1000 kg	2200 lb
Cruise speed	100 m/s	194 kt
Ceiling (operational)	16–26 km	52 500–85 250 ft
Ceiling (design)	24 km	78 700 ft

Range	9773 nm	
Duration 48 hours	@ 18 km	59 000 ft
Max. duration hours	@ 24 km	78 700 ft

Equipment includes various atmospheric monitoring instruments, observation systems and radar.

9.5.3   Northrop Grumman RQ-4A Global Hawk

This aircraft is a high-altitude, long-range, uninhabited aircraft designed to provide military field commanders with high-resolution, near-real time imagery over large geographical areas. It is therefore close to the same operational specification as our design project. At the time of writing, little technical data on the aircraft was available due to US national security.

Wing span	116.16 ft
Overall length	44.33 ft
Overall height	15.16 ft
Max. TO weight	25 600 lb
Payload	2000 lb
Operational height	65 000 + ft
Max. range	13 500 nm
Duration	36 hours

Due to their potential for long endurance flights, powered gliders have also been converted for reconnaissance operations. Two examples are described below.

9.5.4   Grob G520 Strato 1

First flight, as a derivative of the Egrett, was in 1995. This was a proof of concept project as a high altitude 'research' vehicle. It is powered by an Allied Signal turboprop (560 kW). The aircraft has a large area, high aspect ratio wing and a conventional fuselage with rear control surfaces. It carries one pilot and the fuselage accommodates six interchangeable equipment modules. Data is listed below:

Wing span	33 m	108 ft
Overall length	12 m	39 ft
Wing area (ref. gross)	40 sq. m	430 sq. ft
Wing aspect ratio	27.5	
Empty mass	2700 kg	5950 lb
TO mass	4700 kg	10 360 lb
Cruise speed	50 m/s	97 kt
Ceiling	16 km	52 500 ft
Range	1930 nm	
Duration	13 hours	
Structural limits	$3.3/-1.3g$	

9.5.5   Stemme S10VC

This aircraft is also a derivative of an existing powered glider. Its unique feature consists of a propeller designed to fold into the fuselage nose fairing. It first flew in 1990 as a sensor platform for atmospheric research. It is a classically configured, twin-seat, composite-constructed glider with extra equipment pods mounted on each wing. Details are shown below:

Wing span	23 m	75.3 ft
Overall length	8.4 m	27.5 ft

Wing area (ref. gross)	18.7 sq. m	200 sq. ft
Wing aspect ratio	28.2	
Empty mass	670 kg	1477 lb
TO mass	980 kg	2161 lb
Cruise speed	46 m/s	89 kt
Lift/drag ratio 50 @	30 m/s	58 kt

Powered glider derivatives employ a power-off 'glide down' technique to save fuel and extend flight duration.

Another aircraft project to note, because of its unorthodox configuration, is the Boeing project study for a Common Support Aircraft. Though this aircraft is not directly related to our design brief it houses an active phased 360° scanning radar mounted conformally in the unconventional wing structure. This is possible due to the adoption of the 'joined wing' layout. The 40° swept back wings are joined towards their tips to tail mounted, swept forward wings. The radar system is mounted on the wing surfaces to provide nearly all-round scanning. The aircraft, which is designed to fly from/to aircraft carriers, accommodates one pilot and three system operators. Few details of the aircraft are released but it is reported to have a wing-span of 19.3 m (63.3 ft), an overall length of 15.5 m (51 ft), max. mass of 25 535 kg (56 300 lb), max. speed of M0.8 and a patrol speed of M0.38. Although not directly comparable to the design brief under consideration, the aircraft design illustrates how technical innovation can be used to provide an efficient alternative configuration to conventional layouts.

9.6 Design concepts

As the project brief defines a unique mission, the literature search did not reveal any aircraft that could match the requirement. The nearest are seen to be the U-2S, Strato-2 and the Global Hawk.

In order to fly at high speed for quick transfer onto operational stations and to be able to fly at the high Mach number needed for patrol at high altitudes, propeller propulsion systems are not feasible. A turbofan engine with a modest bypass ratio (BPR) that balances low fuel consumption with acceptable thrust loss at high altitude is the preferred choice. An engine with a BPR of between three and five will be compact enough (i.e. a medium size fan diameter) to give a feasible installation on the aircraft. Two engines will be selected to provide improved flight safety, although there will be an increase in engine failures and maintenance.

The fuselage will need to house a variety of different operational equipment/systems. It will also be desirable to be able to quickly change the equipment fits between sorties. This suggests a modular design. Downward and forward sensor viewing will be essential, therefore the fuselage structural frame must consist of a beam stretching fore and aft above the equipment bays. This will allow different modules to be suspended below the beam.

The choice between manned and uninhabited/autonomous operation needs to be carefully considered. The development and flight testing of a new aircraft concept with untried and sophisticated characteristics may be most easily achieved using a human test pilot. Conversely, operating in a hostile environment will demand an unmanned vehicle to avoid the political and diplomatic difficulties that can arise if the aircraft is captured. The best compromise between these requirements is to consider the cockpit to be a 'unique system' module that can be suspended at the front of the fuselage beam structure. As confidence in the aircraft flying characteristics and remote control system performance is gained during flight testing, the cockpit module could be replaced by the autonomous flight system. In service, variants of the aircraft could be either uninhabited or piloted as suits the mission.

Reviewing the layout options for the configuration of the aircraft leads to four possibilities:

1. Conventional (glider type development)
2. Joined wing (Boeing CSA example)
3. Flying wing (USA B2 example)
4. Braced wing (NASA research)

Each of these layouts will be described and considered for the design brief.

9.6.1 Conventional layout (Figure 9.5)

In many ways, this option offers an attractive choice as the wing is aerodynamically efficient and the fuselage acts as a central beam structure which supports the wing, tail surfaces, engines, landing gear, fixed equipment and mission system modules. The aerodynamic and structural features and the associated analyses are well developed. Thus, this option provides low commercial risk with regard to the technological development.

There are two main concerns. First, the long wing-span, and its low height above the ground may make remote piloting difficult during crosswind landings. In fact it would be essential to use a sophisticated auto-land system. The small angle from the main undercarriage contact point to the wing tip will make accurate approach attitude imperative. The aircraft has a low wing loading that will make it very susceptible to gust disturbance. This will make precise landing manoeuvres difficult in all-weather conditions. Touching the wing tip on the ground during landing will inevitably lead to the classical 'ground looping' phenomena.

Second, there may be difficulty in producing the wing thin enough to delay the critical Mach number up to the anticipated operating speed. With a long span, the root bending

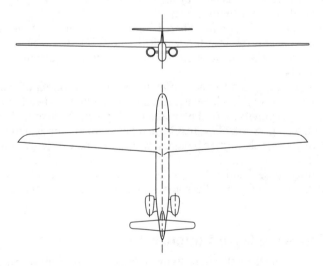

Fig. 9.5 Conventional layout option

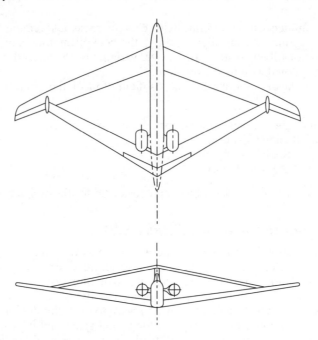

Fig. 9.6 Joined-wing layout option

moments will require a deep section to avoid excessive wing structural mass. Even if it were feasible to design a thin wing, its internal volume would not be sufficient to hold the required fuel load.

9.6.2 Joined wing layout (Figure 9.6)

Although this configuration has been promoted in new civil aircraft projects[5] it has not yet been fully validated in a flying prototype. Even the Boeing CSA project was not taken into the manufacturing phase. There is serious concern about the flow condition at the wing junctions and the interference of the flow field between the two surfaces. These problems could disturb the critical transonic flow conditions and reduce the critical Mach number.

The joined wing is also seen to have difficulty in the positioning of the main undercarriage unit. The wings to fuselage attachment points lie far ahead and well behind the aircraft centre of gravity (cg). The main landing gear unit must be located only slightly behind the cg position. This results in a heavy and complicated fuselage structure. Alternatively a bicycle undercarriage layout could be employed but this is heavy and makes it difficult to rotate the aircraft on take-off and landing.

Notwithstanding these disadvantages, the layout may offer a novel solution to the installation of the otherwise cumbersome antenna required on the aircraft (see the Boeing design).

9.6.3 Flying wing layout (Figure 9.7)

Although several aircraft of the pure flying wing configuration have been proposed and some designs have been flown in the past they have not yet been fully exploited in

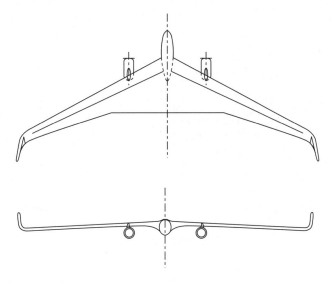

Fig. 9.7 Flying-wing layout option

production aircraft (with the exception of the B-2 stealth bomber). Many enthusiasts for the type have claimed the layout to be aerodynamic and structurally efficient but it seems that such expectations have not yet been realised (the B-2 layout is selected for stealth reasons). The reason may be due to the linking of stable torsional deflections to the aerodynamic forces and the consequential requirement to modify the wing planform and sectional geometry to avoid this problem. The layout is regarded as efficient at one design point but seriously compromised away from this condition. The flying wing configuration was considered in the German study[1] but dismissed on these technical issues.

9.6.4 Braced wing layout (Figure 9.8)

The structural bracing of the wing to the fuselage was a common feature in historic aircraft layouts. This was done to reduce the loads in the wing to match the relatively poor structural properties of the materials used in the construction. The development of stronger and more consistent materials allowed such bracing and the associated drag penalty to be eliminated. The traditional monoplane wing layout has been the preferred choice over the past several decades. As wing aspect ratio is increased, the benefit of bracing becomes more attractive as it significantly reduces wing bending moments. In recent years, some NACA funded research[6] has shown that wing bracing could provide advantages to the design of long-range civil jet transport aircraft. The purely tensile loaded brace reduces the shear, bending and torsion on the wing structure. This correspondingly allows either a thinner wing or a larger aspect ratio to be used on the wing geometry. A thinner wing would allow the wing to be less swept for a design critical Mach number. All of these effects reduce aircraft drag and consequently fuel burn. Positioning the brace attachment to the wing ahead of the sectional structural axis also provides a reacting nose-down moment to stabilise the divergence tendency associated with a swept forward wing planform.

Mounting the wing on the fin structure and adding dihedral to counter the unstable yaw coupling from the swept forward planform places the wing well above the ground

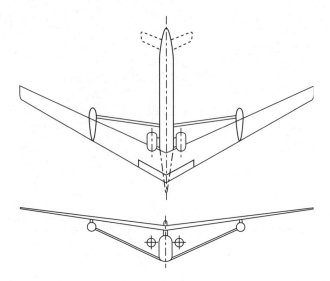

Fig. 9.8 Braced-wing layout option

plane during landing. This provides the aircraft with adequate bank angle to protect against disturbed landing manoeuvres.

The main drawback with the configuration is associated with the novelty of the layout and its potential for technical risk.

9.6.5 Configuration selection

Of the four options, only the conventional and braced wing seems to be worth further consideration. As the German design study[1] selected a conventional layout for their baseline design we will investigate the braced wing layout. This will provide a useful comparison with the previous study.

Having selected the braced wing layout there are several detail design considerations to be made:

1. The engine mountings, fuselage brace attachments and the main undercarriage mounting will be combined into a central fuselage structural framework. This will leave the forward fuselage structure uncluttered and capable of holding the equipment modules as conformal containers below the fuselage structural beam.
2. To avoid the difficulty of attaching the brace to the wing structure, and the possibility of complex airflows at the junction, pylon mounted equipment/fuel pods will be installed on the wing. The brace will be attached to the pod support structure (see Figure 9.9).
3. The brace structure will need to be streamlined and this will provide the opportunity to run equipment service lines or fuel supply pipes directly between the wing and fuselage.
4. It may be possible to use the wing and brace structures to house conformal radar antenna (as proposed by Boeing on their CSA).
5. To reduce trim drag (an important feature on long-range/endurance aircraft) the forward fuselage could support a small canard surface mounted above the equipment modules. Care will need to be taken on the position of this surface relative

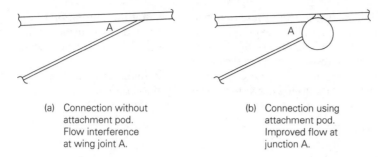

(a) Connection without attachment pod. Flow interference at wing joint A.

(b) Connection using attachment pod. Improved flow at junction A.

Fig. 9.9 Wing to brace interconnection detail

to the engine intakes. Wind tunnel tests will need to be done to finalise the exact geometry.

6. It will be necessary to incorporate wing inboard control surfaces to provide pitch control.
7. Although main wheels will be required, it may be preferable to use skids for the third (nose or tail) unit.

9.7 Initial sizing and layout

While the aircraft is of an unconventional configuration, the initial sizing and layout process will follow the normal procedure. This will involve estimating the aircraft take-off mass, wing loading, some airspeed predictions, wing layout and powerplant sizes. These are described in the following subsections. Finally, all of the component studies are linked together to produce the initial baseline aircraft layout.

9.7.1 Aircraft mass estimation

In order to size the aircraft it is necessary to estimate the maximum take-off mass. The formula below is often used for this purpose (see Chapter 2, section 2.5.1 for the definition of terms):

$$M_{TO} = (M_{UL})/\{1 - (M_E/M_{TO}) - (M_F/M_{TO})\}$$

For the HALE aircraft there are some difficulties that arise from the definitions of aircraft systems to be included in the aircraft empty mass ratio. Many of the systems on the aircraft are directly related to the type of operation. Some of the equipment may be changed to suit the mission (reconnaissance, communication, surveillance, atmospheric research and monitoring). To resolve the lack of knowledge of these systems and the variability with the mission, the equipment mass will be assumed to be 800 kg. This value will be attributed to the 'useful load' in the above equation. At a later stage in the development of the design it will be appropriate to conduct sensitivity analyses around this assumption.

A second difficulty arises due to the expected, unusually large, fuel ratio. An aircraft with a duration of 24 hours is almost unique. Therefore data from other, shorter-range aircraft may be misleading. For this reason it will be essential to check the fuel requirements as soon as the aircraft mass, lift, drag and engine characteristics are known with reasonable accuracy.

The main conclusion from these observations is that the estimation of aircraft maximum mass from the above expression must be treated with suspicion and regarded as tentative. Several iterations of the subsequent analysis will be necessary before confidence in the results can be realised.

Analysis of the technical descriptions of the main aircraft types described in section 9.5 shows the following values for empty mass ratios:

U-2S	0.445
Stratos 1	0.570
Stratos 2	0.500
Stemme	0.680

These contrast with a value of 0.254 for the German EADS project.[1] Our aircraft design brief is closer to the U-2S and EADS aircraft. The U-2S is recognised as being an old aircraft design with 1950/1960's materials and construction methods. Application of modern materials and methods would be expected to reduce the structural mass. The EADS aircraft data may have linked more of the equipment mass to the useful load component than assumed in our case. Without more information, it is difficult to choose between these two values for empty mass ratio, therefore an average figure of 38 per cent will be initially used for our design.

The fuel fraction can be estimated using the Breguet range equation:

$$(M_F/M_{TO}) = (\text{engine cruise sfc}) \cdot [1/(L/D)] \cdot (\text{flight time})$$

Note: engine sfc varies with cruise altitude. For a typical medium bypass ratio turbofan engine, the following relationship is quoted[7]:

$$(\text{sfc})_{\text{altitude}}/(\text{sfc})_{\text{sea level}} = \theta^{0.616}$$

where θ is the ambient air temperature ratio (T_A/T_{SL}). In the stratosphere the ISA temperature is constant at 216.76 K. ISA sea-level temperature is 288.16 K. This makes $\theta = 0.75$.

Hence,

$$(\text{sfc})_{\text{altitude}} = 0.84(\text{sfc})_{\text{sea level}}$$

A medium BPR engine is likely to have a sea-level sfc of 0.55 (lb/lb/hr or N/N/hr). Therefore using the above formulae gives an engine sfc in the stratosphere of 0.46.

With a high aspect ratio wing and slender fuselage the aircraft lift to drag ratio (L/D) in cruise could safely be assumed to be better than the value of 17 which is typical of modern civil airliners. Due to the forward swept wing and the interference arising from the brace structure it will not be possible to achieve the value of 40 which is typical of high performance gliders. Being conservative, we will assume a value of 25 but this will need to be checked and adjusted when detailed drag estimations are available later in the design process.

The duration of the patrol is specified as 24 hours in the design brief. It is unclear if this is to include the time needed to reach the patrol area, so an extra two hours will be added to this time. A design duration of 26 hours will be used in the analysis below:

Hence,

$$(M_F/M_{TO}) = (0.46) \cdot (1/25) \cdot (26) = 0.48$$

We will add 10 per cent for contingencies to give a design value of 0.53.

Using the above values in the initial aircraft take-off mass equation gives:

$$M_{TO} = 800/(1 - 0.38 - 0.53) = 8888 \,\text{kg} \,(19\,600\,\text{lb})$$

To provide for some design flexibility in the subsequent work a design (max.) mass of 9200 kg (20 280 lb) will be assumed.

9.7.2   Fuel volume assessment

The calculation above predicts a fuel mass of $0.53 \times 8888 = 4693\,\text{kg}$ ($10\,350\,\text{lb}$). With a specific mass for aviation fuel of 0.8 (fuel varies between 0.76 and 0.82), this mass will need 5833 litres ($5.866\,\text{m}^3$, $207\,\text{ft}^3$) tankage volume to hold the fuel. When the wing geometry has been defined, a check will be necessary to establish if the volume can be accommodated in the integral wing tanks. If not, the size of other storage tanks to be included in the aircraft layout will need to be determined.

9.7.3   Wing loading analysis

Flying at high altitude where the air is thin requires either a fast airspeed or a high value of lift coefficient (or probably a combination of both) to reduce the required wing area. Maximising these parameters for a chosen altitude sets the value for the maximum wing loading as shown below:

$$L = 0.5\rho V^2 S C_{\mathrm{L}}$$

Substituting $\rho = p/RT$, and using $a = (\gamma RT)^{0.5}$, where (ρ) is air density, (p) is air pressure and (a) is the speed of sound.

Assuming $\gamma = 1.4$ and using $M = $ Mach number $(=V/a)$, gives the equation:

$$L/S = 0.7pM^2 C_{\mathrm{L}}$$

Typical values of the parameter $(M^2 C_{\mathrm{L}})_{\max}$, range from 0.05 for gliders to 0.6 for military jets. Conventional civil transports lie in the range 0.2 to 0.4 (i.e. M0.8 @ $C_{\mathrm{L}} = 0.4$ gives $(M^2 C_{\mathrm{L}}) = 0.24$). Figure 9.10 shows the distribution of maximum wing loading against altitude for various values of $(M^2 C_{\mathrm{L}})_{\max}$. The areas marked for each type represent the common values. Obviously some aircraft are designed to operate away from these regions. Figure 9.11 shows the portion of the previous figure relating to high altitude operations. As discussed in section 9.3, calmer wind conditions are found at

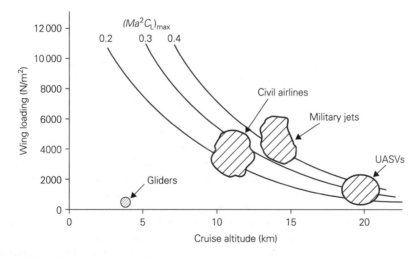

Fig. 9.10 Max. wing loading versus altitude

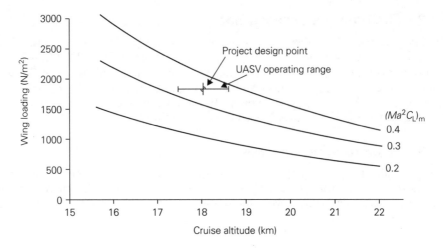

Fig. 9.11 UASV wing loading selection

altitudes around 18 km (59 000 ft). This sets the design point shown in Figure 9.11. Therefore, the selected wing loading is 1800 N/m² (183.5 kg/sq. m, 37.6 lb/sq. ft). For our chosen aircraft design mass of 9200 kg (20 280 lb), this equates to a minimum wing area of 50 m² (537 sq. ft).

9.7.4　Aircraft speed considerations

The aircraft operating envelope is bounded at slow speed by the aircraft stall performance. At high speed, the operating envelope is restricted by the available engine thrust, the rise in transonic wave drag and the effects of the associated buffet on the aircraft structure. The effect of high altitude operation affects both of these speed boundaries. For a given wing area and sectional C_{Lmax} value, the stall speed will increase as air density reduces as defined below:

$$V_{stall} = [L/(0.5\rho S C_{Lmax})]^{0.5}$$

As air temperature reduces with altitude (up to the start of the stratosphere) the speed of sound and thereby the aircraft speed at the onset of transonic flow will reduce. The speed of sound is determined by the relationship:

$$a = a_o \theta^{0.5}$$

where (a_o) is the speed of sound at sea level = 340.29 m/s, 661 kt.

These effects are shown diagrammatically in Figure 9.12.

It is advisable to fly at a speed greater than the stall speed to allow a margin of safety to protect against gusts. This margin will avoid inadvertent stalling and reduce pilot/system control demands. This defines the minimum speed boundary. For many types of aircraft the margin is set by applying a factor of 1.3 to the stall speed in the low speed, approach to landing, phase. High wind speeds may demand an increase in this margin. Although wind speed does not affect the aerodynamic parameters of the aircraft (the aircraft travels with the ambient air and only relative changes are significant) it does alter the perceived (relative to the ground) climb and descent flight

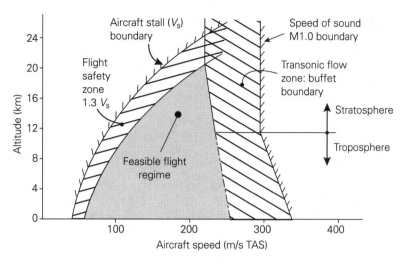

Fig. 9.12 Operating speeds constraints (diagrammatic)

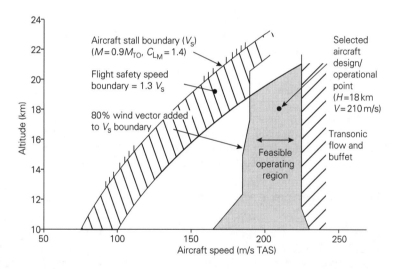

Fig. 9.13 Aircraft speed envelope

paths. This will influence the time needed to get onto, and from, the operating station. The aircraft high-speed boundary is directly affected by the aerodynamic (transonic) characteristics of the aircraft (mainly the wing geometry and pressure interference effects). Smooth aircraft cross-section area shaping, supercritical wing profiles and increased sweep are methods to delay the onset of transonic effects. Civil transports push the boundary to about M0.85 but this increases drag by about 3 to 5 per cent. Reducing the operating speed to less than M0.8 should avoid this penalty.

Figure 9.13 shows the absolute (stall and Mach1.0) boundaries for the aircraft together with the 80 per cent average wind speeds at various heights. The wind speed at altitude is important as it will add or subtract to the ground speed and therefore the search pattern. Note that at about 21 km, the minimum and maximum boundaries are nearly coincident. To fly above this height would require the aircraft to reduce weight, have an increased wing area or an increase in C_{Lmax}, or combinations of these changes. For a given aircraft geometry, a cruise-climb technique, in which height is gained as the aircraft mass reduces with fuel use, could be considered.

The design point selected on the above considerations and the earlier discussion (section 9.3) is:

Operating altitude (initial) $= 18$ km (59 000 ft)

Operating speed $= 210$ m/s (408 kt), representing M0.71 at 18 km

At the above condition the aircraft lift coefficient is 0.604 at the start of patrol (mass $= 0.9 M_{TO}$). This reduces to 0.332 at the end of the patrol (mass $= 1.3 M_{empty}$). At the end of patrol, the aircraft stall speed will have reduced from an initial value of 138 m/s (268 kt) to 102 m/s. This change would allow the aircraft to fly progressively either slower or higher.

The discussion above has concentrated on the cruise performance; it is also necessary to check the approach speed to determine if it is acceptable. Assuming ISA-SL conditions with an aircraft mass on approach of $1.15 M_{empty}$ and a C_{Lmax} of 1.4 (i.e. no flaps):

$$V_s^2 = [1.15 (0.38 \cdot 9200) 9.81]/[0.5 \cdot 1.225 \cdot 50 \cdot 1.4]$$

$$V_s = 30.3 \text{ m/s (59 kt)}$$

If $V_{approach} = 1.3 V_{stall}$, then

$$V_{approach} = 1.3 \times 30.3 = 39.4 \text{ m/s (75.6 kt)}$$

This speed should be slow enough to allow for automatic/remote landing control in the UAV version. For emergency landing at higher weight, consideration may need to be given to the provision of fast fuel dumping.

9.7.5   Wing planform geometry

Selection of the wing planform is the most significant design decision with regard to the aircraft performance. This aircraft will spend most of its time on long-duration patrol missions. It is therefore important to choose the wing geometry to 'optimise' this part of the operating envelope. In this case, drag reduction forms the main basis for the selection of wing characteristics. In the search phase, the aircraft induced drag will form a significant component of drag. Selecting a high aspect ratio for the wing is an effective method of reducing induced drag. On conventional monoplane designs, high values for aspect ratio lead to a substantial increase in wing structural mass. This penalty arises due to the outboard movement of the centre of lift (away from the wing root/bodyside attachment). The bracing structure selected on our design avoids this penalty as part of the outboard lift is reacted by the brace structure. Higher values of wing aspect ratio than normally seen on conventional designs are therefore feasible. For a given wing area, high aspect ratio corresponds to a large span, it may therefore be necessary to impose a limit to ensure that the aircraft is easy to handle on or near the ground. In this respect, an aspect ratio of 25 will be selected. This compares to values in the range 7–9 for civil transports and 20–30 for higher performance gliders.

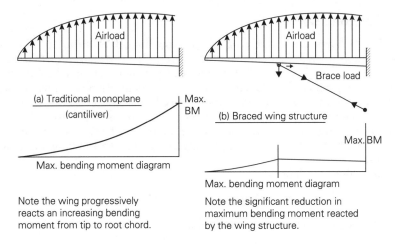

Fig. 9.14 Wing bending-moment diagrams

To delay the onset of transition the wing will be swept forward by 30°. A thin wing (8 per cent) thickness will be adopted. These characteristics should provide a critical Mach number above M0.84 and therefore avoid transonic wave drag penalty and structural buffeting in the cruise/search phases.

The high aspect ratio wing will produce a small chord and correspondingly a low value for the airflow Reynolds number. This will encourage the retention of laminar flow over the wing section. A transition at 70 per cent chord may be possible if the wing profile skins are smooth, continuous (no gaps or junctions) and the surfaces are kept clean. This should be possible with a composite construction and normal military service care.

The reaction force on the wing from the brace will alter the wing bending moment distribution. This will cause an unusual distribution of wing taper. In a traditional unbraced design the maximum bending moment occurs at the wing to fuselage attachment section (see Figure 9.14). This is the position where the deepest wing thickness is required and therefore the widest chord. A straight tapered wing planform with the largest chord at the root is the usual configuration. For the braced wing layout in which the relative stiffness of the wing structure and the brace can be selected, the largest wing bending moment may be at the brace attachment section. To reflect this change of bending moment distribution the wing taper will be unconventional. The largest chord will be at the brace attachment position, as shown in the initial aircraft layout drawing.

The aircraft wing geometry is now defined:

Wing area	50 sq. m	537 sq. ft
Wing aspect ratio	25	
Wing span $(=(S \cdot A)^{0.5})$	35.4 m	116 ft
Wing sweep	30° forward	
Physical span $(=35.4 \cos 30)$	30.6 m	100 ft
Wing thickness ratio	0.08	
Wing max. C_L	1.4	
Wing taper to match brace geometry		
Wing section supercritical cambered		

9.7.6 Engine sizing

As already described, the engine type will be a medium BPR turbofan. The required thrust will be dependent on the climb performance at the cruise altitude. If the climb rate at high altitude is too small then the time required to reach the search height will be too long. Without a specified value for climb rate, we will assume the civil transport criterion of 300 ft/min (1.53 m/s). This assumption will need to be checked when more detailed mass, aerodynamic, propulsion and performance analysis is possible. A sensitivity analysis will be required to show the inter-relationship of this assumption with the commonly specified 100 ft/min service ceiling criterion. The fundamental analysis is shown below:

From basic flight mechanics:

$$(dh/dt) = (T - D) \cdot V/W \tag{9.1}$$

To obtain the required sea level, take-off thrust/weight ratio, equation (9.1) is modified to:

$$(T/W)_{SL} = (\beta/\alpha)[(D/W_o) + (1/V)(dh/dt)]_A \tag{9.2}$$

where subscripts SL is sea level, o is initial (take-off) value and A refers to altitude values. (β) represents the mass fraction at the start of cruise (which, in this case we will assume to be 0.85). (α) represents the engine thrust reduction with altitude (T_A/T_{SL}).

From Eshelby's book[8]: $(T_A/T_{SL}) = \sigma^{0.7}$ in the troposphere (up to 11.02 km) and $\sigma^{1.0}$ in the stratosphere (σ is relative density $= (\rho_{altitude}/\rho_{SL})$).

This assumes a constant engine rating. For this type of aircraft the loss of thrust at this high altitude will be large therefore it is likely that the static sea thrust (T_{SL}) at the climb (or cruise) rating will be suitable to meet the take-off requirement. For this reason, a constant (climb or cruise) rating will be assumed. As our cruise will be in the stratosphere:

$$(T_A/T_{SL}) = (\rho_{11.02}/\rho_{SL})^{0.7} \cdot (\rho_A/\rho_{11.02}) = 0.24$$

The values for ρ (air density) can be found in ISA tables
(in SI units, $\rho_{SL} = 1.225$, $\rho_{11.02} = 0.364$ kg/cu. m)
(in Imp. units, $= 0.002378$, 0.000707 slug/cu. ft)
Aircraft drag at the cruise condition $= 0.5\rho_A V^2 S C_D$
where $V = 210$ m/s (408 kt), $S = 50$ sq. m (537 sq. ft),
and C_D assumed to be 0.022.

Aircraft take-off weight $W_o = 9200 \times 9.81 = 90.25$ kN $= 20\,280$ lb

Equation (9.2) is computed and plotted in Figure 9.15. At our selected design altitude of 18 km (see Figure 9.11) the required thrust to weight ratio (known as the thrust loading) is 0.24 for a 300 ft/min climb ability. The corresponding line at 100 ft/min indicates a service ceiling, with this thrust ratio, of 24 km (78 700 ft). From our previous discussions, this value appears to be satisfactory.

This thrust loading gives a required take-off thrust of:

$$T_o = 0.24 \cdot 9200 \cdot 9.81 = 21.6 \text{ kN} = 4870 \text{ lb}$$

With two engines this equates to 10.8 kN (2434 lb) per engine.

It is necessary to check that this thrust is adequate for safe single-engine take-off in an emergency. The civil aircraft airworthiness requirement sets a climb gradient of 0.024 at 50 ft height and speed V_2 (undercarriage retracted but take-off flaps still deployed).

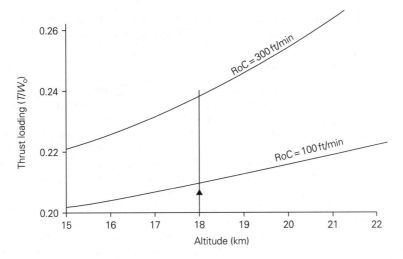

Fig. 9.15 Thrust/weight requirement for climb

Aircraft drag at take-off with one engine inoperative will be affected by an increase due to the flow blockage on the failed engine, extra drag from the asymmetric flight attitude (yaw) and extra trim drag from the control surfaces. To account for these effects we will assume additional drag to increase the aircraft drag coefficient to 0.04.

Assuming an aircraft speed of $1.2V_{stall}$ and a C_{Lmax} of 1.4 at the take-off mass of 9200 kg (20 280 lb) gives:

$$(V_{stall})^2 = (9200 \cdot 9.81)/(0.5 \cdot 1.225 \cdot 50 \cdot 1.4)$$
$$V_{stall} = 45.9 \, \text{m/s} \, (89 \, \text{kt})$$
$$1.2V_{stall} = 55.1 \, \text{m/s} \, (107 \, \text{kt})$$

Aircraft drag (in SI units) $= 0.5 \times 1.225 \times 55.1^2 \times 50 \times 0.040 = 3719 \, \text{N} \, (=836 \, \text{lb})$
Climb gradient $= (T - D)/W = (10\,800 - 3719)/(9200 \cdot 9.81) = 0.078$
Climb rate $= 0.078 \cdot 55.1 = 4.32 \, \text{m/s} \, (850 \, \text{ft/min})$

This result seems to provide acceptable initial, single-engine climb, performance.

The engine will need to provide the power to drive all of the electrical equipment and sensors on the aircraft. For example, when flying for long periods at high altitude it is necessary to warm the electronic and sensor equipment to protect it against the cold ambient temperature. This additional 'load' on the aircraft engine system is likely to be significantly more than on other types of aircraft. It is therefore necessary to install an engine with more thrust than the 10.8 kN predicted above.

From the literature, the Pratt & Whitney of Canada PW530 engine used on the Cessna Citation business jet looks suitable for our aircraft:

Take-off thrust $= 2900 \, \text{lb} \, (12.9 \, \text{kN})$
Fan max. diameter $= 27.3 \, \text{in} \, (0.69 \, \text{m})$
Length $= 60 \, \text{in} \, (1.5 \, \text{m})$
Basic engine weight (mass) $= 632 \, \text{lb} \, (286 \, \text{kg})$
Specific fuel consumption $= 0.55$
By-pass ratio $= 3.3$

This engine will give about 20 per cent extra thrust than required for aircraft performance so should be adequate to meet the aircraft service needs.

9.7.7   Initial aircraft layout

The previous sections have set out the geometrical requirements for the aircraft. It is now possible to produce the first general arrangement drawing (Figure 9.16).

As prescribed, the layout is very unorthodox. Investigating the technical features shows that the configuration is logical. The high mounted wing provides good banking stability when the aircraft is on or near the ground. The high aspect ratio, thin supercritical wing section and swept forward design should reduce drag. The planform taper matches the spanwise loading distribution. The configuration should have good pendulous stability, which will help with low-speed manoeuvrability.

The unobstructed front fuselage provides suitable housing for the observation, reconnaissance and communication systems. These systems are undefined in the project brief but the length and volume provided on the aircraft is consistent with other aircraft of this type. The rear fuselage provides the main structural framework for the attachment of engines, main landing gear, brace connection and the fin/wing mounting. The internal volume in this area provides the main fuel tank. The enclosed volume of the tank is 3 m long × 1.5 m deep × 0.7 m wide, giving a capacity of 3.15 m^3. More

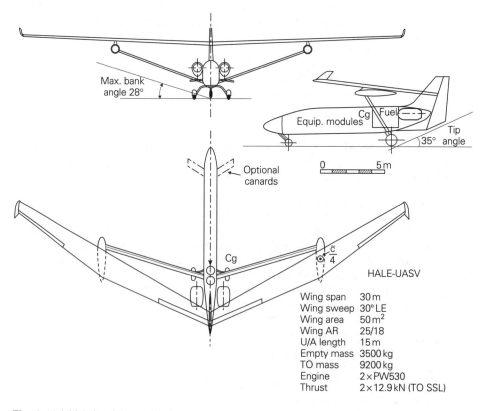

Fig. 9.16 Initial aircraft layout drawing

fuel is housed in the central wing boxes. The capacity of the wing tanks is $0.72 \, m^3$. This combined capacity of the tanks (fuselage and wing) ($3.87 \, m^3$) is substantially smaller than the fuel volume requirement estimated in section 9.7.2. At this stage in the design process no modifications will be made as later calculation of aircraft mass may reduce this early estimate. If it is found later that more fuel is required, the wing mounted 'equipment/brace' pods could offer another $0.64 \, m^3$. However, this would reduce equipment/sensor positioning flexibility. All of the fuel tanks are positioned close to the aircraft centre of gravity (estimated at the wing mean aerodynamic quarter-chord position). This will ensure that fuel used in the mission does not lead to significant increase in trim drag.

The outboard wing control surfaces will act as conventional ailerons. The inboard control surfaces will provide pitch control and aircraft stability. Due to the relatively short tail arm on the aircraft, it may be found necessary to add canard surfaces to the front fuselage to complement the rear controls. Although such an arrangement could reduce aircraft trim drag; the interference of flow over the wing sections may affect the laminar flow condition. The net result could be an aerodynamic inefficiency and a less effective layout. Wind tunnel tests would need to be done to quantify the overall flow condition.

9.7.8   Aircraft data summary

The initial baseline aircraft layout may be summarised as shown in Table 9.1.

Table 9.1

		SI units	Imperial units
Wing	Span	30 m	98 ft
	Aspect ratio	18	
	Sweep	30°	
	Area	50 sq. m	537 sq. ft
Fuselage	Length	15 m	49 ft
	Depth	2 m	6.6 ft
	Width	0.7 m	28 in
Mass	Empty	3500 kg	7717 lb
	Max. TO (design)	9200 kg	20 280 lb
	Payload	800 kg	1760 lb
	Fuel load	4700 kg	19 360 lb
Engine	PW530/545		
	TO thrust	12.9 kN	2900 lb
	Bypass ratio	3.3	
	Cruise sfc	0.54	
	Fan diameter	0.7 m	28 in
Performance	Cruise/patrol	210 m/s	408 kt
		@ 18 km	@ 59 000 ft
	Duration (gross)	26 hrs	
	Approach	40 m/s	76 kt
	TO climb (OEI)	7.8%	

9.8   Initial estimates

With a fully dimensioned general arrangement drawing of the aircraft available it is possible to undertake a more detailed analysis of the aircraft parameters. This will include component mass predictions, aircraft balance, drag and lift estimations in various operational conditions, engine performance estimations and aircraft performance evaluations. The results from these studies will allow us to verify the feasibility of the current layout, and our earlier assumptions, and to make recommendations to improve the design.

9.8.1   Component mass estimations

The geometrical and layout details allow us to estimate the mass of each aircraft component. This will provide an initial aircraft mass statement that we can use to check on our initial empty mass ratio and maximum mass estimates. The new mass predictions will be used in the following performance predictions. It is necessary to estimate each of the mass components in the aircraft mass statement described in Chapter 2, section 2.6.1. These component mass calculations are set out below.

Wing structure

Available wing mass estimation formulae are based on conventional cantilever trapezoidal wing planforms. This presents difficulties in using them to predict our high aspect ratio, braced wing layout. When more details of the wing structural framework are known it will be possible to roughly size the main structural elements and thereby to calculate the mass of the structure. This method will give a reasonable estimate of the wing mass. Until this is possible, we will need to 'improvise'!

Using established wing formula for civil jet airliners results in a mass of about 10 per cent M_{TO} for our geometry. Such formulae are based on much larger aircraft than our design. Therefore, the calculation was repeated using general aviation formulae. This resulted in a prediction of about 18 per cent M_{TO}. This is also regarded as too high and not representative of our aircraft. The high value of the estimate may be due to the sensitivity of the formulae to the high value for aspect ratio. The difficulties that arise from the prediction of aircraft mass for unusual/novel designs are not untypical in advanced project design studies. In the early design stages, all that can be done to overcome these difficulties is to make relatively crude assumptions and to remember to check these as soon as more structural details are available.

Without better guidance, we will average between the two results that have been produced. As the bracing structure will reduce the wing internal loading and as we expect to use high strength composite construction, we will reduce the estimate by 30 per cent as shown below:

Civil aircraft prediction	879 kg	(1938 lb)
GA aircraft prediction	1720 kg	(3597 lb)
Average value	1299 kg	
Less 30%	390 kg	
Predicted wing structure	909 kg	(2004 lb)

Add to this an allowance for surface controls and winglets (10 per cent) = 91 kg
Add 20 kg for each mid-span pod structure = 40 kg

The wing brace structure mass can be estimated by assuming a tube (100 mm diameter × 1 mm thick) and measuring the brace length from the layout drawing (8 m). Note: with these sizes for the brace it may be impossible to avoid the strut buckling from loads in a heavy landing. An aluminium alloy material with a density of 2767 kg/m^3 gives:

> Brace mass (each) = $(\pi \cdot 100 \cdot 1 \cdot 8)$ 2767/$(1000 \cdot 1000)$ = 7 kg
> Add 10 kg (22 lb) for fairing and support structure and add a contingency of 25 per cent:
> Total brace mass (both) = $2 \cdot (7 + 10) \cdot 1.25 = 42$ kg

Hence, total wing mass (including surface controls, pods and brace):

Structure.	909	/	2004
Controls, etc.	91	/	201
Pods	40	/	88
Brace	42	/	92
	1082 kg	/	2385 lb (11.8 M_{TO})

At 11.8 per cent M_{TO} this is slightly higher than modern conventional wing structures but the high aspect ratio and large wing area probably are correctly represented.

Tail surfaces

The mass of the vertical tail is estimated using a typical civil aircraft mass ratio of 28 kg/m^2 (of exposed area). The fin and rudder areas on our aircraft are larger than normal due to the short tail arm and long forward fuselage. Scaling from the aircraft layout drawing gives an area of 6 m^2. Using the same mass ratio as conventional designs predicts the mass at 168 kg (370 lb).

This represents a mass of over 2 per cent M_{TO}. This is larger than normal but reflects the large area. As the wing is mounted on top of the fin structure, a penalty of 10 per cent will be added. The vertical tail mass is therefore estimated as 185 kg (408 lb).

The tailplane/elevator structure (i.e. horizontal tail surfaces) on our aircraft is integrated into the wing. To allow for an increase in structural complexity and for the optional canard control a mass of 1 per cent M_{TO} (=92 kg) will be added to the tail structure mass:

$$\text{Tail mass} = 185 + 92 = 277 \text{ kg (611 lb)}$$

This represents 3 per cent M_{TO}, which is typical of many aircraft

Body structure

The mass of the body is estimated using civil aircraft formulae reduced by 8 per cent to account for the lack of windows, doors and floor. For the body size shown on the drawing, the civil estimate is 808 kg. Therefore, our estimate is 743 kg (1638 lb). This represents 8 per cent M_{TO} which seems reasonable.

The body structure on our aircraft is complicated by a number of special features. These must be taken into account in the estimation:

- add 4 per cent for fuselage mounted engines,
- add 8 per cent for the fuselage brace/undercarriage attachment structure,
- add 10 per cent to allow for the modular fuselage equipment provision.

Hence, body mass = $1.04 \times 1.08 \times 1.10 \times 743 = 883$ kg (1947 lb).

This is 9.6 per cent M_{TO} which is higher than normal but accounts for the complex nature of the fuselage structural framework.

Nacelle mass

Engine nacelle mass is estimated using civil aircraft formulae related to the predicted thrust of 21.6 kN (4856 lb) (i.e. $2 \times 12.9 = 25.8$ kN (5800 lb)). This is acceptable as the installation is comparable to rear mounted engines on civil business jets. The nacelle mass prediction is 147 kg (324 lb) (i.e. 1.65 M_{TO}).

Landing gear

The undercarriage on the aircraft is expected to be straightforward and relatively simple therefore a value of 4.45 per cent M_{TO}, which is typical of light aircraft, is proposed:

$$\text{Landing gear mass} = 0.0445 \times 9200 = 409 \text{ kg (902 lb)}$$

For aircraft balance, it will be assumed that 15 per cent of this mass is attributed to the nose unit (61 kg/135 lb), leaving 348 kg/767 lb at the main unit position.

Flying controls

This item has been included in the wing structural mass estimation.

Propulsion group mass

For large turbofan engines with BPR of 5.0 the basic (dry) mass ratio is predicted from published engine data to be 14.4 kg/kN. This would give a mass of ($14.4 \times 21.6 = 311$ kg/686 lb). Smaller engines with lower BPR would not achieve this value due to the effects of descaling. Data from the suggested engine gives a dry weight for each engine of 632 lb (287 kg). With two engines this gives a total dry-engine mass of 573 kg (1263 lb). There is a substantial difference between these estimations but as the largest one is from an existing engine this will be used. The engine services and systems will increase the dry mass. Typical civil aircraft incur an additional 43 per cent:

$$\text{Propulsion group mass} = 1.43 \times 573 = 820 \text{ kg (1808 lb)}$$

Fixed equipment mass

For conventional aircraft, this mass group would fall within the range 8 to 14 per cent M_{TO}. Our aircraft is not typical as the equipment forms a major subsystem. Observation, monitoring, communication and intelligence gathering equipment will be used on the aircraft on different missions. Versatility of equipment installations will be an essential feature on the aircraft. As discussed earlier, this operational flexibility has been addressed by allowing 800 kg (1764 lb) of equipment mass to be assumed as 'useful load'. However, to support the operational equipment modules the aircraft will need to have some fixed equipment services (e.g. power supplies). It will also require systems to allow the aircraft to function (e.g. hydraulic, electrical, fuel supply, etc.). Some of the systems found on conventional aircraft will not be necessary due to the absence of the cockpit and pilot (e.g. instruments, controls, environmental controls and protection, safety, furnishings). Until more details are available on the systems to be installed we will assume that the fixed equipment accounts for 8 per cent M_{TO} (=736 kg/1623 lb).

Equipment requirements above this figure will be transferred to the previously described (800 kg/1764 lb) 'useful load' component.

Fuel mass

Until a more detailed aerodynamic and performance analysis is done, the previously estimated fuel load of 4693 kg (10 348 lb) will be assumed. As this presents a substantial component to the overall aircraft mass (51 per cent M_{TO}) it is important to carefully estimate the fuel requirements as soon as possible.

9.8.2 Aircraft mass statement and balance

From the sections above it is now possible to compile the detailed aircraft mass statement (see Table 9.2).

The empty mass fraction at 44 per cent is higher than assumed (38 per cent) in the initial sizing. This has increased the aircraft M_{TO} to a value above the 9200 kg (21 168 lb) design mass. A further iteration should have been done to estimate more accurately the component masses and ultimately the M_{TO}. However, as several of the component masses and the fuel mass are based on crude assumptions it is not appropriate to go into such detail at this stage.

The mass statement can be used to determine the position of the aircraft centre of gravity (as described in Chapter 2, section 2.6.2). This will confirm, or otherwise, the assumed longitudinal position of the wing relative to the fuselage as shown on the aircraft layout drawing. The component masses are located around the aircraft structure as shown in Figure 9.17.

These are used to predict the position of the aircraft centre of gravity for different loading conditions. With a datum set at one metre ahead of the aircraft nose the results are:

- at M_{TO}: $x_{cg} = 10.05$ m   33.0 ft   (51 per cent MAC)
- at M_{TO} less body fuel: $x_{cg} = 9.61$ m   31.5 ft   (40 per cent MAC)
- at empty mass: $x_{cg} = 10.04$ m   32.9 ft   (51 per cent MAC)
- at M_{TO} − useful load: $x_{cg} = 10.3$ m   33.8 ft   (58 per cent MAC)

Table 9.2

	kg	lb	% M_{TO}
Wing structure	1082	2386	11.0
Tail structure	277	611	2.8
Body structure	883	1947	9.0
Engine nacelles	147	324	1.5
Landing gear	409	902	4.1
Total structure	2798	6170	28.4
Propulsion group	820	1808	8.3
Fixed equipment	736	1622	7.5
Aircraft empty	4354	9600	44.2
Useful load	800	1764	8.1
Fuel load	4695	10352	47.7
Aircraft M_{TO}	9849	21716	100.0

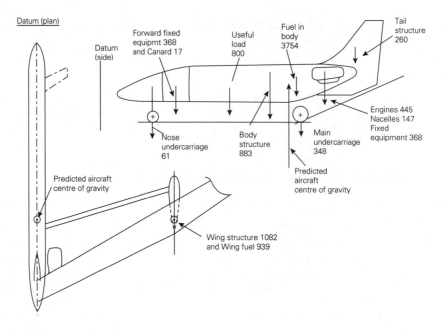

Fig. 9.17 Aircraft balance

The values quoted in parentheses above are the cg positions as percentages of the mean aerodynamic chord (aft of the leading edge). This analysis shows that the wing mean chord position should be moved rearward with respect to the datum. On our design, this is most easily achieved by reducing the sweep angle. Due to the lack of confidence in the component mass estimation at this stage, no changes will be made (yet). It is reassuring to note that even in the present unbalanced configuration the cg range is acceptable and that ballasting to reduce the range does not seem to be necessary.

Although a number of small changes to the aircraft initial layout have been suggested in the mass and balance analysis, it has confirmed the feasibility of the design and provided data for subsequent calculations.

9.8.3   Aircraft drag estimations

The initial drag evaluation will be done using the conventional component drag breakdown and applying the equation below:

$$C_D = C_{Do} + C_{Di} + \Delta C_{Dw}$$

Aircraft cruise speed is set at subtransonic flow conditions. This makes the wave drag component zero. The parasitic drag coefficients (C_{DO}) are evaluated, for each aircraft component, by estimating the terms in the following equation:

$$C_{Do} = \sum (C_{Do})_i = \sum \{(C_f)_i \cdot F_i \cdot Q_i (S_{wet}/S_{ref})\}$$

where C_f = skin friction coefficient
  F_i = form (shape) factor
  Q_i = interference factor
  S_{wet} = component wetted area
  S_{ref} = wing reference area
  S_{ref} = 50 sq. m (537 sq. ft) for our aircraft

Formulae used for the above estimation can be found in most aerodynamic or aircraft design textbooks (e.g. reference 7). Geometrical inputs are scaled from the layout drawing. The results (with a reference area of $50\,m^2/537$ sq. ft) are shown in Table 9.3.

9.8.4   Aircraft lift estimations

To reduce complexity and to avoid drag increases in cruise, the aircraft will be manufactured without conventional flaps. If it is found necessary to increase C_L for landing or take-off, the aileron surfaces could be drooped or a simple leading edge device used. These possibilities will not be considered in the initial layout. Assuming a cambered supercritical wing profile is used, the two-dimensional max. lift coefficient may be 1.65 for our high aspect ratio clean wing.

The three-dimensional value is determined below:

$$(C_{Lmax})_{3D} = 0.9(C_{Lmax})_{2D} \cdot \cos \Lambda$$

Assuming quarter chord sweep $\Lambda = 22°$ gives $(C_{Lmax})_{3D} = 1.4$.
  This confirms our original assumption.

Table 9.3

	Flight cases			
	Cruise	Take-off	OEI climb*	Landing
Airspeed m/s/kt	210/408	37/72	55/107	40/78
		$(0.7V_2)$	(V_2)	
Altitude km/1000 ft	18/59	SL	SL	SL
Mass kg/lb	7820/17 243	9200/20 280	9200/20 280	4976/10 970
C_{Do} ($\times10^4$) fuselage	26.9	26.7	25.2	26.4
wing	58.3	56.1	51.6	55.1
braces	16.8	16.4	15.2	16.2
tail	7.2	7.6	7.2	7.5
nacelles	9.7	9.6	9.0	9.4
C_{Do} total basic	119.4	116.4	108.2	114.6
Add undercarriage	—	104.0	—	104.0
Add trim	2.0	5.0	65.0	5.0
Note: no flaps on this aircraft				
C_{Do} total (incl. contingency)	129.7	237.1	173.1	235.2
C_L	0.592	0.974	0.976	0.996
Induced drag factor	0.022	0.022	0.022	0.022
C_D total ($\times10^4$)	206.9	445.8	381.9	453.5
Lift/Drag (initial cruise)	28.6			
C_L end of cruise ($M = 4976$)	0.376			
Lift/Drag (final cruise)	23.4 (with no height gain)			

*OEI = one engine inoperative at the start of climb, i.e. emergency take-off case.

At the initial cruise speed and height, the design lift coefficient will be 0.59, as shown in Table 9.3.

Much more work would need to be done in designing the best wing section profile. This would entail the application of sophisticated CFD methods that are not practical in the initial design stages. However, the calculations above seem to be reasonable and will provide values for use in the performance calculations that follow.

9.8.5   Aircraft propulsion

The previous initial sizing work identified the required engine parameters and a candidate powerplant. Reference 8 provides formulae to determine engine performance at specified operating conditions (speed and height). The manufacture's quoted values (per engine) for the selected engine at static, sea-level, take-off conditions are:

$$\text{Thrust} = 2900\,\text{lb} \ (12.9\,\text{kN})$$
$$\text{Specific fuel consumption} = 0.55\,\text{lb/lb/hr or N/N/hr}$$

Howe's formulae[9] applied to the cruise condition (M0.7, 18 km) with BPR of 3.3 estimates thrust at:

$$T/T_o = 1[0.88 - (0.016 \cdot 3.3) - (0.3 \cdot 0.7)]0.985^{0.7} = 0.166$$
$$\text{Hence, } T = 12.9 \times 0.166 = 2.15\,\text{kN} \ (48\,\text{lb})$$

Using the same formula, the thrust per engine at the end of take-off (and for OEI) is calculated as $T = 11.37\,\text{kN} \ (2556\,\text{lb})$.

And specific fuel consumption (C):

$$C/C_o = [1 - (0.15 \cdot 3.3)^{0.65}][1 + 0.28(1 + 0.063 \cdot 3.3^2)0.7]\sigma^{0.08}$$
$$\text{Giving, } C = 0.493$$

9.8.6   Aircraft performance estimations

Initial estimates of aircraft performance are based on methods described in most aircraft design textbooks (e.g. references 7 to 10). Point estimates are required to determine the suitability of the aircraft layout to the operational requirements. Three flight phases are investigated:

- field performance,
- climb performance,
- cruise.

Field performance

Although it could be possible to assess the take-off and landing performance using step integration of the aircraft path, it is sufficient in these early stages to use generalised formulae.[10] Four calculations will be made:

(a) stall and operating speeds,
(b) take-off distance,
(c) second segment climb,
(d) landing distance.

(a) As the aircraft wing has been simplified by the avoidance of flaps, the C_{Lmax} is the same for the take-off and landing configurations. The take-off will be calculated at the maximum mass (9200 kg) and the landing at a reduced mass (10 per cent fuel plus full payload) of 4976 kg. An emergency landing calculation will also be done for the aircraft at M_{TO} less 10 per cent fuel. The operating speeds are determined below:

$$V_{stall} = (W/S) (2/(\rho \cdot C_{Lmax}))^{0.5}$$

where $S = 50$ sq. m (537 sq. ft)
$\rho = 1.225 \, kg/m^3$ (0.002378 slug/cu. ft)
$C_{Lmax} = 1.4$

Giving: at take-off, $V_{stall} = 45.9 \, m/s$ (89.1 kt)
take-off speed $V_2 = 1.2 V_{stall} = 55.1 \, m/s$ (107 kt)
at landing, $V_{stall} = 33.7 \, m/s$ (65.4 kt)
landing approach speed $V_A = 1.3 V_{stall} = 43.9 \, m/s$ (85.2 kt)
at emergency landing, $V_{stall} = 44.7 \, m/s$ (86.8 kt)
emergency approach speed $V_A = 1.3 V_{stall} = 58.1 \, m/s$ (112.8 kt)

The normal take-off and approach speeds seem reasonable. As commented on previously, the high speed that is required for the emergency landing case could be reduced if fuel dumping was included in the fuel system.

(b) Take-off distance can be calculated by the formula[10] below (note: the formula in this book is derived in ft-lb units, therefore some conversion will be needed to transform to SI units (see Appendix A)):

$$S_{TO} = 20.9[(W/S)/(\sigma \cdot C_{Lmax} \cdot (T/W)] + 87[(W/S) (1/(\sigma \cdot C_{Lmax})]^{0.5}$$

The two terms in square brackets are for the ground roll (with a rolling friction coefficient of 0.03) and the climb to 50 ft obstacle clearance respectively:

$$(W/S) = 20286/537.5 = 37.74 \, lb/sq. \, ft$$

$$(T/W) = 4856/20286 = 0.239$$

$$\sigma = 1, C_{Lmax} = 1.4$$

$$\text{Hence, } S_{TO} = 2357 + 452 = 2809 \, ft \, (857 \, m)$$

(c) The second segment climb calculation is a check on the ability of the aircraft to climb away from the ground after an engine failure on take-off. The aircraft 'rate of climb' (RoC) is calculated by:

$$RoC = (V/W) (F_N - D)$$

where $V = 1.2 V_{stall}$
F_N = emergency thrust from the remaining engine
D = aircraft drag with the landing gear retracted but with an asymmetric flight attitude to counteract the adverse yaw from the engine thrust/drag

In our case: $V = 55.1 \, m/s$ (107 kt)
$F_N = 11.37 \, kN$ (2556 lb)
$D = (0.5\rho V^2) SC_D = 1853 \times 50 \times 0.03819 = 3538 \, N$ (795 lb)

Hence, $RoC = [55.1/(9200 \cdot 9.81)] (11370 - 3538) = 4.78 \, m/s$ (940 fpm)

We can also calculate the aircraft climb gradient $= \sin^{-1}$ (RoC/V) $= 0.08$ (i.e. 8 per cent which is satisfactory) (note: the minimum value for civil transport aircraft is 2.4 per cent).

(d) The landing distance is calculated using standard formula[10] (in ft-lb units) is:

$$S_L = 118(W/S)/(\sigma \cdot C_{Lmax}) + 400$$

For normal landing: $W/S = 10972/537.5 = 20.4$ lb/sq. ft
For emergency landing: $W/S = 19252/537.5 = 35.8$ lb/sq. ft
With: $\sigma = 1$ and $C_{Lmax} = 1.4$:
Hence: S_L (normal) $= 2119$ ft (646 m)
        S_L (emergency) $= 3417$ ft (1042 m)

It may be desirable to add airbrakes/lift dumpers to the aircraft to increase drag on landing and thereby reduce the landing distances shown above.

Climb performance

The climbing ability of aircraft is a function of the aircraft airspeed, weight and engine setting (for our aircraft the engine power/setting is constant). The best rate of climb for a given weight and thrust will be at the speed for minimum drag. For most operations this speed is too slow. The aircraft is normally flown at a higher speed and often accelerated during climb. This acceleration sacrifices some climbing ability but has the advantage of gaining ground distance and, at the top of climb, matching the climb to cruise speed.

When a full performance estimation is produced it will show the time to climb to specific heights and the associated ground distance covered. With our current degree of knowledge and confidence with the aircraft parameters, such detailed analysis is not appropriate. To predict the climb performance a point analysis will be all that is necessary. This will use the mass, drag and engine thrust data, as described in the previous sections, and an assumed flight speed:

$$\text{RoC} = (V/M \cdot 9.81)\,(F_N - D)$$

The climbing speed profile is assumed to be 250 kt EAS (129 m/s) up to about 30 000 ft (9 km) then Mach 0.7.

Aircraft mass is set at $0.95 M_{TO} = 8740$ kg (19 270 lb) throughout the climb (a simplifying assumption as fuel is continuously burnt). The results of this calculation are shown below and the rate of climb variation plotted in Figure 9.18.

Altitude km	1.4	4.6	9.0	14.0	18.0
ft	4600	15 080	29 500	45 900	59 000
True airspeed m/s	138	163	209	206	206
kt	268	316	406	400	400
Drag N	5362	5486	5552	3457	2930
lb	1205	1233	1248	777	659
Thrust per engine N	8147	6289	4075	2440	1572
lb	1832	1414	916	549	353
RoC (both engs) m/s	17.6	13.5	6.3	3.4	0.51
fpm	3458	2645	1244	674	101

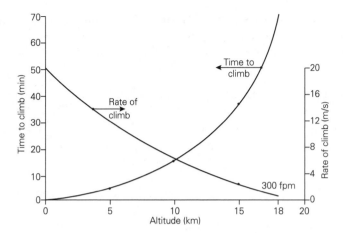

Fig. 9.18 Rate-of-climb and time-to-climb

Time to climb can be roughly calculated using the average values from the RoC graph:

Stage	0–5 km	0–10 km	0–15 km	0–18 km
(Stage to	16 400 ft	32 800 ft	49 200 ft	59 000 ft)
Time min.	5.2	14.5	35.3	68.6

(note that it will take over one hour to climb up to the 18 km operation altitude).

From these calculations it is clear that the cruise altitude of 18 km represents the aircraft service ceiling (normally defined as 100 fpm), at the aircraft conditions assumed. In addition, the calculations show that the time to climb the final 3 km almost doubles the time to reach cruise height. This does not seem to be a sensible operational practice. Either the aircraft weight or drag must be reduced, or the available thrust increased. For example, similar calculations show that when the aircraft mass is reduced to 7500 kg a climb rate of 300 fpm will be possible at 18 km. Alternatively, a different operational practice may be used (e.g. start the mission at a lower altitude and increase this as the aircraft weight is reduced through fuel burn).

Cruise

Several operational strategies can be adopted for the cruise phase. The one to be used in our analysis is to fly the aircraft at a constant angle of attack (constant Mach number). This implies that as the fuel is used and the aircraft becomes lighter, the aircraft gains height. This is known as a cruise-climb profile.

The usual Breguet range equation can be written as:

$$R = (V/c)(L/D) \log_e(M_1/M_2)$$

This gives a maximum value when aircraft speed is 1.316 times the speed for minimum drag. As mentioned above, this speed may be too slow for the aircraft at high altitude where the allowable speed range is narrow. We will cruise at a speed of M0.7. Using the above equation, it is seen to be operationally desirable to start the cruise at a lower altitude than the originally specified 18 km. This confirms the climb result. The aircraft lift to drag ratios for cruise at 15 and 18 km is calculated to be 26.5 at 15 km and 28.6 at 18 km cruise height.

In the cruise phase the true airspeed at M0.7 is 206.5 m/s (400 kt), and cruise sfc = 0.493 per hour. We will calculate the aircraft range taking no account of the fuel used in the climb and descent phases. To correct this assumption we will add 10 per cent to the calculated fuel mass. Hence:

$$\text{Starting mass} = M_1 = M_{TO} = 9200\,\text{kg (20 280 lb)}$$
$$\text{End mass} = M_2 = M_{(\text{operational empty})} + M_{\text{payload}} = 4354 + 800$$
$$= 5154\,\text{kg (11 365 lb)}$$

Hence range, $R = (206.5/[0.493/3600]) \times 26.5 \log_e(9200/5154)$
$$= 23\,154\,\text{km (12 500 nm)}$$

It is also possible to rearrange the Breguet equation to calculate endurance (E):

$$E = (1/c)(L/D)\log_e(M_1/M_2) = 31.14\,\text{hours}$$

This flight time is more than specified; therefore we can determine the mass fraction for the required 24 hour endurance:

$(M_1/M_2) = e^x$, where $x = (E \cdot c)/(L/D) = (24 \cdot 0.493/26.5)$
$(M_1/M_2) = 1.563$
$(M_1/M_2) = M_{TO}/M_2$, which assuming $M_{TO} = M_2 + M_{\text{fuel}}$ gives:
$M_{\text{fuel}} = 0.563$, $M_2 = 2902\,\text{kg (6400 lb)}$, adding 10 per cent to this value gives:
$M_{\text{fuel}} = 3192\,\text{kg (7038 lb)}$

This is substantially less than estimated very crudely in section 9.7.1 (4693 kg). This reduction will make the fuel tankage easier to provide as only 3.99 m^3 will be necessary. We can now estimate the new value of M_{TO} and the mass fractions:

$$M_{TO} = 4354 + 800 + 3192 = 8346\,\text{kg (18 400 lb)}$$
$$M_{\text{fuel}}/M_{TO} = 38.5\,\text{per cent}, M_{(\text{operational empty})}/M_{TO} = 52.2\,\text{per cent}$$

This empty mass fraction is much higher than previously assumed but lies between the Stratos 1 and Stratos 2 values.

We can now determine the engine cruise rating (as a percentage of the max. available) at the operating condition, as shown below.

The aircraft drag at the start and end of the cruise is estimated to be 3236 N and 2086 N (727 lb to 470 lb). The engine thrust at 15 and 18 km is estimated at 4352 N and 3126 N (978 to 703 lb) (both engines operating). Hence the engine rating (aircraft drag divided by available thrust) will be:

75 per cent at 15 km (49 200 ft) at start of cruise
67 per cent at 18 km (59 000 ft) at end of cruise

These seem to be reasonable cruise ratings from maximum and will extend the life of the engine (between overhauls) due to the associated lower operating temperature.

If, however, the aircraft is flown at constant altitude these become:

62 per cent at end of 15 km (49 200 ft) cruise
85 per cent at start of 18 km (59 000 ft) cruise

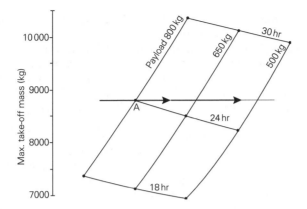

Fig. 9.19 Parameter trade-off study

9.9   Trade-off studies

At this stage in the development of the project, we have sufficient detail to conduct some trade-off studies. Using the engine and aerodynamic equations, it is possible to show the relationship between the two main operational parameters (namely payload and endurance) on the aircraft maximum mass. A classical nine-point carpet plot will be constructed. The endurance will be varied between 18 and 30 hours and the payload reduced from 800 to 500 kg (1764 to 1102 lb). The results are shown in Figure 9.19.

The study shows that the aircraft is relatively insensitive to changes in payload but that endurance is a very influential parameter. A horizontal line drawn across the plot provides an indication of the trade-off between payload and endurance. For example, moving from the design point (A) to a lower payload (500 kg/1102 lb) and substituting extra fuel to replace the lost payload (providing that sufficient tankage is available) allows two extra hours of flight.

It is possible to conduct similar studies to illustrate the effects of varying the following parameters:

- wing area versus aspect ratio,
- cruise altitude versus aircraft maximum mass (M_{TO}),
- system mass versus M_{TO},
- introduction of advanced technologies (e.g. laminar flow, composite materials, etc.),
- variation in mission requirements.

Such studies provide a detailed appreciation of the factors and parameters affecting the aircraft design space. The results of such studies are used to revise the aircraft layout and specification to produce a solution better matched to the design brief.

9.10   Revised baseline layout

The main changes to the initial aircraft layout, from the work done so far, are associated with the provision for adequate lateral (weathercock) stability and control. The long-span, forward-swept wing with winglets, and the long forward fuselage with deep side area, will generate destabilising moments in cross-wind conditions. Balancing these moments is difficult due to the relatively short tail arm. Two modifications are proposed to ease this problem. The forward fuselage length is to be reduced by 2 metres and 'finlets' are to be placed on the wing outboard of the inner wing trailing edge control surfaces. These finlets could be made large enough to double the original fin area if required. Some of the loss of equipment volume resulting from the reduction of the fuselage length could be regained by moving the fuselage fuel tank further back and increasing the amount of fuel held in the wing. These two proposals together should provide sufficient flexibility into the layout to overcome the perceived stability problem.

A second concern relates to the layout of the landing gear. The large wing-span, high aircraft centre of gravity and the narrow main-wheel track combine to make the aircraft potentially unstable in taxi, take-off and landing conditions. The reduced length of the forward fuselage mentioned above will improve the landing gear geometry but this will not be sufficient. It will be necessary to increase the track of the main wheels. This can only be done by adding fuselage sponsons at the main undercarriage mounting positions. Increasing the track to 4 m will provide an overturning angle of about 52° (convention suggests that an angle greater than 60° is unsafe or twitchy in operation). The sponsons will need to be extended fore and aft to provide aerodynamic blending. These extensions will provide extra storage. This new arrangement will also improve the attachment geometry of the braces at the side of the fuselage.

Following the calculations of the component masses and the associated aircraft centre of gravity assessment the wing leading edge sweep will be reduced from 30 to 25°.

The above changes have been included into a revised aircraft general arrangement drawing, see Figure 9.20.

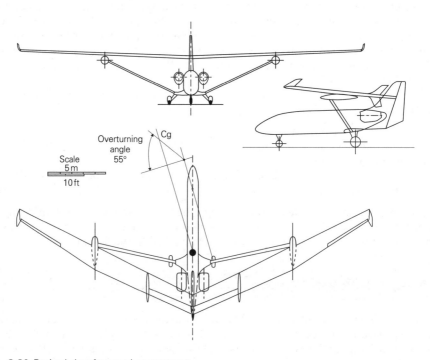

Fig. 9.20 Revised aircraft general arrangement

9.11 Aircraft specification

9.11.1 Aircraft description

Aircraft type: Manned or uninhabited high-altitude, long-endurance, reconnaissance vehicle.

Design features: The novel aircraft layout, with a high aspect ratio, multi-tapered, swept-forward braced wing planform, provides a platform for the mounting of alternative payloads and systems. The wing profile employs a supercritical section. The clean wing is unflapped with outboard ailerons and inboard elevators. The fuselage is configured to allow equipment modules to be quickly changed, giving unique flexibility in operation. One of the forward modules offers the alternative of either a manned cockpit capsule or an autonomous unmanned flight control system. The twin turbofan engines are developments from a similar type currently used on business jets (e.g. P&W of Canada PW-530/545). The tricycle retractable landing gear is of conventional design.

Operational features: The mission profile includes 24 hour flights at 45 000 to 65 000 ft altitude at Mach 0.7. Operation into and from conventional military airfields. Optional detachment of wings and brace structure for rapid deployment to operational theatre.

Structure: Conventional glider-technology composite structural framework with rapid access to interchangeable fuselage equipment modules. Fuel held in integral wing tanks and central fuselage bladder tanks.

Equipment: Space provision for reconnaissance and communication packages to suit variable operational missions.

9.11.2 Aircraft data

		SI units	Imperial units
Dimensions:	Overall length	12.7 m	41.6 ft
	Overall span (incl. winglets)	30.0 m	98.4 ft
	Overall height (incl. winglets)	5.6 m	18.4 ft
	Wing aspect ratio	18	
	Wing taper ratio	0.5 outer, 0.75 inner	
	Wing LE sweep	25° forward	
	Wheelbase	6.6 m	22.6 ft
	Track	4.0 m	13.1 ft
Areas:	Wing (ref.)	50 sq. m	540 sq. ft
	Elevator	4.0 sq. m (total)	43 sq. ft
	Aileron	0.7 sq. m (each side)	7.5 sq. ft
	Fin	2.7 sq. m (upper)	29 sq. ft
		0.5 sq. m (under)	5.4 sq. ft
Mass/Weight	Max. take-off	8665 kg	19 100 lb
	Empty	4354 kg	9600 lb
	Landing	5505 kg	12 140 lb

		SI units	*Imperial units*
	Fuel	3511 kg	7742 lb
	Fuel (US gal)	1200	
	Equipment/useful load	800 kg	1764 lb
Loadings:	Max. wing loading	1637 N/sq. m	331 lb/sq. ft
	Thrust (SSL)/weight	0.315	
Engines (each):	SSL take-off thrust	12.9 kN	2900 lb
	SFC take-off	0.54/hr	
	SFC cruise	0.49/hr	
	Bypass ratio	3.3	
	Dry mass/weight	287 kg	633 lb
Aerodynamic:	C_{Do} (cruise) 50 000 ft @ M0.7	0.01297	
	C_{Do} (landing) SL @ V_A	0.02352	
	C_L (cruise) 50 000 ft @ M0.7	0.60 (initial), 0.38 (final)	
	C_{Lmax}	1.4	
	C_L (landing)	1.0	
	L/D (cruise)	28.6	
Performance:	*Mission*		
	Cruise speed (M0.7 @ 50 000 ft) 207 m/s,		400 kt
	Service ceiling 45 000 ft (initial), 65 000 ft (final)		
	Endurance	26 hours	
	Field		
	Stall speed (V_S)	45.9 m/s	89.1 kt
	Take-off speed (V_2)	51.1 m/s	107 kt
	Approach speed (V_A)	43.9 m/s	85.2 kt
	Take-off distance	857 m	2809 ft
	Second segment climb OEI	0.08	
	Landing distance	646 m	2119 ft
	Climb		
	Rate of climb (initial)	3458 fpm @ 268 kt	
	Time to climb, to 10 km	14.5 min	
	(from TO)　　　to 15 km	35.3 min	
	to 18 km	68.6 min	

9.12　Study review

There are many further design considerations to be studied in the development of this project. The aircraft is a complex combination of advanced technologies in aerodynamics, structures, materials, stability and system integration. This represents a substantial challenge which reflects the nature of future aircraft project work. As aeronautical design matures it will become harder to make significant improvements to current designs. This will force aeronautical engineers to introduce innovation into new designs. The ability to handle the necessary analysis methods to reduce technical risk will form a major feature of future design teams. These teams will include many more specialists from disciplines that have not been traditionally included in aircraft project design. Organising, managing and controlling these teams will demand skills other than those conventionally related to aeronautical engineering. A more 'system-orientated' approach will become the new practice.

As well as dealing with the integration of new technologies and methods, this project has involved the analysis of aircraft operating in the higher atmosphere. In this environment, the stall and buffet flight boundaries begin to converge to make control more difficult. High-speed, high-alpha must be carefully considered to ensure that the aircraft is dynamically stable yet, as in this case, aerodynamically efficient. This combination offers a serious test to the aerodynamic and structural disciplines.

This project has demonstrated the unique features of designing an aircraft to account for:

(a) uninhabited/autonomous missions,
(b) fast and high operation,
(c) system and airframe integration,
(d) the introduction of new technologies and methods.

Not many new projects incorporate such a mixture of challenges to the design team. However, if the difficulties of meeting such demands can be successfully achieved without jeopardising aircraft operational integrity, then we will be in the enviable position of 'pushing the envelope'.[11] Good luck!

References

1 Kampf, K. P., 'Design of an unmanned reconnaissance system', ICAS 2000, Harrogate UK, August 2000.
2 *AIAA Aerospace Design Engineers Guide*, AIAA Publications, ISBN 0-939403-21-5, 1987.
3 *Brassey's World Aircraft & Systems Directory*, Brassey Publications, ISBN 1-57488-063-2.
4 *Jane's All the World's Aircraft*, Jane's Annual Publication, various years. See www.janes.com for list of publications.
5 Lange, R. H., 'Review of unconventional aircraft design concepts', *Journal of Aircraft 25*, **5**: 385–392.
6 Ko, A. *et al.* 'Effects of constraints in multi-disciplinary design of a commercial transport with strut-braced wings'. AIAA/SAE World Aviation Congress 2000/1, paper 5609. See also Gundlach, J. T. *et al.* 'Concept design studies of a strut-braced wing, transonic transport'. *AIAA Journal of Aircraft*, Vol. 137, No. 6, Nov-Dec 2000, pp. 976–983.
7 Jenkinson, L. R. *et al.*, *Civil Jet Aircraft Design*, Butterworth-Heinemann, 2000, ISBN 0-340741-52-X.
8 Eshelby, M. E., *Aircraft Performance, Theory and Practice*, Butterworth-Heinemann, 2001, ISBN 0-340758-97-X.
9 Howe, D., *Aircraft Conceptual Design Synthesis*, Professional Engineering Publishing, UK, ISBN 1-86058-301-6.
10 Nicholai, L. M., *Fundamentals of Aircraft Design*, METS Inc., San Jose, California 95120.
11 Rabinowitz, H., *Pushing the envelope*, Metro Books, 1998 (www.metrobooks.com).

10

Project study: a general aviation amphibian aircraft

10.1   Introduction

Many early aircraft designs were developed for take-off and landing on water. With the absence of readily available level and mowed fields, lakes and even the ocean were looked upon as ideal choices for a place to land or take-off. This allowed operation in a wide range of headings to accommodate wind direction. It also did not require any preparation for landing or take-off other than a quick look to make sure that boats or debris were not in the fight path. This provided a decided advantage over land-based operations where real estate had to be purchased or rented, obstacles (tree stumps and rocks) cleared, and grass cut to a reasonable height or a hardened earth or macadam surface prepared. In emergencies, a lake was also more likely to be clear of obstacles than a farmer's field that might be filled with cattle or bisected by a fence. Hence, many early airplane designers opted for a seaplane configuration. In the event that land operation was sought, an amphibian design offered the capability of water or land operation. In fact, due to the public's lack of confidence in airplane engine reliability, it was not until almost the mid-twentieth century that long, overwater passenger flights (transatlantic, transpacific, Caribbean, etc.) were routinely attempted in anything other than seaplanes or amphibians.

With extensive use of land-based aircraft to transport military personnel during World War II and with improvement in engine reliability, the flying public gained the confidence needed for such aircraft to replace their water-based counterparts. This allowed inland airports to replace coastal sites as ports of entry and exit for overseas flights and the large amphibians and seaplanes of the 1930s and 1940s were retired from service.

In the general aviation (GA) field, seaplanes and amphibians have always occupied a small but important niche in the marketplace, used primarily for operations into and out of remote areas where lakes were more plentiful than airports. Today, most such aircraft tend to be 'floatplanes', aircraft originally designed for land operation to which have been added rather large floats to replace the conventional wheeled undercarriage. Such aircraft are usually considerably slower in flight and more limited in performance than their original designs due to the added weight and drag of the floats. In attempts to get better overall performance, a few specialty aircraft have been designed as amphibians with a hull fuselage. However, the compromises required to allow both land and water operations have still resulted in added weight and complexity, and a lower cruise speed than conventional land-based aircraft designs.

In the following summary of the design process, emphasis will be placed on the factors unique to amphibian aircraft. Consideration of aspects of the process that are common to all aircraft designs will be given more cursory coverage.

10.2   Project brief

The design of a modern, general aviation airplane for operation on both land and water proved an interesting challenge for a group of aerospace engineering students. They wanted to enter their design in the National General Aviation Design Competition sponsored by NASA and the Federal Aviation Administration in the United States in the late 1990s.

In this case, the 'customers' for the aircraft being designed consisted of a group of judges in a design competition and the original 'specifications' for the design were the competition guidelines. Some of these guidelines were rather broad. They included basic goals of promoting the development of designs for aircraft or related systems that would result in the modernization of general aviation programs in the United States.

10.2.1 Aircraft requirements

Specific design guidelines included:

- a payload of four to six passengers/crew,
- single engine (propeller) propulsion,
- a minimum range of 800 to 1000 statute miles (1300 to 1600 km),
- a cruise speed of between 150 and 300 kt (77 to 154 m/s).

The design team set additional general goals which included matching or exceeding the performance capabilities (range, speed, climb rate, take-off and landing distances, etc.) of current, conventional, general aviation aircraft.

10.3 Initial design considerations

The need for waterborne operation places demands on the design of an amphibian aircraft far beyond those encountered in conventional planes. These include the need for a watertight 'hull' (or lower fuselage) and the consideration of buoyancy and center of gravity relationships. These must allow efficient waterborne take-off and landing and provide balance for the craft in low- and zero-speed operations in water.

Wing and engine placement are important decisions in this design process. It is essential to avoid water spray during landing and take-off interfering with the engine. A decision was necessary on the placement of the propulsion unit. Two options are possible, the propeller and engine are either positioned in front of the aircraft and its spray, as is common in floatplanes, or above the wing where the wing and fuselage act as spray barriers. Most modern amphibians have the engine and propeller placed above the wing/fuselage, with some actually mounting the engine in the vertical tail. With this option, attention must be given to the resulting pitching moments caused by engine thrust changes. Placement of the engine above and behind the wing may also result in some interesting weight and balance problems. For both configurations, it is important to be aware of the influence of the propeller wake on aircraft components behind the engine (e.g. vertical fin, rudder, horizontal stabilizer, and the wing). If a tractor configuration is adopted, whereby the propeller is ahead of the wing, the prop-wash has both adverse and beneficial effects on the aerodynamics of the wing. This is especially critical on take-off.

10.4 Design concepts

Comparing existing aircraft, with emphasis on modern amphibian designs, resulted in the selection of a configuration similar to that shown in Figure 10.1. A sleek and relatively simple layout, with both wing and engine mounted on a single strut above the fuselage, was selected. The engine is configured as a pusher propeller. The cruciform tail placed the horizontal stabilizer in the propeller wake, enhancing pitch control during take-off, which is an important factor in take-off from water. Small span sponsons were placed slightly forward of the main wing at the base of the fuselage. This gave the aircraft a 'stagger-wing' appearance. The sponsons provide roll stability in water,

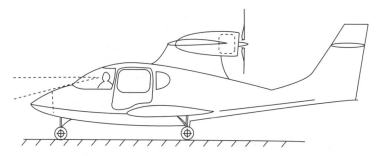

Fig. 10.1 Initial aircraft layout sketch

house the retractable landing gear for use on land, and provide some additional lift. Figure 10.1 shows the initial design layout with all components in their chosen position. This allowed the weight/mass and balance of the aircraft to be calculated.

10.5   Initial layout and sizing

Initial sizing was performed using published methods[1] with data inputs from a comparative study of existing amphibian and conventional single-engine four-place general aviation aircraft. The sizing procedure was tested against existing aircraft and found to be reasonably accurate. This gave the team confidence in the resulting estimate of 1402 kg (3092 lb) take-off gross mass (weight). This is heavier than conventional GA aircraft but is not out of line with current high performance floatplanes carrying four or more people.

10.5.1   Wing selection

The requirements of the General Aviation Design Competition demanded a cruise speed of at least 150 knots and a range of 800–1000 miles. Few current general aviation amphibian aircraft can match these requirements and meeting these criteria would be difficult. It was felt that an excellent engine and a modern, clean-wing design was needed. The Zoche diesel engine was selected to satisfy the first of these requirements. For the second, the NASA LS(1)-0413 airfoil, sometimes known as the GA(W)-2 section, was chosen because of its high lift to drag ratio, reasonable stall, and good pitching moment behavior. NACA 0009 section was selected for the vertical and horizontal tail and the NACA 0009-65 section was used for the sponson. The engine pylon also used a 9 percent thick section.

Based on an assumed design cruise speed of about 90 m/s (175 kt), at an altitude of 2286 m (7500 ft), a wing area of 16.3 m^2 (175 ft^2) was selected. A mean chord of 1.2 m (5 ft) and span of 10.7 m (35.1 ft) gave an aspect ratio of about 7. The use of a wing taper was evaluated but it was determined to be an unnecessary manufacturing complexity. A wing twist (washout angle) of 2° gave near minimum induced drag performance while not making the construction too complex.

Flaps and ailerons were sized using methods based on comparable aircraft designs.[1] However, the desire to avoid deflection of the flow through the pusher propeller led to the study of flaperons (flap/aileron combinations). It was expected that this would allow larger flap spans and improved flap effectiveness at lower angles of deflection than would be required by a more complex and heavier flap. The calculated 2.78 percent reduction in landing speed did not justify this extra complexity. Hence a conventional flap and aileron system was selected with each flap having a 2.95 m (9.66 ft) span and a 0.46 m (18 in) chord. The ailerons have the same chord but 1.3 m (4.25 ft) spans. The 2-D basic airfoil section is known to have a stall lift coefficient in excess of 2.0. The maximum 3-D wing lift coefficients are estimated as 1.7 with no flap deflection, 2.07 with 15° flap deflection, and 2.44 with 30° flap deflection. The 30° flap deflection gives a landing speed of 25.2 m/s (49 kt).

10.5.2 Engine selection

The choice of an engine for this aircraft was influenced by the need for sufficient power to provide performance comparable to conventional GA aircraft, reasonable cost, and low maintenance requirements. More unique requirements included the desire to operate on fuels other than conventional GA fuels (to allow operation in remote locations) and the desire to meet modern emissions requirements. A dozen or more commercially available conventional aircraft piston engines along with several turboprop and turbo-shaft designs from several manufacturers were evaluated. Most of the turboprop and turbo-shaft engines provided superb power to weight ratios and the promise of low cost maintenance and relatively low fuel cost; however, their initial price was several times that of their piston counterparts. The piston engines provided a proven product for efficient operation at lower cruise speeds but were comparatively heavy and most required fuel which is not universally available.

The engine selected was a modern diesel engine designed by the Zoche Aero-Diesel Company of Germany. The ZO-02A radial diesel engine promised operation at up to 300 horsepower with a mass (weight) of 123 kg (271 lb). The engine accepts a wide range of fuels including diesel, JP-4, JP-5, JetA, and even ordinary kerosene, with lower pollutant emission than comparable piston engines. The manufacturer also promised a 30 percent reduction in specific fuel consumption compared to conventional piston engines. The engine also promised a number of advanced features which would provide much easier starting and lower vibration than typically found with diesel engines. The selection of a diesel engine for this design may appear somewhat unconventional but diesels have been used in aircraft applications since the 1920s and today several advanced diesel designs are being evaluated for use in new general aviation applications.

The Hartzell Propeller Company was asked to recommend a propeller which would match the desired performance characteristics of the aircraft to the power output of the Zoche engine. Hartzell recommended a 1.78 meter (70 inch) diameter, three blade, composite, variable pitch, reversing propeller design based on the Hartzell HC-C2YR-1RLF/FL6890.

10.5.3 Hull design

The unique requirement for an amphibian aircraft is its need to take off and land on water. This operation must also include the ability to maneuver on water at low speed and to be both statically and dynamically stable.

Developing a hull for an amphibian requires the aircraft designer to become acquainted with somewhat different terminology than that usually associated with modern airplanes. The fuselage width becomes the 'beam' and the 'waterline' commonly used as a reference in aircraft design drawings takes on a more realistic meaning. Relevant hull (fuselage underside) dimensions now include the 'maximum beam', the 'step height', the forebody and afterbody 'keel angles', and the 'sternpost angle'. These and other terms are illustrated in the Figure 10.2.

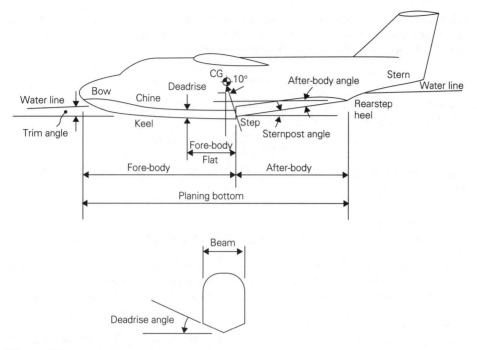

Fig. 10.2 Seaplane hull geometry (reference 2)

Because of the relative rarity of amphibious aircraft in today's marketplace, coverage of the design requirements for this type of aircraft in modern design texts is often omitted. Two exceptions are the texts of Darroll Stinton[2] and the slightly older work of David Thurston.[3] The reader is referred to these excellent references for a complete coverage of this subject. Using these texts, along with NACA TN 2503,[4] and a 1989 Dornier report,[5] a hull shape appropriate to the proposed four person amphibian aircraft was designed.

The amphibian aircraft must meet five water-related criteria:

- it must be buoyant,
- it must be statically stable when sitting in water,
- it must be dynamically stable when moving through water,
- it should be shaped to minimize water spray impingement on the aircraft during waterborne operation, and
- it must be shaped to allow hydrodynamic lift during take-off, and in so doing, to counter the suction force between the water and the hull.

In addition to these criteria, it must have sufficient power to overcome the high drag of a waterborne take-off and have enough pitch control force to counter the moments imposed by the hydrodynamic forces. Many plots and equations have been developed in references 3, 4, and 5 to aid the designer in selecting appropriate hull shapes and dimensions and in determining the location of the waterline under various loading conditions.

The process began by estimating the MTOM (TOGW) and the 'beam' of the aircraft. The selected beam width of 1.3 m (4.25 ft) was based on cabin ergonomic requirements.

The aircraft forebody length (step to bow), at 4.75 m (15.6 ft), was determined using the equation below and a graph relating the hull forebody length to beam ratio, to the gross load coefficient ($C_{\Delta 0}$) from published criteria[3]:

$$C_{\Delta 0} = \Delta_0/(wb^3)$$

where b is the maximum beam of the chine, w is the specific mass (weight) of water, and Δ_0 is the displacement of the aircraft.

$C_{\Delta 0}$ was determined to be 0.626

Using the same criteria[3] and the equation below, the spray coefficient, K, was calculated:

$$K = \Delta_0/(wbL_f^2) = 0.0465$$

L_f is the length of the forebody.

The magnitudes of both $C_{\Delta 0}$ and K point to a light spray design which will have stable landings.

Based on published recommendations, the afterbody length was determined[5] to be 114 percent of the forebody. A step is needed in the hull profile to introduce a layer of air between the water and the hull. This breaks the inherent suction force during take-off. The dimensions of this and the sternpost angle, σ (the angle between the forebody and afterbody), were found from Thurston.[3]

A simple transverse step was sized to be 10 percent of the maximum beam (0.155 m or 6 in). This should provide adequate hull ventilation and quick transition from the displacement to the planing modes on the take-off run.

The aft 1.9 m (6.24 ft) of the 4.76 m (15.6 ft) forebody is referred to as the 'forebody flat' and is designed to reduce porpoising. The keel line of the flat is inclined at 2° to improve planing effects. The 'deadrise' angle, the angle between the vee-shaped hull bottom and the horizontal, was selected at 20°. This is a compromise between the need for efficient planing and to reduce impact forces on landing.

The afterbody keel angle was determined to be 6.6°, giving a sternpost angle of 8°. The dead-rise warping of the afterbody increases linearly from 20° at the step to 40° at the stern.

The resulting lengths and angles, along with the calculated static waterlines, are presented in Figure 10.3. The waterlines are those for the static aircraft in fresh water for both the empty and fully loaded cases.

10.5.4 Sponson design

For static stability, and to a lesser degree dynamic stability, there is a need for additional surfaces to provide balance in roll when the aircraft is afloat. The static roll stability depends on the relative locations of the center of gravity and the center of buoyancy of the aircraft. The relatively high center of gravity location on an aircraft necessitates devices such as widely spaced floats or sponsons to maintain positive stability in low-speed water operation. Having made the decision not to design a floatplane, the choice was between sponsons or floats placed somewhere on the wings. Sponsons offered the promise of lower drag and added lift in flight when compared to wing mounted floats unless a heavy and complex retractable float system was designed. Wing tip floats, used on many amphibians, would have required long struts (about 6 feet) and would have resulted in high drag and high in-flight twist moments on the wings unless they were retractable. The choice of sponsons rather than floats was also based on an ability to

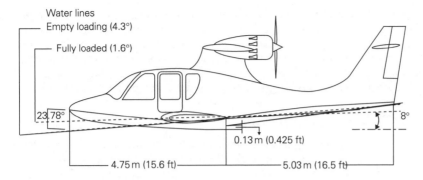

Fig. 10.3 Hull dimensions and static waterlines

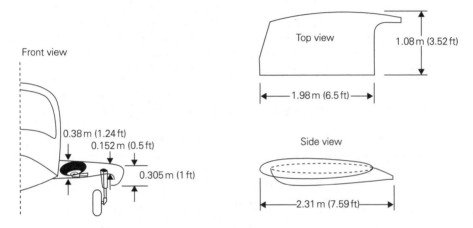

Fig. 10.4 Sponson geometry

provide a location for the main landing gear with sufficient lateral spacing to ensure stability in conventional land operations.

The sponson dimensions, shown in Figure 10.4, resulted from buoyancy and stability calculations and from the desire to use them for passenger egress, for the main gear location, and for spray suppression in take-off. The sponsons used an NACA 0009-65 section, a modified 0009 with the maximum thickness moved aft to accommodate the landing gear and tire. A 'Finch' wing tip was employed to increase the displacement and to provide a slight aerodynamic performance boost. The sponson leading edge was swept 16.7°. The sponson section profile is mounted at an angle to give a slight positive angle of attack in the take-off run for any loading situation and a near zero lift at cruise incidence.

10.5.5 Other water operation considerations

To aid in control at low speeds in water, a small, retractable water rudder is designed to be deployed from the aircraft afterbody. This is not extended during take-off or landing but, when deployed, is coupled to the control cable system to operate in co-ordination with the aircraft rudder.

To meet FAR requirements for water operations mooring hooks must be provided. These were designed to be temporarily mounted to the fuselage for water-based use. Seat cushions which are approved flotation devices and an anchor must also be provided, and oars must also be available for use in case of engine failure while in water.

Operation in conditions above 'sea state two' is not recommended. If the craft is to be parked or docked for prolonged time it should be taxied onto land.

10.5.6   Other design factors

Due to the nature of the design competition for which this aircraft was being developed, considerable attention was given in the design process to cockpit layout, pilot and passenger ergonomics, ice detection and elimination systems, and manufacturing requirements. Indeed, the unique structure of the aircraft was designed with a strong emphasis on ease of manufacturing. Extensive use of composite materials was employed with carbon fiber used with aluminum for all structurally critical areas. Inexpensive to fabricate, blown chopped fiberglass is used in low stress areas of the hull and fuselage. A complete plan for a manufacturing plant, assembly procedures, tooling requirements, quality assurance inspection procedures, and even a full analysis of production times and personnel requirements, was included in the final design report.[9]

10.6   Initial estimates

Once the amphibious features of the aircraft were developed it was possible to perform a relatively conventional analysis of the vehicle's aerodynamic behavior and flight performance and to design a satisfactory structural layout.

10.6.1   Aerodynamic estimates

Based on calculations of wetted areas and using average skin friction coefficients, the C_{D0} for the entire aircraft, referenced to the gross wing planform area, was estimated to be 0.0227. A well-known vortex lattice code developed at NASA[6,7] was used to calculate the wing and sponson aerodynamics. This was also used to determine the wing spanwise loading, under various flight conditions, for structural analysis. The Oswald efficiency factor for the wing was calculated to be 0.88. The drag polar based on these calculations is:

$$C_D = 0.0227 + 0.05154 C_L^2$$

10.6.2   Mass and balance

Using the initial sizing provided a starting point and some guesses for aircraft geometry. A statistical group weights method[1] for general aviation aircraft was used to more accurately determine component masses. The aircraft was designed with a constant 1.53 m (60.2 in) chord wing and its quarter chord is located 4.425 m (174.22 in) from the nose of the aircraft. The final mass statement showed a preliminary design estimate for MTOM (TOGW) of 1311 kg (2890 lb). This was slightly less than the initial estimate. Much of this improvement was due to the selection of a lightweight, diesel engine to power the aircraft. Table 10.1 lists the determined component masses (weights) and their positions (fuselage station and waterline) are quoted relative to the nose of the aircraft and the nominal ground plane. The fuel weight is based on 80 US gallons capacity. The component mass locations are shown in Figure 10.5.

Table 10.1 Component masses (weights) and locations

Item	Description	Mass weight (lb) kg	Fuselage stn. m (in)	Waterline m (in)
	Structures group			
1	Wing	117.9 (260)	4.65 (183.0)	2.94 (116.0)
2	Sponsons	20.0 (44)	3.64 (143.4)	0.77 (30.5)
3	Horizontal tail	9.5 (21)	9.56 (376.4)	2.95 (116.0)
4	Vertical tail	10.9 (24)	9.40 (370.0)	2.61 (103.0)
5	Fuselage	132.0 (291)	4.40 (173.3)	1.65 (65.0)
6	Nacelle	17.7 (39)	5.09 (200.4)	2.95 (116.0)
7	Main landing gear	47.6 (105)	7.07 (278.2)	0.77 (30.5)
8	Nose landing gear	19.5 (43)	0.76 (30.0)	1.02 (40.0)
	Total structures	375.1 (827)		
	Propulsion group			
9	Engine	122.9 (271)	5.77 (227.0)	2.95 (116.0)
10	Propeller	22.7 (50)	6.27 (247.0)	2.95 (116.0)
11	Fuel system	15.0 (33)	4.78 (188.3)	2.41 (94.7)
	Total propulsion	160.6 (354)		
	Fixed equipment			
12	Avionics, electronics	35.8 (79)	1.48 (58.4)	1.48 (58.4)
13	Electrical system	21.3 (47)	4.34 (170.8)	1.72 (67.7)
14	Battery	11.3 (25)	5.49 (216.0)	1.48 (58.4)
15	Hydraulic system	1.4 (3)	4.34 (170.8)	1.48 (58.4)
16	Flight controls	13.2 (29)	5.08 (200.0)	1.65 (65.0)
17	Furnishings	84.4 (186)	2.84 (112.0)	1.31 (51.4)
18	Anchor, mooring lines	13.6 (30)	3.92 (154.4)	1.20 (47.4)
	Total fixed equipment	181.0 (399)		
	Useful load			
19	Passengers	308.4 (680)	2.84 (112.0)	1.31 (51.4)
20	Baggage	68.0 (150)	3.91 (154.0)	1.20 (47.4)
21	Fuel (usable and reserve)	217.7 (480)	4.11 (162.0)	1.78 (70.0)
	Total useful load	594.2 (1310)		
	MTOM (TOGW)	1310.9 (2890)	3.99 (157.1)	1.82 (71.7)

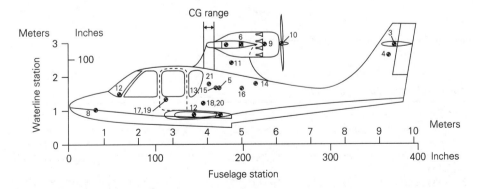

Fig. 10.5 Location of component masses

A weight and balance analysis based on the above information yielded the in-flight CG excursion diagram shown in Figure 10.6. This figure shows the envelope of possible loading configurations of the design. In this figure, the empty mass (weight) is defined as the aircraft equipped but empty. The operating masses (weights) includes the weight of the pilot and unusable fuel. This represents the minimum mass (weight) at which the airplane can fly. For this design, this represents the aft limit of CG placement. The forward-most position of the center of gravity occurs with a pilot, three passengers, baggage, and with minimal fuel. Figure 10.6 also shows the calculated control limits of the aircraft. The positive static limit allows 5 percent positive static stability. The take-off rotation limit defines the extreme position allowed for rotation at take-off with full horizontal stabilizer deflection.

The aircraft moments of inertia are needed for determination of the vehicle stability derivatives. These were calculated[1] using the component masses (weights) shown previously and summarized in Table 10.2.

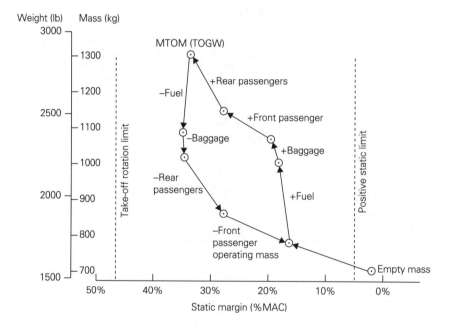

Fig. 10.6 Aircraft center of gravity diagram

Table 10.2 Aircraft moments of inertia

Loading	Mass (weight) kg (lb)	I_{xx} kg-m^2 (sl-ft^2)	I_{yy} kg-m^2 (sl-ft^2)	I_{zz} kg-m^2 (sl-ft^2)
Operating mass (weight)	796.1 (1755)	1726.7 (1273.23)	2534.3 (1868.78)	3216.6 (2371.85)
MTOM (TOGW)	1310.9 (2890)	1902.8 (1403.24)	3143.9 (2318.25)	3725.7 (2747.28)

10.6.3 Performance estimations

The available engine and propeller data plots of thrust and power were developed. These, together with the aircraft drag, are presented in Figures 10.7 and 10.8.

The aerodynamic data and the engine fuel flow data were used in standard aircraft performance and Breguet range equations. Based on Hartzell propeller performance data,

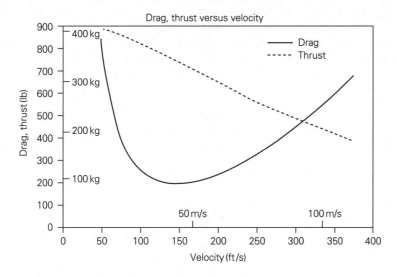

Fig. 10.7 Drag and thrust against aircraft forward speed (SL)

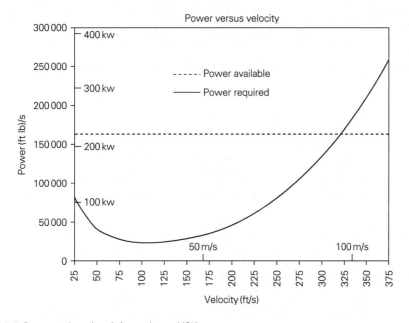

Fig. 10.8 Power against aircraft forward speed (SL)

the propeller propulsive efficiency was set at 87 per cent. The specific fuel consumption, at 75 percent power setting, is quoted by the manufacturer for the Zoche engine at 0.375 lb/hp-hr. These were used in the estimation of range and endurance.

Using the drag estimate given above, the cruise speed, at 75 percent power at 2286 m (7500 ft) altitude, was calculated to be 85 m/s (165 kt). This gave a range of 2253 km (1400 statue miles). This is well above that of most conventional general aviation aircraft at a normal cruise speed and is largely attributed to the very low specific fuel consumption quoted for the Zoche engine. Maximum range was calculated to be almost 3862 km (2400 statue miles) at a relatively slow, 48 m/s (95 kt) cruise speed.

It is obvious at this point that the assumed 80 US gallon fuel tank is larger than needed for most flights. A reduction to 50 US gallon tank would be possible. This would result in lower aircraft mass and improved performance in many areas. However, it was decided to retain the large tank and to stress in marketing the long range of the aircraft. This feature is unique among current amphibian GA aircraft.

Take-off and landing calculations were also made using standard performance equations. Using the assumption that the speed at lift-off would be at 1.2 times the stall speed, the take-off ground run from a paved runway at sea-level conditions was found to be 308 m (1009 ft). At Denver on a hot day (5000 ft altitude at 100°F) the ground run increases to 539 m (1738 ft). Landing with full flaps deployed at sea level was found to require a ground run of 179 m (586 ft).

Waterborne take-off is much more difficult to estimate due to the need to conduct towing tank tests of the planing hull design to determine its aircraft drag in the water. Based on published water/ground take-off ratios for existing general aviation amphibians of similar size it was estimated that about 550 m (1800 ft) would be required at sea level. On landing, due to the effects of water drag the aircraft will require less distance after touchdown than for a conventional runway-based landing. As with the take-off performance, no estimates are possible without further hull drag data.

The maximum cruise speed of the aircraft was calculated to be 94 m/s (182 kt) using propeller thrust data for the Hartzell propeller and the 300 hp (223 kW) power available from the Zoche engine. Using this same engine/propeller data the best rate of climb was found to be almost 15.25 m/s (3000 fpm) at sea level at a speed of 33.4 m/s (65 kt).

A summary of the aircraft performance estimation is shown in Table 10.3.

Table 10.3 Aircraft performance summary

Cruise speed at 75% power at 7500 ft	165 kt (85 m/s)
Maximum cruise speed (sea level)	182 kt (94 m/s)
Stall speed (sea level, no flaps)	52 kt (27 m/s)
Take-off speed (sea level, no flaps)	62 kt (32 m/s)
Fuel capacity	80 US gal
Range at 75% power at 7500 ft	1218 nm (2253 km)
Max. range	2087 nm (3862 km)
Take-off ground run at sea level	1009 ft (308 m)
Landing ground run at SL (full flaps)	586 ft (179 m)
Estimated water take-off (SL)	1800 ft (550 m)
Max. rate of climb (SL)	3000 ft/min (15.25 m/s)
Speed of max. rate of climb	65 kt (33.5 m/s)

10.6.4   Stability and control

The aircraft control surfaces were initially sized using conventional design methods[1] and by comparison with current land and amphibious general aviation aircraft. All controls are conventional with the possible exception of the horizontal stabilizer on which a fully movable stabilizer (stabilator) was used instead of a stabilizer/elevator combination. This was done to provide the pitch control needed for water take-off and landing where hydrodynamic forces have a major influence on aircraft pitch. To further enhance the stabilator effectiveness in take-off, a cruciform tail was employed placing the stabilator directly in the wake of the propeller.

The calculated longitudinal and lateral static and dynamic stability derivatives were calculated using established methods.[8] Complete results of the stability and control study were presented in the project report, Reference 9.

10.6.5   Structural details

The basic structural layout of the aircraft is illustrated in Figure 10.9. The structural framework consisted of three distinct sections:

- the forward fuselage/hull,
- the aft fuselage/hull, and
- the wing/engine mount/undercarriage support section.

Each of these sections required a different design philosophy but all were designed to transfer loads smoothly between the mating components.

The aft fuselage section is the simplest of the three structural components. It incorporates a fuselage shear web. This extends from the engine support bulkhead frame to the stern post beneath the tail. Positioned along this web are transverse frames which support the aft fuselage skin and transfer bending loads to the web. The non-load-carrying chine deck and cabin floor create several watertight compartments in this part of the fuselage. These structural elements are constructed as a composite of Nomex foam core sandwiched between from 3 to 15 layers of woven carbon fiber.

The forward fuselage is built up from a planing hull on a central box beam going from the forward-most cabin frame to the engine support bulkhead. Two upper longerons supplement the box beam in carrying forward fuselage bending loads. Transverse frames encircle the structure to transfer loads, enclose the cabin, and to support cabin seating, instruments and controls. Two main forward frames also support the sponsons, wing pylon, and the wing forward spar. All of the forward section frames also serve as watertight bulkheads in the forward fuselage hull area.

The third component is the most complex part of the structure. It consists of the wing and engine support section. The wing structural box is built from two straight spars, running continuously between each wing tip, and top and bottom profile skins. The engine firewall and support structure is mounted to the aft wing spar. This, in turn, is supported by two main fuselage frames. Like the other two structural sections, all frames, longerons, spars, ribs, and the wing skin are constructed of a composite of woven carbon fiber sandwiching a Nomex core.

The engine is attached to its support frame through an aluminum firewall and support bracing.

The retractable main gear is attached to the frame that provides the aft spar for the sponsons. The retractable nose wheel is attached to the forward bulkhead. The

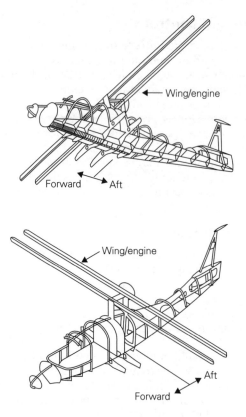

Fig. 10.9 Structural framework

compartments into which the wheels retract are not watertight but are designed to drain any water ingress.

The outer fuselage and hull skins are all molded of a bi-directional woven fiberglass composite reinforced with blown chopped fiberglass using techniques common in the boating industry.

The entire structure is designed to meet FAR 23 requirements. The aircraft $V\text{-}n$ diagram is presented in Figure 10.10.

10.7   Baseline layout

The resulting design, shown in Figure 10.11, was selected as a finalist in the 1997 NASA/FAA General Aviation Design Competition and was awarded third prize.

This sleek design attracted considerable attention when the student design team presented it at a NASA forum at the 1997 Experimental Aviation Association annual air show in Oshkosh, Wisconsin. Many requests for the team's design report were also received from people in several different parts of the world. It was obvious that there continues to be a high level of interest in a high performance general aviation amphibian aircraft and that there may be a market for such a vehicle if it was produced by an established aircraft manufacturer.

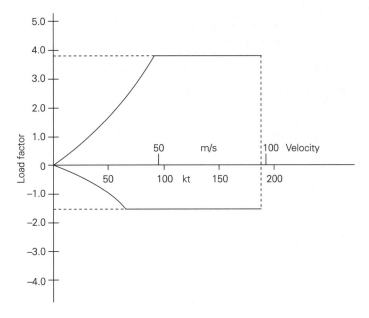

Fig. 10.10 Structural flight envelope

10.8   Revised baseline layout

It is obvious that this design is not optimized in several ways. Early in the process an 80 US gallon fuel tank was selected because this provided fuel capacity similar to long-range versions of existing general aviation aircraft such as the Cessna 182 (Skylane). It was later determined that the selected engine had a very low specific fuel consumption and that a satisfactory range could easily be attained with much less fuel. Cutting the fuel in half would reduce the mass (weight) of the loaded aircraft by some 109 kg (240 lb). This, in turn, could lead to a lighter structure, lower landing loads, etc., all of which would result in a lighter, smaller, and less expensive aircraft.

Another interesting student design project could begin with this design and rework it for a 40 or 50 US gallon fuel tank and the same payload to see how costs could be reduced or performance improved. Alternatively, the reduction in fuel mass could be used to increase payload.

10.9   Further work

An interesting added element in this design project was the construction and limited testing of a wind tunnel model of the aircraft. A 1.22 m (4 ft) span model of the design was constructed of wood, plastic foam, fiberglass, and plaster and tested in the Virginia Tech Stability Wind Tunnel test section 1.83 m × 1.83 m (6 ft × 6 ft).

The model fuselage was made using a wood center line profile section fitted with evenly distributed wood bulkheads attached along its length. Spaces between the bulkheads were filled with plastic foam material and these were cut to a rough surface shape using a hot wire stretched between the bulkheads. The resulting fuselage was coated

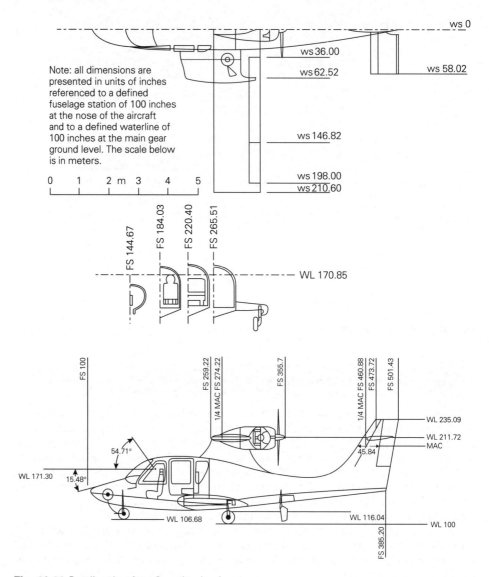

Fig. 10.11 Baseline aircraft configuration drawing

with a plaster-like compound and sanded smooth to the desired shape. The engine nacelle was constructed in a similar fashion.

The model wings, tail, sponsons, and engine support were cut from plastic foam to their desired airfoil profiles using a hot wire traversed over wing section, profile templates. These were reinforced with carbon fiber spars and covered with fiberglass, giving very rigid aerodynamic surfaces.

After assembly, finishing, and painting, the model was mounted on a six component strain gage strut balance in such a way as to allow testing over a reasonable range of angle of attack in the wind tunnel. Two types of tests were conducted in the tunnel.

Fig. 10.12 Aircraft model in the wind tunnel

In one test small lengths of yarn (tufts) were attached to the surface of the model and wing and the flow behavior was observed from a negative angle of attack of about 5° up through the wing stall. In the other test the model was adjusted in angle of attack from 0° to 16° in 2° increments. The lift, drag and pitching moment were measured. Figure 10.12 shows a photograph of the model in the wind tunnel (the model has been painted black to allow better observation of the 'tuft' tests).

It should be noted that student-built wind tunnel models rarely yield research quality test results. This is due to limitations in test Reynolds numbers and the less-than-precise airfoil and fuselage contours. Surface roughness, errors in attaching wings and control surfaces at correct angles, and other factors combine to influence measured forces and moments. Among other deficiencies, drag will usually be higher than expected and stall will often occur at lower than expected angle of attack. On the other hand, such tests can confirm that the design exhibits no regions of serious flow separation over the aft portion of the fuselage. They can also reveal patterns of stall on the wing and tail which may confirm control effectiveness in stall. Most importantly, the tests can show students that their design is actually capable of producing forces and moments which affirm that it would fly. This is considered as the primary conclusion reached as a result of these tests. For this project the flow visualization tests demonstrated that the design had attached flow over the rear part of the fuselage through stall and that when the wing was stalled the flow over the tail was still attached. The force and moment test results, shown in Figure 10.13, revealed very reasonable looking data plots for lift, drag, and pitching moment coefficient within the limitations mentioned above.

Wind tunnel testing of a student design is not always possible due to time constraints and facility availability but it is to be recommended where the schedule and facilities allow. Building and testing a model provides a valuable 'hands-on' experience as part of the design exercise as well as a satisfying 'closure' to the process, even if the test results do not fully confirm the predicted airplane performance. An added benefit is having a model of the design which can be painted for later display and used to make photographs like the one in Figure 10.14.

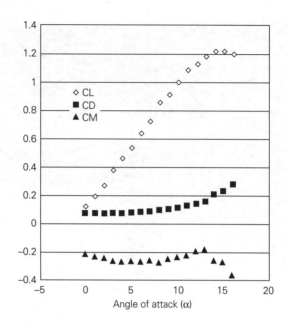

Fig. 10.13 Wind tunnel test results

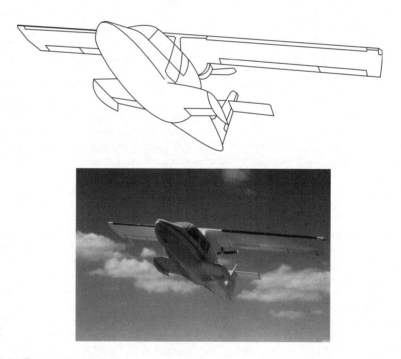

Fig. 10.14 The completed aircraft

10.10   Study review

This project proved an interesting academic exercise for a class in aircraft design. The study of an amphibious aircraft provided a challenge for students who had not previously studied waterborne flight operations. They found that most of the background material related to amphibious aircraft design and operation was published over a half-century ago. Indeed, one of the most interesting challenges of the project was to attempt to merge the semi-empirical charts and equations of an earlier era with modern computer analytical techniques. The use of a diesel engine also represented a return to an old technology which may hold significant promise when blended with modern technology.

This design is obviously a first iteration in a process which could be taken much further, with the principal driver of its future evolution being the revision of the fuel capacity. As mentioned in section 10.8, an initial assumption was made that an 80 US gallon fuel tank would be reasonable based on a comparison with existing four-place, long-range, general aviation airplanes. The final analysis showed that with the selected modern diesel engine the chosen fuel capacity was about twice that needed for the initially desired flight range. This suggests two directions for continued work on the design. One new study could begin with an analysis of the changes that could be made in the aircraft starting with an assumption of a 40 US gallon fuel capacity and the present payload. A second study could assume that half the existing fuel load can be replaced with 109 kg (240 lb) of additional payload. This study would need to redesign the fuselage to provide space for the larger payload.

It would be desirable to conduct a more detailed analysis of the take-off and landing capabilities of the aircraft in water. Take-off and landing distances on water were estimated based on a comparison with existing amphibious aircraft. It would be interesting and instructive to build a model for hydrodynamic testing in a towing tank. A suggested follow-up study would involve the construction of a small-scale, radio controlled, flying model of this design as a 'proof-of-concept' project. Such an experiment would provide the designers with a 'hands-on' experience of many of the aerodynamic/hydrodynamic, stability and control problems unique to amphibious aircraft.

The amphibious aircraft represents only a small niche in the current, general aviation market. This study appeared to show that it is possible to design and build a reasonably priced amphibian with flight performance superior to some current land-based and amphibious designs. If, as this study indicates, such a vehicle could be built and sold for a price comparable to single-engine, land-based aircraft, there seems little reason why such a design could not live up to the title given by its student designers, of 'general aviation's sport utility vehicle'.

References

1 Raymer, Daniel, P., *Aircraft Design: A Conceptual Approach*, American Institute of Aeronautics and Astronautics, Washington DC, 1992.
2 Stinton, Darrol, *The Anatomy of the Aeroplane, Second Edition*, Blackwell Science, London, 1998.
3 Thurston, David, B., *Design for Flying*, McGraw Hill, New York, 1978.
4 Hugli, W. C. Jr and Axt, W. C., 'Hydrodynamic investigation of a series of hull models suitable for small flying boats and amphibians', NACA TN 2503, November 1951.
5 Dathe, Ingo, 'Hydrodynamic characteristics of seaplanes as affected by hull shape parameters', DORNIER Engineering Report, 1989.

6 Lamar, J. E. and Frink, N. T., 'Experimental and analytic study of the longitudinal aerodynamic characteristics of analytically and empirically designed strake-wing configurations at subcritical speeds', NACA TP 1803, June 1981.
7 Lamar, J. E. and Herbert, H. E., 'Production version of the extended NASA-Langley Vortex Lattice FORTRAN computer code, volume I – Users Guide', NASA TM 83303, April 1982.
8 Roskam, Jan, *Airplane Flight Dynamics and Automatic Flight Controls*, Roskam Aviation and Engineering Corp., 1979.
9 'The VenTure, a design report submitted to the NASA/FAA General Aviation Design Competition', Dept. of Aerospace and Ocean Engineering, Virginia Tech, Blacksburg, VA, USA, May 1997.

11

Design organisation and presentation

The early chapters described how the preliminary design phase is organised. The case studies have illustrated several different ways that the design process can be applied to the design of different types of aircraft. From these studies, you will have realised that each aircraft project is unique. However, the process that is followed is relatively constant. By concentrating on methodology and organisation, it is sometimes easy to forget the main purpose of the work. Remember that conceptual design is largely concerned with the innovative aspects of engineering. Mechanically applying prescribed sequences of analysis, without careful thought about the real nature of the problem, will lead to weak and shallow work. To help you to concentrate on the fundamental design problem a checklist has been compiled. This is described in section 11.1 below.

In many academic design courses, aircraft projects are undertaken by groups of students. This is intentionally arranged by the course designers to provide experience in 'team working'. For most of the studying that you have done previously, you will have been expected to have worked alone (or at most in pairs). When working in industry you will be expected to be a contributor to a team effort, therefore previous experience at working in groups is seen as an important part of your professional training. Team working is not just a matter of joining together to complete the project. There are several differences between working in teams and alone. These must be understood if the group work is to be successful. Each member of the group will bring along his or her own skills, abilities, expectations, and prejudices. The quality of the final project will involve the merging of such diversity into a successful team effort. This will not be achieved without careful management and organisation of the group. Section 11.2 below describes some of the potential pitfalls in team working and offers some advice on how these may be avoided.

At the conclusion of the preliminary design course, and probably at times before this, you will be required to produce technical reports and presentations describing your, and/or the team's, work. These may form part of an overall assessment procedure. It is important to find out when these are required so that you can plan your work schedule for the project. For example, the first group report may coincide with the time that a decision has been taken on the preferred aircraft configuration. If the reports and presentations are to be assessed it is important that you do not underperform. Experience has shown that students' enthusiasm for the design work has sometimes left too little time to plan and prepare for these assessments. Sections 11.3 (on writing technical reports) and 11.4 (on making a presentation) have been included to help you to avoid some of the deficiencies that have been seen in previous student work. These sections have been written as 'stand-alone' guides.

11.1   Student's checklist

The checklists below can be used as a kind of 'route map' through the maze of often conflicting considerations that must be assessed in order to reach the final aircraft layout. The list may be regarded as a summary of the descriptions in this book. The earlier chapters (1 and 2) should be referred to if more detail is required. The individual case studies (4 to 10) may be used as a guide to the issues raised in the checklists.

The preliminary design process requires the sequential attention to six questions followed by eight technical work tasks. The process concludes (task 8) with the writing of a technical report describing all the details of the baseline configuration that are known at that point in the development of the design. This may form part of the final group report. In this case, you may need also to refer to section 11.3.

11.1.1   Initial questions

1. What is the problem to be addressed? (This should be answered by writing a report outlining all the main issues raised by the problem, the problem constraints and the overall assessment criteria.)
2. What information is available to assist the definition of the problem and subsequent analysis? (The answer to this question involves a literature search to find technical reports and articles relevant to the project.)
3. What data on previous and competitor aircraft designs can be found and how can this data be manipulated for use in subsequent technical analysis? (This involves the generation of a database of information on existing appropriate aircraft and the plotting of useful parameters, e.g. thrust to weight and wing loading.)
4. Do we need to take into account any specific issues regarding engines, systems and airworthiness and operational features? (It would be very unusual for the answer to this question to be 'no'. List all the issues that you are aware of at this point in the design process.)
5. With the preceding knowledge, can you write a complete aircraft operational requirement? (If not, more work will be needed on questions 1 to 4!)
6. What aircraft concepts might be considered for the design solution? (This requires a freethinking (brainstorming) period in which all your dreams (about aircraft!) can be considered. Do not worry about feasibility at this stage, as this will be considered in the first of the technical tasks that follow.)

11.1.2   Technical tasks

1. From the available configurational options, decide which one(s) might be worth investigating further. (Start by throwing away all the 'weird' concepts but be careful not to 'throw the baby out with the bath water'.) Try to select only one, or at most, two different configurations. Remember that eventually a decision on the 'preferred design' will have to be made and all the effort given to the discarded layout(s) will be wasted so be as forceful as you can but be impartial.
2. Size the baseline aircraft configuration(s) using data from previous designs and some simple initial sizing methods.

3. Estimate, in more detail, the mass, aerodynamic (lift and drag) and performance characteristics of the baseline aircraft using the geometry from the initial layout drawing.
4. Conduct a constraint analysis to identify the validity of the selected design point (configuration) and to determine the sensitivity of the constraints imposed on the problem. Plot the positions of the existing aircraft on this chart to confirm the viability of the specified constraints.
5. In view of the previous analysis, redefine or reconfigure the baseline design to produce a new three-view general arrangement drawing considering aspects including those relating to non-technical issues.
6. Re-estimate, using more detailed methods, the mass, aerodynamic and performance characteristics of the new layout and then conduct parametric/trade studies to fix all the detail geometric parameters (e.g. aspect ratio, wing area, engine size, etc.).
7. Produce a final drawing of the aircraft, taking into account all the previous studies, and then conduct a SWOT (strength, weaknesses, opportunities and threats) analysis on your final design.
8. Finally, produce a report describing, in detail, all the known features of the baseline aircraft. Include a full performance estimate in the report.

Figure 11.1 illustrates the above procedure as a block diagram.

Although the design method is shown as a sequential process in Figure 11.1, it is often possible to overlap some of the stages, especially when you are working in a group of students. For example, while some students are sorting out the problem definition, others can start collecting data (i.e. information retrieval). In addition, the design process is not as smooth as implied in Figure 11.1. It is often necessary to iterate between some of the steps to produce a better answer. For example, the constraint analysis may suggest that a significant change in the aircraft configuration is required, or that a supercritical constraint is relaxed. In such cases, it would be wise to return to the previous stages to obtain some more accurate estimations for the new configuration, or to review the aircraft specification. It would be necessary then to recalculate the constraint diagram to see if the design lies in a feasible design region.

Checklists are only a means of logically triggering your thoughts about a problem to be solved. They are not designed, or intended, to be a replacement for your own careful consideration of the problem. Many design projects involve significant individual facets that may not have been mentioned or included in the general lists above. Bear in mind that the design you produce is totally your responsibility. Do not rely solely on a mechanical, externally produced set of prompts to get the best result. Think about the problem for yourself, as broadly as you can, in the time you have available and do not be afraid to consider outrageous options with all the other ideas you have.

11.2   Teamworking

While some aircraft design courses are taught with an individual design format where each student works on his or her own design, it has become commonplace for most design courses to employ team projects. This is usually done in an attempt to emulate the design environment one would encounter in industry. Working in groups or teams requires attention to the development of management, organisational, communication and interpersonal skills. As such, 'teaming' places unique demands on the design process. Nevertheless, most experts agree that, despite the added difficulties,

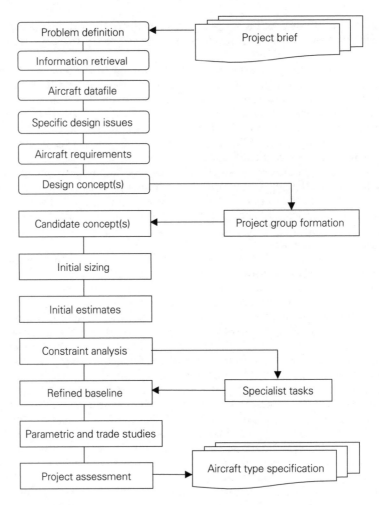

Fig. 11.1 Student's checklist

teamworking results in a better design and provides the student with more relevant industrial experience.

There are many advantages to the team approach to design:

- Teaming brings together the strengths of a diverse group of people.
- Team consensus usually results in a better product.
- Everyone on the team has a backup person to check his or her work.
- Sharing the task makes the job easier and more fun.
- Teaming allows each member to use his or her talents in a more focused way.

Teaming also has disadvantages:

- The success of the team may be limited by its weakest members.
- Some members may tend to procrastinate, delaying the work of the entire team.
- Teaming makes it easier to blame problems on the least liked or least productive member of the team.

- Personalities can get in the way of progress.
- The most energetic team members usually do more than their fair share of the work.

The first challenge faced in a team design environment is the selection of team members. Students often enter a design course with a desire to be a part of a team of students that they have chosen, a team composed of long-standing friends. However, this is not a good simulation of industrial team formation. In practice, those selected for membership on a design or product development team may have never met before. Indeed, many modern industry teams are multinational in membership and members may never meet except through Internet, telephone, and teleconference communication.

Forming a team from a group of friends is often a recipe for disaster. Team members are likely to find that the social skills and personality interactions which made them friends do not necessarily serve them well in accomplishing the goals of a design team. The laid back personality which makes a person fun to be with on Saturday nights at a bar may prove exasperating when one is expecting that friend to finish his or her critical part of the design analysis. Consequently, it is common for 'self-selected' design teams to have serious problems as members fail to measure up to the expectations of others on the team. Both long-standing friendships and teamwork suffer.

Many experienced design course instructors prefer to form teams by randomly selecting members from class rolls rather than letting them 'self-select'. This often results in fewer teamwork problems since members bring fewer expectations of their teammates to the group and often work a little harder to create a team environment.

Some experts recommend using various versions of personality profile testing in selecting team members. Alternatively, they suggest personal profiling of existing team members to enable them to work together optimally. This recognises that each team member has unique capabilities to bring to the work of the team. It is meant to encourage members to interact in such a way as to make the best use of their individual talents. While it might be interesting to analyse student design teams in this way, if only to study team interactions and dynamics, it is a luxury and distraction, which time schedules on most university design classes do not permit.

11.2.1 Team development

It is important for team members to understand that building a good team takes an effort by all team members and that conflict along the way is inevitable. Most teaming experts recognise several distinct stages in the team development process.

When a team begins its work, interactions among team members are tentative and polite as they struggle to find their own place on the team and to identify the strengths and weaknesses of other members. These formative stages of teamwork involve building team interdependencies through an initial exchange of information, through task exploration and identification of common goals.

The second stage of team interaction usually involves some conflict as members begin to disagree over procedures and direction. There is a tendency of team members to criticise each other's ideas as they strive to assert their own approaches to the solution of the problem. Team members often respond to criticism emotionally and hostility may develop between members. Coalitions begin to form within the team and team polarisation is often the result. It is in this stage of team formation that poor meeting attendance or lack of participation in team discussions by some team members can lead to hostile reactions from others. These problems are a normal part of the team formation process and they must be recognised as such if the team is to progress further.

The period of potential conflict is usually followed by a growth of cohesiveness and unity as team members begin to establish their roles and relationships. An agreement on team working relationships, standards, and procedures can result in a realisation that the group is beginning to think and work as a unit.

Following the establishment of team cohesiveness, the group is ready to proceed in a spirit of mutual co-operation to make the needed design decisions and to solve the assigned problem. This brings an increased sense of task orientation with an emphasis on team performance and achievement. With a little luck and a lot of patience and perseverance, this stage of team performance will be reached before the end of the academic term and the deadline for project completion.

11.2.2 Team member responsibilities

Many books and articles have been written on the traits needed for successful teaming and for being an effective team member. All agree on certain rules of teamwork requiring that each team member:

- is equally responsible for the progress and success of the team,
- must attend all meetings and be there on time,
- must carry out assigned tasks on schedule,
- must listen to and show respect for the views of others,
- can criticise ideas but not team members,
- needs to give, and expect to receive, constructive feedback,
- needs to resolve conflicts constructively,
- should always strive for 'win/win' resolution of conflicts,
- must pay attention in team meetings and refrain from wasting time in mindless discussion of irrelevant matters,
- needs to ask questions when clarification of what is happening is required.

11.2.3 Team leadership requirements

Selection of a team leader is one of the most important tasks facing a design team. Team leaders are often chosen for the wrong reasons. Personal popularity or past academic success is not necessarily the best basis for selection of a leader. Effective leadership embodies many traits beyond being the life of the party or the class valedictorian. A good team leader will be able to:

- motivate and encourage the team,
- keep the team organised and on schedule,
- keep team meetings on the agenda,
- make sure everyone's ideas are heard and evaluated,
- keep all team members on their assigned tasks,
- encourage and maintain effective individual and team communication,
- keep team information resources up to date,
- utilise the talents of individual team members effectively,
- do his or her share of the team's work while helping others when needed to keep the team on schedule.

It is often wise to take a little time to observe the team members in action before selecting a team leader. Where time and the number of team members permit, it may be useful to begin the design process by dividing the group into two or three subgroups. Each subgroup is asked to propose one or more candidate design concepts. Working in these subgroups, the team members are able to identify and assess potential leaders for the later combined team.

An effective team may have other 'leaders'. These are members who formally or informally serve in various roles to keep the team effective, and on task. While the team leader can fill some of these roles, various members of the team may assume others. Some of the tasks are shown below:

- Team gatekeeper: Makes sure all team members are heard by keeping dominant members in check and encouraging less assertive members.
- Team checker: Makes sure everyone understands what is happening and what everyone else is doing.
- Team recorder: The team record keeper, or most commonly called secretary, makes a record of all team ideas and decisions and makes sure the team is not continually rehashing old issues.
- Devil's advocate: Makes sure that opposing ideas are considered.
- Team motivator: Gives everyone positive recognition for their contributions.

Sometimes, despite the best efforts of the team and 'management' (the academic supervisor?), the selected team leader or leaders simply do not perform in the best interests of the team. While it may be rare for a 'forced' change in team leadership to be necessary, this option should always be kept open.

It should also be noted that, particularly with larger teams, co-chairpersons might be appropriate. Splitting the role of team leader may be a very effective way to utilise the different strengths of two members, as long as the co-leaders can work well together.

11.2.4 Team operating principles

There are many 'rules' for effective team operation and success. Almost every major company has developed its own set of rules or recommendations for effective teamwork. The following principles for design teaming are based on a set of ideals used by a major international aircraft company.

A successful aircraft design team must:

- have a compelling vision,
- have clear performance goals,
- have a single plan,
- recognise that 'data sets us free' (i.e. that technical claims must be supported by calculations or test results),
- have no secrets (about the design project!) among team members,
- allow complaints by members but require them to offer alternative, constructive solutions,
- listen to each other and help each other,
- have emotional resilience, be able to bounce back after difficulty or criticism,
- have fun and make the journey 'working together'.

11.2.5 Brainstorming

One final trait of a good team that bears mention is the ability to 'brainstorm'. Brainstorming is often thought of as a very informal way of suggesting sometimes 'off-the-wall' ideas for solving problems. However, effective brainstorming can result from a planned team activity with a set of well-defined rules. An idea central to successful brainstorming is that all ideas are equally acceptable without regard to how ridiculous they may first appear. Criticism of ideas is forbidden during the session. Because of its freewheeling nature, many people are not comfortable in a brainstorming session. Others, who thrive in such an environment, fail to see the need to document its results.

Brainstorming is an effective way to generate a high volume of ideas in a non-analytical way and to stimulate creative thought and interaction where one idea leads to another. This is a way to maximise individual, team member involvement and commitment and to document the scope of knowledge of the team quickly.

The following are offered as general guidelines for effective brainstorming:

- Define and write out the topic.
- Silently generate a list of ideas, then share these and add to them as a team.
- Record all information as it is given.
- Do not criticise either ideas or the people presenting them.
- Build on the ideas of others.
- Maximise the quantity of ideas.
- Welcome wild ideas as these can stimulate new directions.

Brainstorming is probably most useful at the beginning of the design process but it can play a useful role at any point where the design team has reached an obstacle in the road to success. It is important to realise that brainstorming offers a means of exploring a wide range of design options without having to thoroughly analyse all aspects of each suggestion. It is a harvesting process, one that gathers all possible approaches, no matter how wild-eyed, to solve the design problem. All ideas are given an initial assessment and the most promising approaches to the solution of the problem are identified. These are then subjected to analyses that are more detailed and complete.

The brainstorming approach recognises that there is simply not enough time to completely analyse every possible solution to a design problem and to reach a reasonable conclusion. It is a way of recognising that there is value in every suggestion even when not all of them may merit detailed analysis. With effective use of brainstorming, a design team can take a quick look at a variety of good ideas and combine them into even better designs before taking the time to perform an in-depth evaluation of the concepts. This can save considerable time in the design process. It will also broaden the scope of the search and the approaches to the solution of the problem. Be brave and have a go at brainstorming. It will lead to you finding your best design and will be fun (a win/win process).

11.3 Managing design meetings

A team meeting does not consist of a group of students getting together to talk about what they have done or anything they feel they would like to raise at that time. This would be, at best, a waste of time and, at worst, would not lead to a successful project.

In order to reach a rational conclusion to the design task it is necessary to organise and plan the work that is required to develop the aircraft configuration. To achieve this without causing unnecessary conflict and confusion between members of the group, it is desirable that all members understand and are willing to work within and adopt a formalised procedure. This implies a stylised format to each meeting. The notes below set out a suggested form of such meetings. These procedures are not specific to aircraft design; they can be adopted for all professional group work.

11.3.1 Prior to the meeting

It is essential that, before anyone comes to the meeting, the purpose and procedure to be followed are made public. The 'Meeting Agenda' is the notice that is published to provide this information. It is essentially a list, in chronological order, of the topics to be raised and discussed at the meeting. These act as a reminder to those students who will be the main participants to come prepared to talk about their subject and to distribute any supporting data. It is also a call to other members of the group who might want to raise issues on the agenda items. They should come to the meeting with all the information they want to raise and have copies, if appropriate, for the other members of the group.

The agenda should be compiled by the chairperson and circulated at least 24 hours prior to the start of the meeting. An efficient chairperson would normally have spoken to the group members who will speak on the principal subjects. To ensure that there is no difficulty with presenting the main items, this discussion should have been done before issuing the notice.

A suggested format of the meeting notice is shown below. The heading is followed by a general description of the meeting. For example:

Aircraft Preliminary Design Project
The regular weekly aircraft design group meeting*
to be held on 25/12/04 at 1400 hr in the departmental seminar room.
(*Or any other special/extraordinary meeting as required)

AGENDA

Appointment of new meeting secretary (if appropriate).

1. Apologies for absence.
2. Approval of previous minutes (usually the last meeting).
3. Matters arising from the minutes.
4. Main agenda items:
    4.1 The title and one or two sentences of explanation for each item,
    4.2 giving the essence of the discussion and if appropriate the decisions likely to be taken.
    Etc.
5. Any other business.
6. Date, time and venue of next meeting.

11.3.2 Minutes of the meeting

During the meeting the secretary should be making notes of the main points raised, any future action to be taken and any decisions reached. It is important to record the nature of the responsibility (action) placed on any member of the group and the date when they must supply a response to the group. These notes must be written up by the secretary in the form of minutes of the meeting (see below). These will be the official account of the discussions and decisions taken by the group. It is good practice to show the minutes to the chairperson to confirm that they are 'a true and accurate record' of the meeting. This should be done before publishing them to the rest of the group.

It is important that the minutes are prepared in advance of the next meeting. As design meetings on academic courses may be frequent (e.g. weekly) a deadline no later than 72 hours following the end of the meeting is reasonable.

The format of minutes is standardised as shown below:

Aircraft Preliminary Design Group Project
M I N U T E S
of the regular aircraft design group meeting
held on 25/12/04* at 1400 hr
in the department seminar room.
(*Note, no rest for Christmas for our team!)

Present: Neil Armstrong   Louis Bleriot   Richard Branson
Octave Chanute   Glenn Curtiss   Henry Farman
Hermann Goering   Amy Johnson   Chas Lindbergh
Freddy Laker   Henry Royce   Igor Sikorsky

(i.e. a list in alphabetical order of those people present at the meeting).

1. **Apologies for absence** (List all the members of the group who are not present and record their reason for non-attendance, if known. (Note: it is regarded as bad form not to let the chairperson know that you will not be attending the meeting.)

2. **Minutes of previous meeting(s)** (Normally states 'Approved without correction' or 'Approved with the corrections detailed below'.)
   2.1 List of corrections to minutes
   2.2 etc.

3. **Matters arising from the minutes**
   3.1 A matter will only arise from earlier minutes if an action has been
   3.2 placed on a previous topic. The action will have been allocated to a
   3.3 specified person in the group by name. He will be required to report
   3.4 to the meeting on the action. The secretary will summarise the
   3.5 discussion and decisions and if further action is required this will
   3.6 be recorded to a specified member of the group and 'actioned' for a
        future meeting.
   Etc.

4. **Agenda items**
   4.1 Agenda items may arise from previous discussions on the development
   4.2 of the aircraft, or when a specialist has been asked to report on his/her
   4.3 subject to the group. Regular presentations from each specialist should
   4.4 be scheduled to keep the rest of the group informed about progress
   4.5 on the work and to inform on the detailed investigations into the
        aircraft configuration.
   Etc.
   (It is common practice to place the 'actioned' item clearly away from the main text to ensure that it is not 'lost' within the rest of the document, namely:
      ACTION: Richard Branson to provide air tickets by next meeting.

5. **Any other business**
   5.1 This section may only be used for non-substantive items which have
   5.2 not required the group to be informed of the detail prior to the meeting.

5.3 It may be used to present web addresses and data sources, considerations for future main agenda items, or short reports requesting extra information or data (that could be available within 24 hr) from the group members.

Etc.

6. **Details of the next meeting of the group** (This records the decisions taken at the end of the meeting on the date, time, venue of the next meeting and any subsidiary meetings that have been arranged.)

Meeting ended at 1845 hr!

Name of secretary
Date of issue

11.3.3   Dispersed meetings

Most of the meetings that are held for academic work will involve group members meeting face to face in their college department. Some design courses run a group project in which it is not possible for the team to meet in this way. For example, some members may be in different colleges or universities, or even in a different country. In this case, the meeting format may involve teleconferencing or some other form of electronic communication. Such meetings may be fun at first due to the unusual nature of the interaction, but there are several safeguards that must be followed in order to avoid confusion. Some of these difficulties arise from the remoteness of the contributors, some from the instant response of the communication system, and some from cultural differences. For example, it is sometimes not possible to pick up non-aural clues when members are making suggestions (was he being sarcastic or serious?). Everyone has experienced sending an email response too quickly and then regretting the consequences. Even time zone differences and course programme variations can lead to confusion and disruption. All of these difficulties are surmountable with a little care and patience in the management of the meetings. It is even more essential to follow the conduct of meetings described above for such meetings.

11.4   Writing technical reports

The ability to write a clearly understood, concise, and accurate technical report is regarded as a principal requirement of a professional engineer. You will be judged in your career by your skill at producing high-quality and readable reports of your work. You probably have all the ability necessary to do this but our experience has shown that students need to be informed of the procedures and characteristics that are common to technical writing. The notes below describe the features found in a good technical report:

- It has a neat presentation in both appearance and language.
- It is easy for readers to 'navigate' the contents to find the information they seek. This requires careful attention to page numbering, references in the text to each figure

and appendix (these need to be numbered) and clear identification of the source of any external material pertinent to the content of the report.

- It is written in a language that is impersonal (mainly in the third person). A technical report is not a narrative. Therefore, there is a difference in the language you use between a letter home and a report to your supervisor!
- Although the specific points mentioned above need to be taken into account the main criterion of a well written report (to the intended readership) is that it is – *interesting*.

You are not the first person to write such a report. There are many good examples available for you to follow. Textbooks, company reports, research papers, and good quality aeronautical journals provide adequate evidence of good practice.

The most annoying faults that are found in student work relate to avoidable 'typos', missing labels (e.g. figure and table numbers), out of sequence pages, duplicate pages and references, and figures that cannot be found. Mostly these deficiencies arise due to the shortage of time allowed to adequately proof-read the final version. The handing-in time should not be assumed coincident with the final printing and collation of the report!

The sections below bring together advice on the production of the report.

11.4.1 Planning the report

Before starting to write the report, make sure that you can answer the following questions:

- What is the objective of producing the report? What use will the readers make of the report? What material has to be included in the report? Have you been made aware of any criteria to be used if the report is to be assessed?
- Who will read the report? Are you aware of what they are expecting to see in the report? Have they informed you of the purpose to which they will use the report? What experience in the subject matter of the report do they possess? What prior knowledge do they have that is pertinent to the content of the report?
- How much time is available for you to write the report? What length should it be? Are there any formatting criteria that must be met?

11.4.2 Organising the report

This involves listing the topics that you want to include. The report must have a logical progression. It starts with an 'introduction' section. This is followed by the 'main body'. The report is completed by a 'termination' section. The topics should be grouped and allocated to one of the sections. These groups should be titled with appropriate headings. A smooth transition between groups of text in the report must be provided.

The start of the report establishes the relevance of the work and puts your findings into context by identifying other relevant studies/information. This is usually followed by an overview of the structure of the report. This describes the progression of the work and any subsidiary information. Do not assume that the reader is familiar with your design problem even if the report is to be read only by your design instructor.

The 'main body' contains the detail design work covered in the report. The nature of this section will vary depending on the purpose of the report. For example, it could contain the methods of analysis used in the design work, explanations of the theories used and any assumptions that have had to be made. It will describe the results that have been found and how these results can be interpreted relative to the study objectives. Obviously, this could involve a substantial amount of text and several figures. To make it easier to read it is advisable to subdivide the section (e.g. mass estimation, aerodynamic analysis, propulsion, performance, etc.). Avoid unnecessary padding (e.g. extractions from textbooks and the Internet). Try to be concise but make sure that the reader, who is not as familiar with the work as you, can understand what you have done.

To finish the report you should summarise the main design features, make observations on the results, draw conclusions (this may involve making relevant comparisons with competitor and other aircraft) and finally make some recommendations. No new material should be brought into the termination section.

Several 'administration' sections sandwich the three main textural sections. These include:

(a) prior to the introduction section:
- title page
- abstract or executive summary
- contents list
- notation and list of symbols
(b) after the termination section:
- list of references
- bibliography
- figures
- tables
- appendices

A typical report layout is shown in Figure 11.2.

11.4.3 Writing the report

The sequence in which the report is written will not be in the same order as it is collated in the finished version. For example, the abstract/executive summary is usually the last section to be written. Write each section separately but ensure that the style and format are consistent throughout the report. If you are working in a group, it is possible for separate individual members to write different parts. This will require decisions to be taken on the appearance of the report. When using word processing software, this is achieved most easily by designing a 'format template' that each writer uses.

Ensure that the report is readable by following the suggestions below:

- Avoid conversational language (especially colloquia, jargon and slang).
- Avoid the use of undefined acronyms and technical terms unfamiliar to the reader.
- Use an impersonal style (e.g. 'it was found that...', not 'we found...').
- Keep sentences short and to the point, the language style precise, and the report concise.
- Use paragraphs to group descriptions, reasoning, and thoughts together. Do not use them just to space out the text.
- Try to avoid repetition except when summarising your previous findings.
- Avoid using 'fancy' fonts and too many textural, format changes.
- Use a 'serif' text font (e.g. Times) for the main text passages, as this is easier to scan-read. Use a 'san serif' font (e.g. Arial) for headings.
- Avoid font sizes of less than 10pt for A4 or American 8.5 in × 11 in paper.
- Avoid colour if the report is to be photocopied or printed.

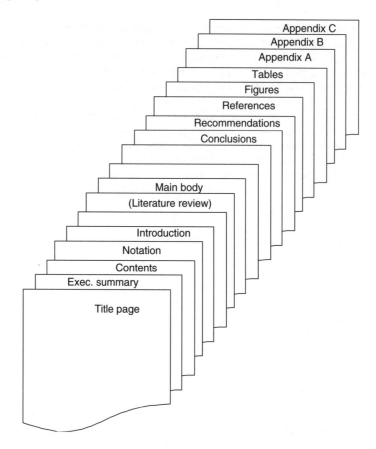

Fig. 11.2 The full report

Before you start the report decide the hierarchical structure of your headings. Select progressively fewer prominent styles and sizes for lower order headings.

11.4.4 Referencing

Using text, figures, diagrams, data, photographs (etc.) from sources other than those produced by you or your group (e.g. textbooks, industrial reports, technical journals, the Internet, etc.) and not correctly indicating the origin is called:

PLAGIARISM

Apart from this being against the law (e.g. infringement of copyright), it is regarded as professional cheating. It could lead to your career suffering and possibly result in exclusion from your professional body. Referencing and displaying the source of external information is your only 'insurance policy' against an accusation of plagiarism.

Correctly referencing any material used in the study and quoted in your report adds validity to your work. It also indicates that you are familiar with the area of study and have spent time and effort researching the field. Assessors will penalise you heavily if you, intentionally or not, omit to quote sources (the Internet included). They are likely to give you credit if you demonstrate your knowledge of the subject area.

There are several different methods of showing the source of external material. For some reports the method of referencing is prescribed by the recipient. When including a substantial extract in the report (more than a few words), it is common practice to put it in quotation marks and/or change the text style (often to italic) to distinguish it from the original text. If you are using extensive extracts from an external source, it is advisable to obtain written permission from the author or the copyright holder. They may require you to fully credit them at the point of insertion.

When referring to external work, it is usual to add an indicator (a number) in the text and to provide a full reference to the original text in a reference section towards the end of the report. Alternatively, the full reference can be added as a footnote on the page containing the indicator. The full reference must contain sufficient detail to allow the original to be located in a library or on the Internet. Most textbooks (including this one) and other technical reports will provide you with examples of the correct formats for writing the full reference. For visual material (diagrams, graphs, photographs) it is normal just to quote the source in the caption but if several are used from one source it is necessary to apply the full referencing method.

11.4.5 Use of figures, tables and appendices

Without a link in the text, a figure, table or appendix should not be included in the report. Figures, tables, and appendices should be numbered separately and sequentially (e.g. Figure 9.5, Table 2.3, Appendix 3). The numbering order should coincide with the order they are introduced into the text. Each figure should be titled and listed in the contents section of the report.

The word 'figure' is used for any visual representation. Do not use words like 'sketch', 'graph', 'plot', 'photograph', 'plate', 'drawing', 'illustration', 'diagram' as a caption; they are all termed figures. All figures and the text appearing on them must be readable from the bottom edge (portrait) or the right-hand side (landscape). Remember to leave space around the figure to allow for binding. When electronically inserting figures or scanned material, avoid distorting the image. If the figure is to be reduced for insertion in the text, make sure that the original is sufficiently bold to avoid unreadable text and lines when made smaller.

Be aware that some standard templates available on widely used drawing and graphing software are not suitable for direct insertion into technical reports. Drawings generated by several popular CAD programs, which look very good on the computer screen, sometimes virtually disappear when inserted into standard word processing applications. Colours that appear brilliant against another colour background in a computer graphics presentation will fade into oblivion or become indistinguishable from other hues in print. Graphs will need modifying as suggested below, especially if the report is to be printed in monochrome:

- Avoid the use of colour on the graph, particularly as a background to the plotting area.
- Distinguish lines either by using different line types (solid, chain, dashed) or by applying different line markers.
- Indicate suppressed zeros on the axes.
- Select a text format that is consistent with the report style.
- Make sure that the text is not too small (or too large).
- Apply gridlines to both axes, or neither.

- Ensure that the axes have number divisions that can easily be measured. The reader may want to scale a point from your graph so provide a scale that you would like to use if you needed to do this.
- Quote units on each axis and provide annotation for the line markers.
- Title the graph in the same format as used for other figures (normally placed below the graph).
- Preview the figure or table before printing to make sure all the details are visible and clear.

In most software packages it is possible to create a template that matches the format specification of your report.

Large amounts of detailed analysis and/or data can be removed from the main text and put into separate appendices. This avoids an interruption to the 'flow' of the text. In such cases, the report must be capable of being understood, without reference to the detail in the appendices. This may mean that the main results from the detailed work will need to be added to the main text. The appendices should be regarded as supplementary to the report. They are there to allow the reader to clarify any confusion that arises from the results in the main text.

11.4.6   Group reports

When working as a design team, individual members will be 'expert' in different specialised subject areas (e.g. structures, aerodynamics, flight simulation, etc.). This may mean that the work done in each specialisation is reported by the team member responsible for that subject. This approach may lead to some difficulties if the production of the report is not carefully controlled and managed. Here are some guidelines to avoid such problems:

- It is important to decide, within the limitations of the length of the report, what subjects/topics should be included. Each specialist may feel that his or her work is more significant than other members' work. It will be necessary to agree what is the best overall composition of the report to meet the objective. Compromises will need to be agreed before individuals start writing their input (see comments below about individual assessment reports).
- Although it may be common practice to compose the table of contents after the report has been written, it is worthwhile drafting this at the start. This helps to succinctly define the nature of the report. It can act as a list of sections to be written and as an action list that defines the names of authors of various sections and when their contributions are required.
- One of the best ways of selecting the content of each section is to identify and agree what 'key' figures should be included.
- As mentioned above, the group must decide the design of the report style and set this into a template that each writer will use.
- When several different people are independently writing technical analyses, it is important that the nomenclature is agreed (e.g. will M be used for mass, moment, or Mach number, etc.?).
- The group report will not be satisfactory if it is composed from 'cut and paste' extracts from individual specialist reports. The group report must be created as a totally separate entity. This will require the specialists to précis their more extensive individual reports. Only the main findings and conclusions will be required for the group report.

- Assign a subgroup who is not too involved with the production of the report to act as proof-readers and editors. For this task to be effective, they will need time to thoroughly review the finished report. This will mean that time must be allowed in the production schedule for this to be done and to allow any changes that they recommend being included in the final version.
- The figures to be included must be in a standard format. This may be most easily achieved if a subgroup of students, who are not involved in writing the text, is made responsible for this work.

Throughout the design process it will have been necessary to progressively improve the detailed analysis of the aircraft to gain confidence in your predictions. In group work, this will have involved individual members of the team performing analysis in various specialist areas. The group will require verbal and written reports from these 'experts' as the design unfolds and decisions are taken on the aircraft configuration. These individual reports will form the basis of specialist reports to be submitted at agreed milestones in the project development. These reports will contain figures and data that will be required for the final group report. These figures and data will form the basis for the descriptions to be included in the main body of the final group report.

11.4.7 Review of the report

When the report is complete, it is worth answering the following questions to make sure that you have not overlooked anything that is important:

- Is the report presentable? Are you pleased with the way it looks? Does it look 'professional'? How does it compare with the reports that you used as examples of good practice?
- Does it accurately present your understanding of the problem and your abilities in finding a solution to the original problem?
- Does it show an imaginative and fresh approach to the problem?
- Does it match your original intentions?
- Is it understandable to the people who will be reading it?
- Will it meet the expectations of the assessors? (*Do you know what these are?*) For example:
  - Does it demonstrate a thorough understanding of the problem (e.g. aircraft operational requirements, external threats to a successful outcome, risk assessment, and potential for development)?
  - Have you clearly described the technical approach that you have adopted?
  - Have all the theories and data used in the design been correctly applied and validated against known information?
  - Have you identified the critical technical problem areas and offered potential solutions to these difficulties?
  - Have you covered all the factors that the assessors will be expecting?

Figure 11.3 shows an example of a contents list taken from a student final, aircraft design, group report.

This was the format of a report to the NASA 'University Design Competition' in 2002. The competition required a short (20 page) report supplemented with a series of detailed specialist reports (in appendices). The aircraft was a personal (two-place) very short take-off and landing vehicle with a new type of engine and twin ducted fans.

TABLE OF CONTENTS
Team members
Abstract
Executive summary
    1. Project description
    2. Aircraft specification
        2.1 Introduction
        2.2 Performance
            2.2.1 Takeoff and Landing
            2.2.2 Cruise Performance
            2.2.3 Climb and Turn Performance
            2.2.4 Performance Comparison
        2.3 Propulsion
            2.3.1 Rand-Cam Engine
            2.3.2 Engine Installation
            2.3.3 Ducted Fans
        2.4 Aerodynamics
        2.5 Stability and Control
        2.6 Structures
            2.6.1 Structural Design Aspects
            2.6.2 Structural Overview
            2.6.3 Landing Gear
        2.7 Avionics and Systems
            2.7.1 Avionics
            2.7.2 Systems
        2.8 Ergonomics/Human Factors
        2.9 Manufacturing
        2.10 Cost
        2.11 Practical Applications
    3. Small Aircraft transportation systems (SATS)
    4. Student work schedule
Appendix A: Competition information   Appendix B: Aircraft performance
Appendix C: Propulsion   Appendix D: Aerodynamic analysis
Appendix E: Stability & control   Appendix F: Weight & balance
Appendix G: Structures   Appendix H: Avionics & systems
Appendix I: Cockpit layout   Appendix J: Manufacturing issues
Appendix K: Cost analysis   Appendix L: Design selection process

Fig. 11.3 Example content list (group design report)

11.5　Making a technical presentation

There is a fundamental difference between making a technical presentation and giving an after-dinner speech! Although in both a friendly and relaxed approach is expected, it is not advisable to be too flippant or risqué in a technical talk. Since your audience will be looking at you, it is important that your appearance should be in keeping with the importance of the occasion. In a team presentation, the members should agree in advance on a dress standard. Be well groomed and smartly dressed but do not take this to an extreme, as this will also be distracting. The intention is to appear professional and to display confidence in your subject. Try not to shuffle papers prior and during your speech. Look towards the audience as much as possible so that they feel you want to involve them in your experience. Try to avoid standing between your audience and the projection screen or between the projector and screen.

The sections below give advice on the preparation and presentation of the talk.

11.5.1   Planning the presentation

There are four questions that you should answer before starting to prepare the talk:

- What is/are the reason/s for making the presentation? For example:
  - to present information
  - to display your abilities
  - to explain your methods of study and results
  - to explain your decisions
  - to persuade people
  - to sell your ideas, or yourself
- Who will be in the audience and what experience do they have of the work you are presenting? For example:
  - What is their technical knowledge?
  - What are their motives for being at the talk?
  - What are they expecting to gain from the talk?
  - Do they have any specific requirements of which you know?
- How much time is available for the presentation?
  - Knowing this, you can divide the time into segments to suit the material to be presented.
  - Do not attempt to get too much detail into a short presentation.
  - Make sure that you do not overshoot your time.
  - Do not try to get too many slides into your presentation (see later advice for timing of each slide).
- How many separate topics can be covered in the allowable time?
  - Too many and each topic will appear to be shallow.
  - Too few and some important detail will need to be left out which will make the presentation appear to be shallow.

11.5.2   Organising the presentation

Organising the presentation will require you to:

- Provide a logical progression to the talk.
- Divide the talk into clear segments, each with a specific objective.
- Start by establishing the relevance of the talk to the audience.
- Make sure that, in the main body of the talk, you have smooth and logical transitions between each segment.
- Finish with a summary of your work making recommendations that lead from your results.

The opening and closing sections are the most important parts. Initially you need to get the attention of your listeners and finally leave them with something to remember your work. In the opening set the style of your talk by engaging the audience with eye contact, relaxed body language and airs of confidence in what you are about to tell them. Speak clearly and not too fast. Make sure you can be heard and understood at the back of the room. At the end of the talk, bring your audience into your work by making comparisons with which they will be familiar. Describe future considerations and at the end throw down a challenge. Finally, do not be afraid to ask if they have any questions.

11.5.3   Use of equipment

In our modern high-tech environment, you need to make a careful choice of equipment to use in your presentation. Your choices will usually include:

- Overhead projector (OHP).
- Slide projector.
- Computer software (e.g. Powerpoint) and projection.
- Video/DVD.

Students sometimes feel tempted to use several different methods in their presentation in order to 'jazz up' their talk and demonstrate their abilities at handling equipment. Experience has shown that this is a very hazardous strategy. The favourite habitation of gremlins is in visual demonstration equipment. They know their mischief here can cause the most chaos. The best advice is to keep the presentation as simple as possible and to have a back-up strategy available in case of equipment failure.

Here are some tips:

- Back up all your files and take these with you to the presentation venue.
- If you are taking your own equipment to the venue, know what you will do if/when it fails.
- If you are intending to use equipment supplied at the venue, ensure that it is compatible with your files, tapes, etc. Note that European and US video systems are different, PC and Apple operating systems are different, and not all computers have the same versions of standard operating systems and software.
- Well before you are scheduled to give your talk go to the presentation room to install your equipment, to try out the provided equipment and to practise your talk.
- Always have a 'back-up' set of overhead slides for use should computer presentation fail.

If you intend to change the method of presentation during your talk (e.g. showing a computer simulation of your aircraft flying), this will take extra time from that allocated for your presentation. It will also interrupt the audience concentration in your talk. The value of adding the item to the talk must be considered against these disturbances. For aircraft design presentations, it is not often necessary to introduce dynamic (video) clips. If available, these may be held in reserve to show to interested participants during the question time or later. On the other hand, for a long presentation (i.e. more than 50 minutes) changing the presentation method will help keep the audience awake. A technique along these lines is simply to stop using visual equipment for a short time (e.g. turn off the OHP). This will refocus the attention of the audience onto the speaker. However, be careful not to use these tricks too often.

As mentioned above, the best presentations are intentionally kept simple. This strategy also applies to the use and design of the visuals. Here are some more tips:

(a) OHP:
- Know the purpose of displaying each slide in your talk.
- Make sure that all the slides are in the same format so that they appear as a set.
- Keep the slide 'crisp'. For scanning or reading pages of text a serif font, like Times, is good but for displayed text a san-serif font, like Arial, is easier to read

from the screen. Definitely never use 'fancy' fonts and styles, as they are much more difficult for the audience to read. It may also convey the wrong impression of your intelligence to the audience!

- Do not put too many words on each line of the slide. Six to eight is recommended depending on font size. The maximum and minimum point size will depend on the equipment to be used and the size of the presentation room. If possible, choose the best size by making a few trial slides and projecting them before spending too much time on the final set.
- Put a heading on each slide (often in a different colour and font size to the text) and only use six to eight lines of text or bullet points below.
- Avoid using too many colours, different fonts, text styles and sizes (remember simple is best).
- Avoid the use of block colour unless you have a good printer on which to produce the slides.
- Although cartoons and clipart help to brighten up the presentation try not to use them too often as they become distracting.

Aircraft design presentations will need some graphs, tables, diagrams and drawings to be displayed. Keep these as uncomplicated as possible. Always allow time to explain the information on these slides to the audience. For example, if you are showing an aircraft general arrangement drawing it is important to allow enough time for the audience to appreciate the detail. Make sure that the people at the back can read the projected slide. If these requirements are difficult to meet it may be better to have the aircraft drawing printed and displayed separately. It is especially important to take the time to explain graphs and charts. Remember that this may be the first time your audience has seen them and they may not be able to understand your figures without a brief description of the axes or scales used.

(b) Computer-based presentations. The advice above is also applicable to computer application software (e.g. Powerpoint); however, there are extra pitfalls to be avoided:
- Do not use distracting background template designs.
- Do not use animation unless it is used to aid the presentation (e.g. progressively displaying points as the verbal explanations are given).
- Do not use sound effects.

11.5.4 Management of the presentation

- Keep the pace of the presentation leisurely.
- Although you may be nervous, try to speak slowly and leave enough time intervals between topics so that the listeners have time to digest your information.
- Plan the number of slides in the talk to give 1.5 to 2.5 minutes' display time for each. Remember to extend this time for the more complex slides. An absolute limit should probably be one slide per minute.
- Make sure that you look towards the audience. Eye contact is important. Never look at the projector screen, as this will break your link with the audience.
- Never read the text on the slides. The slide is not there as your prompt. It is intended to guide the audience through your presentation.
- If you feel that you may 'dry up' and feel that you would be more confident if you had a prompt, use small cards held in the palm of one hand as a 'security blanket'. Do not fiddle with these cards throughout the talk as this will be distracting to the audience.

- Diagrams are often a much clearer way of describing complex issues to the audience than text but you will still need to explain the slide carefully.

For aircraft design work involving teams of students, it is not advisable for each member of a large group to be involved with speaking. It is distracting to change the speaker too frequently. Limit the number of speakers to two or three. It is an advantage to have one of the speakers making the introduction and putting the work into context and for the same person to return towards the end to summarise the work and conclude the presentation. Remember to make contingency plans in the group in case one of the speakers is absent. The non-speaking members of the group can be involved with the preparation of the slides and other display material.

At the end of the presentation, it is common practice to ask the audience if they have any questions. It is at this point that the non-speaking members join the presenters to field the questions as they may relate to detail that has not been possible to put into the talk for lack of time. The appropriate expert will provide the answer to the question. For this reason alone, it is important that all team members are available at the presentation. It is often worthwhile to prepare a few extra slides relating to potentially contentious topics and keeping these in reserve to help with the explanations.

11.5.5 Review of the presentation

Like all good performances, it is essential to have at least one rehearsal before the big day. Get a small audience to listen to the rehearsal. This may largely be composed of the non-speaking group members (and their friends). They must critically review the presentation. The questions below will help in this process:

- Was the opening strong enough and did it introduce the presentation so that the audience knew what to expect? Did it grab your attention and make you look forward to the talk?
- Did the presentation have a logical flow?
- Did the speakers convey the points clearly and with authority?
- Was there a memorable closing to the presentation?
- Overall, was the presentation 'professional', 'credible' and 'fluent'?
- Could you hear everything?
- Could you see everything?

Get the reviewers to comment on the 'visuals':

- Did the visuals support and supplement the talk without dominating it?
- Were they all relevant, readable, understandable and clear?
- Were they displayed for long enough?
- Did they have a 'professional quality' and 'eye appeal'?

Get the reviewer to comment on the speakers:

- Were they sincere and enthusiastic or were they flippant and dull?
- Did they maintain eye contact with the audience?
- Did they have clear and up-beat voices or did they mumble?
- Did they have distracting body language or mannerisms (like turning away from the audience, swaying, walking about or shuffling papers)?
- Did they read too much from their notes?

The most important question to be answered by the reviewers is:

'Was the presentation interesting?'

Finally, there is often a fear among student presenters that some in their audience, particularly faculty assessing their work or students on other design teams, will attempt to ask questions that will 'trip them up'. This is usually not true. Questioners are usually seeking to improve their understanding of the presentation or to clarify some point they missed. Questions are seldom asked simply to make the presenters look bad.

When asked a question, keep calm and ask for clarification from your questioner if needed. Try to answer honestly. If you don't know the answer, say so and/or ask someone else on the team if he or she can provide an answer. Never try to make up an answer because this will almost inevitably lead you into deeper trouble. Think of questions as simply a further chance to make clear all the good work you have done on your project. Avoid paranoia and enjoy yourself.

11.6　Design course structure and student assessment

Design courses are intended to act as an integration of the various disciplines studied individually in other parts of an engineering education programme. It is this blending of the many technical and management topics that gives the design course its unique flavour and sets the fundamental aims and objectives. A vital aspect of the course is the consideration of the broader aspects of engineering. To this end, an overall system approach to design is often followed in the course structure. The various specialised disciplines (e.g. aerodynamics, propulsion, mass and balance, performance, configuration, structures, flight dynamics, cost, manufacture, serviceability, airworthiness, environmental, financial, commercial, political, social, etc.) form the substructure for the integrated/system approach. The many, often conflicting, requirements from the specialist areas need to be incorporated and resolved into a feasible design solution. To do this requires careful teamworking. Therefore, the overall management and organisational procedures of the design process are essential elements of an aircraft design course.

Course organisers will define their own aims and objectives and students must be aware of these. Typical examples of both of these are shown below.

11.6.1　Course aims
- To introduce students to the process of designing to a specific set of requirements.
- To introduce and provide experience in the practice of group working.
- To enable the student group to produce and present a feasible design solution to a given operational requirement, taking into account all aspects to the design.
- To broaden the students' perspective of professional engineering practice.
- To prepare students for a career in engineering technical management.

11.6.2   Course objectives

On completion of the course the student should:

- Appreciate the overall design process.
- Have experience in assisting in the development of an aircraft design from initial conception to a preliminary design solution.
- Have experience in teamworking.
- Have contributed to both oral and written presentations of the group work.
- Appreciate the requirements of good management, organisational and communicative skills in working with integrated project teams.
- Understand how the above experiences are related to commercial engineering practice.

11.6.3   Course structure

The 'aims and objectives' that are specified for a design course can be achieved in many different ways. These may range from a highly structured lecture/tutorial system to an unsupervised student 'free for all' approach. Usually something between these extremes will be found to work best. As for all systems, the course designers must clearly identify the course requirements. To meet these, the separate parts of the course must be arranged to link into a successful entity. For example, to identify individual student abilities in a final assessment process, it may be necessary to combine individual studies with the overall group work. In such a case, the course structure may consist of interlinked design activities. The structure below is an example of a course designed to provide individual and group assessments.

Example

After presenting the group with the design brief, the individual students separately consider the problem and propose a personal layout option. A reported description of this option may be assessed as an individual piece of coursework. The options from the full group members are then considered by the whole group to decide the best layout to adopt. This may be one of the individual designs but it more likely will be a composite design which takes the best component ideas from all of the concepts available. Industry calls this process 'technical transfusion'. The group then analyses the new layout to produce the 'baseline' aircraft layout. This may be presented as the first group report (written and oral) and assessed. In the next stage, the group allocates individual specialist tasks to specific students. They apply their effort to analysing the baseline layout with respect to their speciality. After this has been done, some recommendations on modifications to the baseline layout will be made.

Description of the final design (written and oral) will form the basis of the final group assessment. In this example, there are four assessment components to the course:

1. Individual conceptual design report (30 per cent).
2. Initial group 'baseline' report (20 per cent).
3. Individual specialist analysis report (30 per cent).
4. Final group report and presentation (20 per cent).

Typical assessment weightings for each piece of coursework are shown above but these may be varied to suit specific regulations. Detailed criteria for the assessment of each piece of coursework is described below.

11.6.4 Assessment criteria

A feature of aircraft design work is the necessity to make reasoned choices between alternative options. It will be necessary to review, and possibly revise, these decisions as more detail and knowledge of the aircraft and its requirements become available. A vital thread running throughout a design course is the need to make defendable decisions. It is important to identify and record the process by which such decisions are made. This will enable an audit trail to be generated by which development of the project can be assessed.

The example above shows that four reports are required during the design study. Students tend to underestimate the importance of good report writing. They underestimate the significance placed on such reports in industry. One of the important aspects of design project work is the reliance on accurate reporting, both written and aural. In section 11.4 above, we offer some detailed guidance to this topic. Assessment marks will usually be influenced by the students' ability to produce a competent professional report. In this book we make a strong plea that you do not start the design process before you understand the requirements of the problem. The same strategy should be applied to the production of any report. Students should be aware of the overall objective of the report and the assessment criteria that will be applied by their faculty. For the example given above, the following are descriptions of the assessment process.

Report 1: Individual 'conceptual design'
This report is based on the individual work completed as the first task. The objective of this report is to record the work that you did in arriving at your preferred design layout. This report will include the analysis of the given design brief, information collection and analysis, design options and your selection procedures to choose your favourite design. It must be possible to clearly identify your design thinking and your design layout in as much detail as you can manage in the time you dedicate to this work.

Report 2: Group 'initial baseline configuration'
This is a group effort. As such, it will require decisions by the design team to allocate responsibilities to individuals and subgroups for various sections of the report. What is required is a clear description of the process used by the group to settle on their preferred baseline layout. This will be followed by a detailed technical description of the aircraft. This part of the report might be regarded as an initial draft of the aircraft type specification. The value of the report will be judged on the clear understanding of your aircraft and the confidence in the design detail.

Report 3: Individual 'specialist report'
A separate individual report is required for each specialisation (e.g. aerodynamics, structures, configuration, propulsion, flight control and stability, etc.). The report will outline the detail analysis that the specialist has conducted on the design. Each report may exhibit a different format, as suits the specialist task. Advice will be given throughout the period of study, on the nature of each task. The report must exhibit both your own effort and understanding of the subject and the application of this knowledge to the aircraft layout. It may be appropriate to conclude your report with recommendations to the group on changes that would benefit the design (from your own specialist perspective).

Report 4: Group final baseline aircraft
Part of the last coursework assessment will be based on the group presentation given at the end of the course. Associated with this presentation will be the submission of the group design report. This will be, in effect, a full type specification of the

aircraft. This is not just a collection of parts of each of the specialist reports but a separate, self-contained and complete description of the aircraft. The objective is to enable a reader to fully understand the final configuration and technical parameters of the aircraft. An introduction may be appropriate to describe any changes that were introduced to the final layout from the previous group report.

11.6.5   Peer review

One of the frequently voiced criticisms of student group or teamwork in an educational environment is the difficulty of accurately assessing the individual student's work. This criticism is aimed mainly at courses in which the overall group assessment is credited to each student equally. In such cases, the concern is raised about the undervaluation of students who have contributed more than average to the group activity, against the undeserved advantage gained by weak and lazy members of the group. Unless the group supervisor has had a close involvement with the students, throughout the whole project, it is impossible to distinguish individual effort and ability with the accuracy and fidelity required for examination purposes. To avoid this difficulty it may be necessary to introduce a component of the assessment that is based on a 'peer review'.

There are several different types of peer review. One of the simplest consists of giving the group 100 points and asking them to share them between group members according to individual effort and contributions to the overall group effort. A more sophisticated method requires each member of the group to submit an assessment of each of the other members' contributions to the group work. In this case, care must be taken to eliminate the possibility of 'block voting' by friends! Obviously, there are numerous other possible forms of peer review. The selection of a suitable method will be related to the course structure, local assessment procedures and general academic regulations.

The way that the results from the peer review are used in the overall assessment process is a variable that must be decided by the course supervisor. In some courses, it is necessary to inform the group at the start of the course that a peer review will be undertaken as a part of the overall assessment method towards the end of their work.

11.7   Naming your aircraft

Ask any parents and they will be pleased to tell you how much enjoyment they had in choosing the names of their children. The same pleasure can be had in deciding the name of your new aircraft. It must be suitable, mean something to you and the team, recognisable and not offensive to the sensibilities to others. Nearly all aircraft have been 'christened' by their designers in the past. From the 'Flyer' to the 'Typhoon', aircraft are recognised by their name. It is preferable if the name links to some characteristic of the aircraft (like 'Concorde', an early joint venture between UK and France). A great deal of fun and friendly discussion can be had in selecting a suitable name for your project. For example, the Deep Interdictions Strike Aircraft (Chapter 8) was named by a student team 'CHIMERA'. This, it appears, is the name given to a mythical fire-breathing monster with the head of a lion, the body of a goat, and the tail of a serpent. As the course they were attending was titled 'Aircraft Synthesis', it seems a very appropriate name. However, another generalised definition of the word is quoted as 'a wild and unrealistic dream or notion'. You would have to ask the team which definition stimulated the adoption of the name. The amphibious aircraft (Chapter 10) was named by the Virginia Tech team as 'VenTure' to identify the university (VT) at which it was created and to identify the operational freedom of the specification. The roadable aircraft presented in Chapter 7 was named the 'Pegasus' after the well-known mythological flying horse. Choosing a name for your 'baby' does not need to be left until the end of the project. Have a go and let your imagination take flight, right from the start.

Footnote

Remember that even though much effort and thought is necessary in preliminary design work, it is often remembered as one of the most enjoyable parts of your academic work. After graduating it is unlikely that you will be given the opportunity to get involved with this type of activity again. So go-ahead and enjoy yourself while you can.

Have fun in all your work but remember to be professional at the same time.

Appendix A Units and conversion factors

In UK and European college courses and industries, aeronautical calculations will be conducted in SI units (Systeme International d'Unites). This system has the basic units of kilogram, metre, second, kelvin, and derived units of newton, joule, pascal, watt, etc. Most USA companies and universities and some older UK reports will be in measurements of foot, pound and second and Rankine (known as Imperial or British System of units). Some European industries still use the Old Metric System (not to be confused with the SI system) which uses the metre, kilogram force, second and kelvin. To work as an aeronautical engineer, you will need to understand all these systems of measurement. Unfortunately, it is common practice for a variety of different units to be used in most types of aircraft analysis. This will mean that you will need to understand, and be able to convert, units in different systems.

Coupled to the confusion that can arise from converting between different systems of units are the 'special' units used in the aircraft industries. These include the aptly named slug, and the use of nautical terms for speed (knot) and distance (nautical mile). Operational parameters (e.g. air traffic control and some flight instruments) are often required when interpreting or confirming flight data (e.g. altitude is often quoted in feet or hundreds of feet, e.g. flight level 330), rates of climb in feet per minute, weight in pounds (or kilograms force), engine thrust in pounds, speed in nautical miles per hour (or kilometres per hour) and pressure in millibars.

Definition of aircraft speed can be particularly confusing, even if it is quoted in knots. Pilots like to use the speed shown on their cockpit instruments (known as indicated airspeed, IAS). Flight test engineers like to use the indicated airspeed corrected for measuring inaccuracies (known as calibrated airspeed, CAS). Design engineers need speed related to the forces on the aircraft (referred to as true airspeed, TAS, or in a different system, equivalent airspeed, EAS). The inter-relationships between TAS, EAS, CAS and IAS are dependent on the properties of the ambient and local air mass, the design of the flight instruments and their interaction with the airflow around the aircraft. For example, flying at sea level on a standard (ISA) day, CAS = TAS. Most classical aerodynamic and aircraft performance textbooks define the mathematical relationships involved in making the conversions between the different airspeed definitions.

Derived units

Isaac Newton defined the laws of motion that are used as the basis for the linking of mass and acceleration to force. He proved that force exerted by an object in motion is *proportional* to mass times acceleration. As engineers, we have devised systems of measurement to allow Newton's law to be redefined as Force = Mass × Acceleration. This equation is only true for a consistent system of units, for example:

In SI units: Newton (force) = Kilogram (mass) × Metre/second^2 (acceleration)

In Imperial: Pound (force) = Slug* (mass) × Feet/second^2 (acceleration)

   * See below for the definition of the slug.

There are several other derived units shown in the list of conversions below.

Funny units

***Nautical mile* (nm):** The international nautical mile is 1852 m exactly. The Imperial nautical mile is 6080 ft but aviators and other navigators have sometimes used 6000 ft (2000 yards) as a crude approximation.

***Knot* (kt):** is (nm/hour) = 0.514 m/s = 1.852 km/hr = 1.688 ft/s = 1.1508 mph.

***G (gee)* (g):** is the gravitational acceleration at ground level (average) = 32.2 ft/s^2 = 9.81 m/s^2. Gee is used to divide aircraft accelerations to relate them to steady flight conditions (i.e. the aircraft is pulling 6 g).

***Slug*:** is a contrived unit used in the Imperial system to avoid multiplying by gravitational acceleration. It is a mass unit which provides a force of 1 pound when it is subjected to an acceleration of 1 ft/s^2. It is often used for specifying air density in aerodynamic equations in ft:lb:s units (e.g. air density at SL.ISA (denoted as ρ_o) = 0.002378 slug/ft^3 which is equivalent to 1.225 kg/m^3).

(Note: gravitational acceleration at ground level 'g' (average) is 32.2 ft/s^2 in Imperial units, so a slug is effectively 32.2 pounds mass!)

***British thermal unit* (Btu):** The heat required to raise the temperature of one pound of water through 1°F. (Note: 1 Btu = 1055 joule.)

***Horsepower* (hp):** An artificial measure of power = 550 ft. lb/s = 3300 ft lb/min = 746 watts.

***Bar* (bar):** A measurement of pressure = 106 dyne/cm^2 (often quoted in millibars = 10 kN/m^2). (Note: standard atmosphere (atm) is sometime quoted for pressure measurements = 10.01325 bar = 101 325 N/m^2.)

***Imperial gallon* (Imp. gal):** The volume of 10 lb of water at 62°F = 277.4 in^3 = 4.546 litre. (Note: the US gallon (US gal) is a smaller volume than the Imp. gallon, equaling only 231 in^3 = 3.785 litre. Therefore, **1 US gal = 0.83267 Imp. gal**. (Be careful when interpreting aircraft fuel burn and tankage volumes from unspecified gallons.))

***Ton*:** A measure of weight = 2240 pounds = 1016 kg (f) (approx. 1000 kg). (Note: in the US 1 ton = 2000 lb and 1 tonne = 2240 lb.)

***Thou*:** One thousandth of an inch = 0.001 in. (1 mm = 40 thou approx.)

***Flight level*:** is a derived unit of altitude used in air traffic control. One hundred feet altitude is the basic unit, therefore flight level 330 is flying at a height of 33 000 ft.

***Drag count*:** is used as a measure for the change in drag coefficient (ΔC_D) = 0.0001. (Note: drag count is not a direct measure of drag as it is associated with a reference area. It is only valid for use as a relative assessment of change for designs with the same reference area (e.g. trade studies).)

Conversions (exact conversions can be found in British Standards BS350/2856)

Multiply	by	to get
Inch (in)	25.40	millimetres
Feet (ft)	0.3048	metres
Feet (ft)	3.048×10^{-4}	kilometres
Statute mile (mi)	1.609	kilometres
Nautical mile (nm)	1.852	kilometres
Nautical mile (nm)	1.1508	statute miles
Square foot (ft^2)	9.290×10^{-2}	square metres
Cubic foot (ft^3)	28.317 (2.832×10^{-2})	litres (cubic metres)
Cubic inch (m^3)	1.639×10^{-5}	cubic metres
US gallon (US gal)	3.78542	litres
Imp. gallon (Imp. gal)	4.546	litres
Foot/second (ft/s)	0.3048	m/s
(ft/s)	1.097	km/h
(ft/s)	0.6818	mph
Knot (nm/hr) (kt)	1.689	ft/s
(kt)	1.151	mph
(kt)	1.852	km/h
(kt)	0.5151	m/s
Mile/hour (mph)	1.467	ft/s
(mph)	1.609	km/h
(mph)	0.8684	kt
(mph)	0.4471	m/s
Slug (slug)	14.59	kg
Pound (lbf)	4.448	N
Pound/Sq. in (lbf/in^2)	6895	N/m^2
Pound/Sq. ft (lbf/ft^2)	47.88	N/m^2
Slug/cubic ft (slug/ft^3)	515.4	kg/m^3
Foot pound (ft. lbf)	1.356	Nm (joules)
Foot lb/s (ft. lb/s)	1.356	Joules/s (watts)
Horsepower (hp)	550	ft.lb/s
(hp)	33 000	ft.lb/min
(hp)	745.7	watts

The following data has been compiled to help you to understand some of the above units and to provide assistance in converting to and from different systems of measurement. We start by defining units derived from the basic Imperial System.

Density: $\text{Slug/ft}^3 = 515.4\,\text{kg/m}^3$ ($1\,\text{lb}_m/\text{ft}^3 = 16.02\,\text{kg/m}^3$)

Force: Pound (lbf) $= 4.448\,\text{N}$

Work: $\text{Slug ft}^2/\text{s} = 1.356\,\text{Nm}$

Power: $\text{Slug ft}^2/\text{s}^3 = 1.356\,\text{Nm/s}$

Pressure: $\text{Slug/ft s}^2 = 47.88\,\text{N/m}^2$ ($1\,\text{lb}_f/\text{in}^2$ (psi) $= 6895\,\text{N/m}^2$)

Gas constant: ft lbf/slug $°R = 0.1672\,\text{Nm/kg}°\text{K}$

Coeff. of viscosity: $\text{Slug/ft s} = 47.88\,\text{kg/ms}$

Kinematic viscosity: $\text{ft}^2/\text{s} = 9.290 \times 10^2\,\text{m}^2/\text{s}$

Specific fuel consumption: (jet a/c) $\text{lb}_m/\text{hr lb thrust} = 0.283 \times 10^{-4}\,\text{kg/Ns}$

Some useful constants (standard values)

Gravitational force at SL $= 9.80665\,\text{m/s}^2 = 32.174\,\text{ft/s}^2$

Air pressure at SL (p_o) $= 760\,\text{mm Hg} = 29.92\,\text{m.Hg} = 1.01325 \times 10^5\,\text{N/m}^2$

                                    $= 2116.22\,\text{lb/ft}^2$

Air temperature at SL (T_o) $= 15.0°\text{C}$ (conversions ⇓)

                                    $= 288.15°\text{K}$ ($°\text{K} = °\text{C} + 273.19$)

                                       (where $-273.19°\text{C}$ is absolute zero temperature)

                                    $= 59.0°\text{F}$ ($°\text{C} = (°\text{F} - 32)\,(5/9)$)

                                    $= 518.67°\text{R}$ ($°\text{R} = °\text{F} + 459.7$)

                                       (where $-459.7°\text{F}$ is absolute zero temperature)

        (where: $°\text{C}$ is degrees in Centigrade, $°\text{K} = $ Kelvin, $°\text{F} = $ Fahrenheit, $°\text{R} = $ Rankine)

Air density at SL (ρ_o) $= 1.22492\,\text{kg/m}^3 = 0.002378\,\text{slug/ft}^3$

Air coeff. of viscosity SL (μ_o) $= 1.7894 \times 10^{-5}\,\text{kg/ms} = 1.2024 \times 10^{-5}\,\text{lb}_m/\text{fts}$

Air kinematic viscosity SL (ν_o) $= 1.4607 \times 10^{-5}\,\text{m}^2/\text{s} = 1.5723 \times 10^{-4}\,\text{ft}^2/\text{s}$

Specific gravity at $0°\text{C}$ (lb_m/ft^3, kg/m^3):

Water $= 1.000$ (62.43, 1000)   Seawater $= 1.025$ (63.99, 1025)

Jet fuel JP1 $= 0.800$ (49.9, 800)   JP3 $= 0.775$ (48.4, 775)

JP4 $= 0.785$ (49.0, 785)   JP5 $= 0.817$ (51.0, 817)

Kerosene $= 0.820$ (51.2, 820)   Gasoline $= 0.720$ (44.9, 720)

Alcohol $= 0.801$ (50.0, 801)

Appendix B
Design data sources

As part of the design process you will need to become aware of information about the particular project that you are working on. There are many sources of such information. This section of the book is intended to direct you to the most commonly used references. Of course, the sources quoted below are in no way the only ones that may be of use. They are intended to provide a start to your searches. Some of these have been referenced in the project studies described in Chapters 4 to 10.

Technical books (in alphabetical order)

Brandt, S. A., Stiles, R. J., Bertin, J. J., Whitford, R., *Introduction to Aeronautics: A Design Perspective*, AIAA Education Series, 1997, ISBN 1 56347 250 3
A book based on the lecture courses for new recruits to the US Air Force Academy and as such is a good introductory text. The case studies (Wright Flyer, DC-3, and F-16) are more concerned with historical development than as examples of technical analysis. The sections on 'Performance and constant analysis' and 'Sizing' are worth some study. As might be expected, the text is mainly illustrated with military examples.

Eshelby, M. E., *Aircraft Performance – Theory and Practice*, Butterworth-Heinemann, UK, April 2000, ISBN 0 340 75897 X
There are many good textbooks available on aircraft performance but this one is relatively new and has the merit of linking fundamental analysis to operational and airworthiness requirements. It is particularly good for civil aircraft design.

Fielding, J. P., *Introduction to Aircraft Design*, Cambridge University Press, UK, Sept. 1999, ISBN 0521 65722 9
John Fielding is the Professor of Aircraft Design at the College of Aeronautics, Cranfield, England. This book is mainly aimed at the detailed design phase which follows the preliminary design studies. It contains a lot of valuable information that students will find useful in the layout stage of their design work. The book will help students to get a complete overview of all types of aircraft and aircraft project management.

Howe, D., *Aircraft Conceptual Design Synthesis*, Professional Engineering Pub. Ltd, UK, October 2000, ISBN 1 86058 301 6
Professor Howe taught many students at Cranfield College of Aeronautics (later university) and this book reflects his extensive understanding of his experience. It is a very useful text covering detail design and synthesis in a readable format. It contains a case study of the design of a feeder liner and is accompanied with a PC CDROM with Excel spreadsheet programs. The book concludes with four design examples (aerobatic piston-engined trainer, twin-turboprop airliner, reconnaissance UAV and a supersonic STOVL fighter).

Huenecke, K., *Modern Combat Aircraft Design*, Airlife Publishing Ltd, UK, 1987, ISBN 1 85310 002 1
Aimed obviously at fast military aircraft design in which high performance has to be coupled with advanced technology. Students will be able to use this book to gain an insight into the operation and capability requirements of fighter aircraft.

Jenkinson, L. R., Simpkin, P., Rhodes, D., *Civil Jet Aircraft Design*, Butterworth-Heinemann, UK, and AIAA Education Series, August 1999, ISBN 0 340 74152 X
This book describes in detail the conceptual design of civil transport aircraft powered with turbofan engines. It contains detailed explanations of the aircraft performance analysis and engine integration, together with sections on environmental and cost issues. Four case studies illustrate the application of the methods to civil aircraft design. Detailed data on aircraft, engines and airports can be found in the web pages associated with this book.

Mattingly, J. D. *et al.*, *Aircraft Engine Design*, AIAA Education Series, 1987, ISBN 0 930 403 23 1
Along with G. C. Oates' book in the same series this is the best book on engine design for performance and engine/aircraft integration. The format of the book is specifically tailored to the teaching of design and the development of the design process as applied to propulsion aspects. There is an excellent section on the use of constraint analysis in aircraft design.

Nicolai, L. M., *Fundamentals of Aircraft Design*, METS, Inc., San Jose, California 95120, 1984
Although somewhat old (original version 1975) this book does contain all the necessary methods and a lot of useful data. Unfortunately there are no example projects included but the book has stood the test of time and is still widely appreciated.

Raymer, D. P., *Aircraft Design: A Conceptual Approach*, third edition AIAA Education Series, 1999, ISBN 1 56347 281 0
An excellent book, describing in detail the fundamental principles underpinning the initial design of aircraft. This book is the most popular aircraft design book used in undergraduate studies. It is particularly good for military aircraft projects but also useful across the whole range. Two examples (a light sports plane and a military fighter) are included.

Roskam, J., *Airplane Design*, DARC, Kansas 66044, USA
This is a collection of eight volumes (parts) dealing with separate topics that comprise aircraft preliminary design. It is particularly useful for students as it contains detailed design data that can be used in undergraduate design courses. Although the full set is expensive for personal purchase, it is held in most academic libraries.

Stinton, D., *The Anatomy of the Aeroplane*, second edition, 1998, Blackwell Science Ltd, ISBN 0-632-04029-7
This recently updated version of a very successful earlier book is an excellent introduction to aeronautics. It explains, without too many formulae, how aircraft work and what are the operational and flying characteristics. It presents a good overview of aircraft and aircraft design.

Stinton, D., *The Design of the Aeroplane*, Blackwell Science Ltd, 1983, ISBN 0-632-01877-1
In this book, Darrol Stinton presents the formula and detail design data for aircraft design. It has good information on the light aircraft sector and many unusual configurations. It also contains useful information over the full spectrum of aeronautics. Some useful design data for project work is included.

Stinton, D., *Flying Qualities and Flight Testing of the Aeroplane*, Blackwell Science Ltd, 1996, ISBN 0-632 02121 7
In his third book, Darrol Stinton concentrates on both designing aircraft for good flying quality and meeting appropriate airworthiness testing requirements. Although dealing with manually controlled flight systems on light aircraft many of the fundamental aspects can be applied to larger types.

Torenbeek, E., *Synthesis of Subsonic Airplane Design*, Delft University Press, 1981
Historically this is perhaps the most used handbook in academic aircraft studies. Providing allowance is made for the 20 years since it was first written, it contains much useful data which is still valid. No example aircraft projects are given and it is mainly based on civil transport aircraft design yet it is still widely respected.

Janes Information Systems: website www.janes.com
This company publishes many types of reports and books on aircraft subjects. The best known is *Janes' All the World's Aircraft* which is published annually and contains technical information on most types of aircraft. They also publish books such as Whitford, R., *Design for Air Combat* (1989).

Reference books

Useful reference books for aircraft projects include *Janes' All the World's Aircraft*, mentioned above, and Brassey's *World Aircraft and Systems Directory*.

Detailed analysis methods at a professional level can be found in Engineering Science Data Unit (ESDU) Data Sheets. There are several volumes dealing with various specialised sections. Their web site is www.esdu.com but many universities' departments, or libraries, are subscribers to their services which means that students can access their publications on-line.

Research papers

Research publications can be found from conferences and journals from professional bodies including The Royal Aeronautical Society, The Institution of Mechanical Engineering, The American Institution for Aeronautics and Astronautics, and The Society of Automotive Engineers.

Journals and articles

General articles can be found in magazines such as *Flight International* and *Aviation Week*. These also regularly (annually and bi-annually) produce surveys containing technical data on groups of aircraft. These are very useful in the early stages of aircraft project work.

The Internet

Apart from the addresses recorded in the previous chapters of this book, many useful Internet pages can be found from manufacturers, academic institutions, research bodies, and enthusiast groups. Be aware that in an industry as dynamic as aerospace, some of the addresses may have changed since they were last accessed and it is certain that new sites will have been created. For this reason, addresses of web sites are not recorded here. When starting a project it is best to make your own new search for addresses specific to the project on which you are working.

Some publishers, and other companies, have 'search engines' that are specifically aimed at scientific information. Elsevier Science, the group of companies that publish this book outside Europe, make the SCIRUS system available. Dedicated search engines are more efficient at finding specific information and they avoid companies trying to sell you things.

Index